SCANNING PROBE MICROSCOPY – PHYSICAL PROPERTY CHARACTERIZATION AT NANOSCALE

Edited by **Vijay Nalladega**

Scanning Probe Microscopy – Physical Property Characterization at Nanoscale
Edited by Vijay Nalladega

Published by InTech
Janeza Trdine 9, 51000 Rijeka, Croatia

Publishing Process Manager Oliver Kurelic

Technical Editor Teodora Smiljanic

Cover Designer InTech Design Team

Additional hard copies can be obtained from orders@intechopen.com

Scanning Probe Microscopy – Physical Property Characterization at Nanoscale,
Edited by Vijay Nalladega
p. cm.
ISBN 978-953-51-0576-3

For EU product safety concerns:
IN TECH d.o.o., Prolaz Marije Krucifikse Kozulić 3, 51000 Rijeka, Croatia,
info@intechopen.com or visit our website at intechopen.com.

Meet the editor

Dr Vijayaraghava Nalladega is a Research Engineer in the Structural Integrity Division of University of Dayton Research Institute. He has PhD and MS degrees in Mechanical Engineering from University of Dayton, respectively. The primary focus of his research work is materials characterization using atomic force microscopy and scanning probe microscopy techniques. His research also involves the design and development of novel scanning probe microscopy techniques for imaging multiple physical properties of materials at the nanoscale, as well as development of AFM based nanoscale nondestructive evaluation (NDE) techniques. Dr. Nalladega's other research interests include ultrasonic NDE, nonlinear ultrasonics, and design of interdigital transducer sensors for structural health monitoring (SHM) applications.

Contents

Preface

The invention of scanning tunneling microscope (STM) by Binnig and his colleagues in 1982 opened up the possibility of imaging material surfaces with spatial resolution much superior to the conventional microscopy techniques. The STM is the first instrument capable of directly obtaining three-dimensional images of solid surfaces with atomic resolution. Even though STM is capable of achieving atomic resolution, it can only be used on electrical conductors. This limitation has led to the invention of atomic force microscope (AFM) by Binnig and his co-workers in 1986. These techniques have the characteristic that their resolution is not determined by the wavelength that is used for the interaction as in conventional microscopy but rather by the size of the interacting probe scanned over the sample surface. Thus the resolution that is achieved using these techniques is far superior to the wavelengths involved. Although the initial applications focused on near-atomic resolution surface topography measurements, the AFM has been used extensively to measure and image surface physical properties. Several new microscopic techniques, based on AFM have been developed to measure properties such as elastic modulus, magnetic, electrical and thermal properties in the nanometer regime. These instruments, commonly known as scanning probe microscopes (SPM) have opened up new vistas in many interdisciplinary research areas with wide-range applications across multiple disciplines.

The book presents selected original research works on the application of scanning probe microscopy techniques for the characterization of physical properties of materials at the nanoscale. The chapters in this book are arranged into three sections. Section 1, Instrumentation Development, describes two novel SPM techniques for physical property characterization of materials. One of the techniques is based on the combination of atomic force microscopy and a quartz tuning fork sensor for the analysis of topography and mechanical properties of materials with different elastic stiffness including biomolecules with potential applications in nanotribology. The other chapter describes the design and development of an AFM based eddy current force microscopy for the characterization of electrical properties of metals, composites and nanocomposites without the requirement of a bias voltage. The application of the technique to study magneto-elastic and electromagnetic properties as well as for nondestructive evaluation (NDE) at the nanoscale is presented.

Section 2, Surface Morphology, deals with the application of AFM techniques to study the topography features of polymer films and membranes. The section has four chapters. The first chapter in this section discusses the nanoscale features of the structure of polyamide imide membranes of different morphology studied using an AFM. The second chapter in the section deals with the statistical analysis of the surface topography images of the surface of a poly (*p*-phenylene vinylene (PPV) to quantify the topology and to identify the type of the surface. Thickness, mechanical modification, and photo-blanch effects on surface topography of the PPV film are discussed using first-order and second-order statistical analysis. Spintronic devices, composed of alternating magnetic and nonmagnetic multilayer structures, are one of the key components of the data storage. The interfacial roughness is an important parameter for the proper operating of the device. In the third chapter of this section, AFM is used to study the morphological properties and interfacial roughness in the multilayers stacks of spintronic devices. The analysis of the topography is used to study the growth mechanisms of the thin film as a function of growth parameters. Furthermore, magnetic force microscopy (MFM) is used to study the micro-magnetic properties of magnetic thin films and patterned magnetic objects used in high temperature superconductors. The optimal structure and magnetic properties of the thin films are studied for efficient operation of the devices. TiO_2 thin films are used in optical coating for visible and near infrared optics and electrical devices. The structure of the thin films strongly affects the optical and structural properties of the thin films. The last chapter in this section discusses the characterization of surface morphology, roughness of TiO_2 thin films using AFM.

Finally, Section 3, Characterization of Mechanical Properties, discusses the application of AFM for the evaluation of nanoscale mechanical properties. Atomic force acoustic microscopy (AFAM) is an AFM based technique to image nanoscale elastic property variations using ultrasonic waves. The evaluation of elastic and nano-wear properties of SiO_2-PMMA and hybrid coatings using AFAM and nanoindentation techniques is described in the first chapter of this section. Grain boundary sliding is one of the important deformation mechanisms during the process of creep in metals. The study of grain boundary sliding during ambient temperature creep in a hexagonal close-packed metal, zinc, using AFM is presented in the second chapter in this section. The role of grain boundary in the deformation mechanism during the ambient temperature creep is analyzed using the grain boundary sliding evaluated by an AFM. An electron back-scattered diffraction pattern is used to observe the grain boundary structure. Ultrathin lubricant films are being used as a protective layer against wear and corrosion at the magnetic head-disk interface in hard disk drives. The third chapter in the section discusses the tribological and nanomechanical properties and lubricant-surface interactions in ultrathin films used in hard disk drives as measured by AFM and nanoindentation techniques. The microtribological properties of polymer-carbon nanotubes (CNT) nanocomposite thin films studied using AFM and frictional force microscopy is presented in the last chapter of the section.

The topics presented in this book reflect the strong interdisciplinary character of the research in scanning probe microscopy and its application for the characterization of physical properties at the nanoscale. The book gives a unique opportunity to study and understand the possible future trends in the research in the field of scanning probe microscopy and its ever-increasing applications in materials science and engineering.

Vijayaraghava Nalladega
University of Dayton Research Institute, Dayton OH, USA

Section 1

Instrumentation Development

1

Multiple Material Property Characterization Using Induced Currents and Atomic Force Microscopy

Vijay Nalladega[1], Shamachary Sathish[1], Kumar V. Jata[2] and Mark P. Blodgett[2]
[1]*Structural Integrity Division, University of Dayton Research Institute, Dayton, OH*
[2]*Air Force Research Laboratory, Wright-Patterson Air Force Base, Dayton*
USA

1. Introduction

The invention of atomic force microscope (AFM) by Binnig and his co-workers (Binnig et al., 1986) has led to the imaging of conducting and insulating surfaces with nanometer scale resolution. The AFM measures very small forces (less than nN) between a cantilever-tip and the sample surface. When the tip is brought near the surface, the interaction forces between the tip and the sample cause the cantilever to deflect.A topographic image of the surface is obtained by raster scanning the tip across the sample surface and using the interaction force as a parameter for a feedback electronics system which maintains the force at a constant set value. Since the invention of the AFM, it has become a popular tool for surface characterization and is now routinely used in many industries and academic research labs with applications in several research areas.

The initial applications of the AFM were focused on high resolution surface topography imaging of materials. Though it provides high resolution topography images, it cannot provide physical property information. This has led to the development of AFM methods designed to image simultaneously physical properties with topography.Tapping mode AFM, magnetic force microscopy (MFM) (Hartmann, 1999), electric force microscopy (Bluhm et al., 1997; Nyffenegger et al., 1997) are some of the examples. Further imaging modes were developed later to image elastic stiffness (Burnham et al., 1995; Dinelli et al., 1999; Nalladega et al., 2008; Rabe & Arnold, 1994; Yamanaka et al., 1994), surface potential (Nonnenmacher et al., 1991), thermal conductivity (Gu et al., 2002), dielectric properties (Stern et al., 1988), and optical properties (Betzig et al., 1991). The families of instruments based on AFM are known as scanning probe microscopes (SPM). All SPM techniques are based on the same principle, i.e., scanning a probe in the near-field across the sample surface. The techniques of SPM differ only in the selective detection of different sample-probe interactions among the many kinds of interactions between the probe and the sample. For example, if an electrical potential difference is externally applied, electrostatic interactions can be imaged by utilizing a conductive probe. Similarly, a magnetic probe is used to image magnetostatic interactions between the magnetic probe and ferromagnetic surface. The unique combination of nanoscale resolution and broad applicability has led to the proliferation of SPM techniques into virtually all areas of nanometer-scale science and technology.

To measure electrical properties using AFM, a bias voltage is applied between a conducting probe and the sample and the resulting electrical interactions (electrostatic forces, electric currents, resistance, and capacitance etc.) are measured. Depending on the type of the electrical interaction, different electrical property can be measured. Various techniques have been developed based on measuring these interactions to image electrical properties. Electrostatic force microscopy (Bluhm et al., 1997; Nyffenegger et al., 1997), conducting AFM (Oh & Nemanich, 2002; Olbrich et al., 1998), tunneling AFM (Gautier et al, 2004; Ruskell et al., 1996), scanning capacitance microscopy (Matey & Blanc, 1985; Williams, 1999), surface potential imaging (Weaver & Abraham, 1991), and piezoresponse force microscopy (Franke et al., 1994; Gruverman et al., 1996) are some of the widely used AFM techniques for studying electrical properties. To obtain an image of the electrical property, the probe measures the interactions at each location by moving from one discrete location to the next across the scan area. Therefore, these techniques are quite time consuming. Moreover, a bias voltage is always applied between the sample and the tip requiring a conducting tip to perform the measurements. In addition, some of these techniques require a physical contact between the tip and sample.

Electrical properties can also be measured using electromagnetic induction. When a conductor is placed in a time varying magnetic field, currents are induced in the conductor by the magnetic field. These currents are known as eddy currents. Since currents are induced in the conductor, no physical contact between the source and the conductor is needed. Several techniques have been developed based on eddy currents to develop electrodeless methods to measure electrical properties of materials. In these methods, the sample is placed in the field of a coil excited using an AC source. The time-varying magnetic field induces currents in the sample. The induced currents produce a magnetic field opposing the primary field, which changes the impedance of the coil. The electrical impedance can be used to determine the resistivity of the sample.

In addition to the measurement of electrical conductivity, eddy currents are also used in nondestructive evaluation (NDE) of defects in materials. It is well known that defects in a material modify the flow of induced currents in the vicinity of the defect. Consequently, the electrical conductivity around the defect is also different. This fact has been effectively utilized for NDE applications (Libby, 1971). In a typical eddy current testing method, a coil is located as near as possible to the sample being tested and is excited with a time-varying magnetic field at a given frequency. When the coil is scanned across a defect, the impedance of the coil is modified. Therefore, by monitoring the changes in impedance of the coil, it is possible to detect defects in the material. This methodology has been used for NDE applications as well as for the measurement of electrical and magnetic properties under various environmental conditions. It is possible to generate electrical conductivity images by scanning the coil in a raster pattern (Kirby & Lareau, 1997). The spatial resolution in eddy current imaging is dependent on the diameter of the coil and the best spatial resolution is about 50 μm (Karpen et al., 1999). Eddy current methods are sensitive to small changes in electrical and magnetic properties. Thus, small changes in the properties can be detected. However, eddy current methods are essentially comparison methods and it is not possible to get absolute values of electrical conductivity. The electrical conductivity is always given in terms of conductivity of a calibrated standard.

The invention of AFM has enabled the development of eddy current microscopy techniques with better spatial resolution than that of conventional eddy current imaging systems. In magnetic force microscopy (MFM), a magnetic probe is oscillated above a magnetic surface. The oscillating magnetic probe generates eddy currents. This concept was used in the development of an MFM based eddy current microscopy (Hoffmann et al., 1998). This technique was used to image local variations in electrical conductivity of a sample consisting of TiC precipitates in Al_2O_3 matrix with nanometer scale resolution. However, since the magnetic field of an MFM tip is small, this technique is not suitable to image small variations in conductivity.The sensitivity of this technique was improved by using large magnetic fields from a tip made ofpermanent magnet (Lantz et al., 2001). This resulted in increased sensitivity but reduced the spatial resolution down to hundreds of nanometers.

From the above discussion it is evident that it is difficult to achieve both high resolution and high sensitivity to local variations in electrical conductivity using eddy current microscopy by MFM. To improve the sensitivity, a flexible cantilever capable of detecting small variations in the forces can be employed. However, in MFM techniques, a stiffer cantilever, vibrated at its resonant frequency, is used in order to make the cantilever sensitive only to the long-range magnetic forces. But by using a stiffer cantilever, it is difficult to measure small forces generated due to very small variations in the electrical conductivity. The magnetic tips used in MFM have small magnetic field strength. Therefore, the eddy current density that can be induced in the sample material is limited.

The above considerations led to the development of a new high-resolution, non-contact electrical conductivity imaging technique. The technique, called scanning eddy current force microscopy (SECFM), combines the principles of eddy currents and AFM to achieve high spatial resolution and high sensitivity to variations in electrical conductivity on nanoscale. The technique is based on a simple principle- detecting the magnetic forces due to the interactions between a magnetic probe and the magnetic field generated by eddy currents in a conducting sample. To achieve higher sensitivity, a small electromagnetic coil is excited near the sample and eddy currents are generated in the sample. Further sensitivity is achieved by employing soft cantilevers (0.1 N/m) to detect small changes in electrical conductivity. The magnetic field due to eddy currents interacts with the static magnetic field of the probe resulting in magnetic forces. The magnitude of the magnetic forces generated is directly proportional to the electrical conductivity of the sample. The deflection of the cantilever due to the forces is measured and analyzed by a custom-built electronic instrumentation to generate surface topography and electrical conductivity images simultaneously. Since currents are induced, bias voltage is not required between the probe and the sample thus removing the need of conducting tips. The electrical conductivity images are obtained in non-contact fashion. The new instrument has a spatial resolution of 20-25 nm. The instrument is used to characterize electrical properties of different materials. The contrast mechanisms in different materials are explained based on the variation of the magnetic forces caused by eddy currents in different materials. In addition to the electrical properties, we also show that by doing small modifications to the system, it is possible to characterize magnetic, magneto-elastic properties. The advantages, limitations and possible applications of the instrument in materials characterization and nano NDE are discussed.

2. Theory and development of SECFM

The central element of an AFM is the force sensor. For maximum sensitivity to electrical conductivity variations, the sensor should detect small forces generated by the eddy currents. Therefore, it is necessary to select a cantilever spring constant capable of measuring small forces. In order to do so, however, magnetic forces between the probe and eddy currents should be known first. Therefore, a theoretical model is first used to calculate the eddy current forces in a typical metal. The model is used to describe electrodynamic interactions between eddy currents and the probe. Based on the calculated forces, a suitable cantilever is selected.

2.1 Theory

Eddy current fields are considered to be quasi-static fields. Quasi-static condition requires that the wavelength λ of the field is much greater than the dimensions of the conductor (Landau & Lifshitz, 1960). The magnetic field **H** generated by the eddy currents in a nonmagnetic conductor is

$$\nabla^2 \vec{H} = \sigma\mu \frac{\partial \vec{H}}{\partial t} \tag{1}$$

where σ is the electrical conductivity, $\mu = \mu_0\mu_r$, μ_0 is the magnetic permeability of the free space, and μ_r is the relative permeability. In a variable field of frequency ω, all quantities depend on the time through a factor $e^{j\omega t}$. The magnetic field intensity, therefore can be written as

$$H(t)=H^* e^{j\omega t} \tag{2}$$

Substituting Eq. (2) into Eq. (1)

$$\nabla^2 H = j\omega\sigma\mu H \tag{3}$$

or

$$=k^2 H \tag{4}$$

where $k^2=j\omega\sigma\mu$ and $j^2=-1$. The constant k is connected with the penetration depth of an electromagnetic wave. The eddy currents tend to flow near the surface of the conductor. The eddy current density in a conductor is strongest near the source of the field and exponentially decreases with increasing thickness of the conductor. This effect is known as skin effect and is dependent on many factors such as electrical conductivity, frequency of the source, and magnetic properties (Libby, 1971).

2.1.1 Magnetic field due to eddy currents in a conductor

Figure 1 shows the schematic of a non-magnetic electrically conducting sample placed in the field of an electromagnetic field with a magnetic probe above its surface. The electromagnetic field is excited by a small cylindrical coil of diameter a, with N number of turns. The diameter of the coil is smaller compared to the lateral dimensions of the sample. The thickness of the sample, t, is very small compared to the diameter of the coil. The time-varying AC signal through the coil produces a uniform magnetic field within the diameter of the coil. The normal component of the magnetic field is designated as $\mathbf{B_0}$ as shown in Fig. 1. The oscillating normal component of the magnetic field produces eddy currents in the conductor. The magnetic field generated by the eddy currents in the sample is assumed to be uniform within the diameter of the coil.

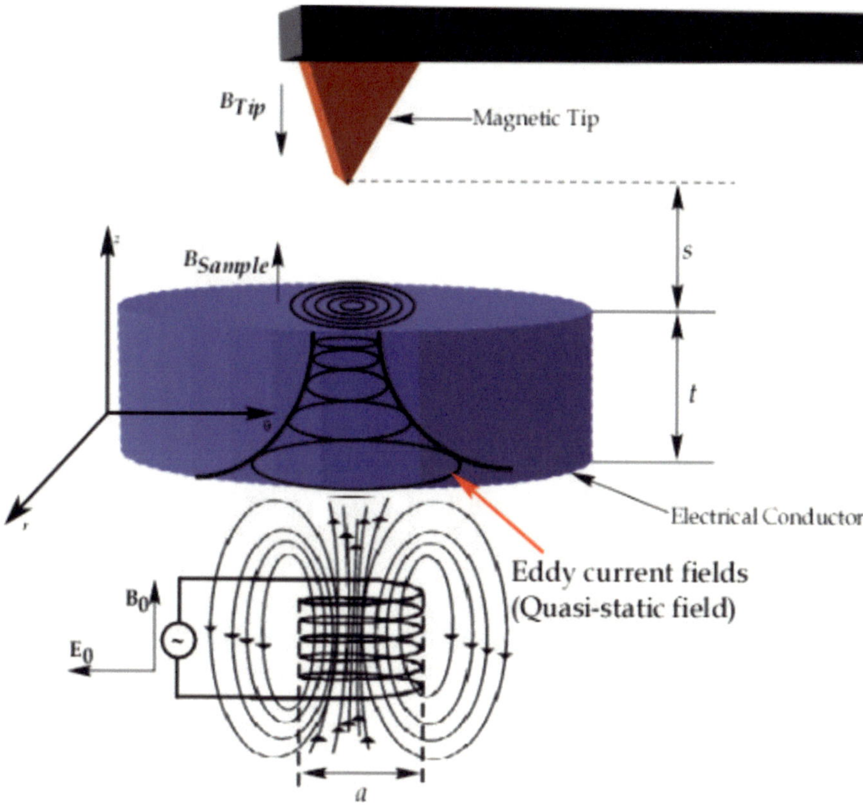

Fig. 1. Schematic of a non-magnetic electrical conductor placed in an oscillating electromagnetic field. A magnetic tip attached to a cantilever is positioned above the sample.

The relationship between eddy current density**J** and magnetic field**H** is given by

$$\nabla \times H = J \tag{5}$$

$$\nabla \times J = -(j\omega\mu_0\sigma H_z + j\omega\sigma B_0)e_z \tag{6}$$

where H_z is the normal component of the magnetic field due to eddy currents and e_z a unit vector in the z-direction. The electrical conductivity is assumed to be constant along the thickness of the sample. In cylindrical coordinates, the eddy current density **J** can be represented by a scalar potential *u(r)*(Poltz, 1983) as

$$J = \frac{1}{t}(\nabla \times u(r))e_z \tag{7}$$

Substituting Eq. (7) in Eq.(5),

$$\nabla \times \left(H - \frac{1}{t} u(r) \right) e_z = 0 \tag{8}$$

The normal component of the magnetic field due to eddy currents H_z is written as

$$H_z = \frac{1}{t} u(r) \tag{9}$$

In the experiments, we used a magnetic tip with a diameter *d*, positioned above the sample to measure the magnetic interactions generated by the eddy currents. Therefore, the magnetic interactions occur over a region equal to the diameter of the tip. Therefore, the scalar potential *u(r)* should be evaluated over the region equivalent to the tip diameter. Substituting Eq. (7) in Eq. (6), the scalar potential function evaluated in $0 \leq r < a$ is written as

$$\frac{\partial^2 u}{\partial r^2} + \frac{1}{r}\frac{\partial u}{\partial r} = j\omega\mu_0 \sigma u(r) + j\omega\sigma t B_0 \tag{10}$$

The solution (Krakowski, 1982) to the above equation is

$$u(r) = \frac{tB_0}{\mu_0}\left(\frac{I_0(kr)}{I_0(ka)} - 1 \right), 0 \leq r \leq a \tag{11}$$

where $k = \sqrt{\omega\mu_0\sigma e^{\left(j\frac{\pi}{4}\right)}}$, $I_0(kr)$ and $I_0(ka)$ are zeroth order Bessel function. The constant *k* is related to the penetration depth of the electromagnetic waves into the sample and is an important factor considered in eddy current testing. The constant *k* can be written in terms of penetration depth, δ as

$$k = \frac{1+j}{\delta} \tag{12}$$

$$\delta = \sqrt{\frac{2}{\omega\sigma\mu_0}} \tag{13}$$

Using the scalar potential function, the normal component of the secondary magnetic field H_z can be calculated using Eq. (9). This magnetic field interacts with the static magnetic field of the tip.

2.1.2 Magnetic field of the tip

A pyramidal shaped magnetic coated tip attached is used as a force sensor in our experiments. Let **M** be the magnetization of the magnetic tip. Let the magnetic field generated by a magnetization **M**of the tip (Hirsekorn et al., 1999) is given by H_{tip}. Then,

$$H_{tip}(s) = \frac{1}{4\pi} \oiiint d^3 r_i \left(\frac{3M(s - r_i}{\left| s - r_i \right|^5} (s - r_i - \frac{M}{\left| s - r_i \right|^3} \right) \tag{14}$$

where r_i is the location within the magnetic coating of volume V, d is the thickness of the magnetic coating of the tip, and s is distance between the tip and sample surface. The tip can be modeled using either monopole or dipole approximation (Hirsekorn et al., 1999). Since the dimensions of the tip are large compared to the distance between the tip and sample, a monopole approximation is used. In this case, H_{tip} can be written as (Hirsekorn et al., 1999)

$$H_{tipM} = -\frac{q}{4\pi}\frac{r_i}{s^3} \tag{15}$$

where q is the monopole moment of the tip magnetized along the z-axis and given by

$$q = \frac{MV}{l} \tag{16}$$

where l is the length of the tip.

The eddy current forces can be determined once the magnetic field strengths of both secondary magnetic field due to eddy currents and the magnetic tip are known. The eddy current force as defined in this work is the difference in the magnetic force measured by the tip before and after the introduction of the sample. When there is no sample between the coil and the tip, the interaction is between the magnetic fields of the coil ($\mathbf{B_0}$) and the tip ($\mathbf{B_{Tip}}$). When a conductor is introduced between the tip and coil, the eddy currents screen the magnetic field and decrease the force on the tip. The difference between the two forces is the eddy current force.

The eddy current force for a typical metal ($\sigma = 10^7$ $(\Omega m)^{-1}$) is calculated based on the theoretical model.The frequency of excitation is taken to be 100 kHz. The coil is taken with 100 turns of copper wire with 6 mm diameter. The magnetic field, B_0 when a current of 86 mA flows through is approximately 17 kA/m. The thickness and volume of the magnetic coating are taken as 60 nm and 4.2 X 10^{-19} m^3 respectively. The magnetization of the coating, **M** is 114 kA/m (Wadas & Hug, 1992).The theoretical eddy current forceis calculated to be 50 pN at a separation of 100 nm between probe and tip.

2.2 Scanning eddy current force microscope

Figure 2 shows a schematic diagram of the experimental setup used for electrical conductivity imaging. A Digital Instruments Dimension 3000 was modified for the purpose of electrical conductivity imaging (Nalladega et al., 2008b). The maximum scan area of the scanner in this system is 100 μm. Magnetic tips used in MFM have a spring constant of greater than 2 N/m. Based on the theoretical calculation, the spring constant of a cantilevers should be less than 0.5 N/m. A magnetic tip-cantilever with spring constant of 0.1 N/m (Veeco Probes, Model MSNC-MFM) was used as the probe The cantilever is a V-shaped cantilever made of Si_3N_4 with a resonant frequency of 25 kHz with a length of 153 μm and width of 44 μm. The tip is coated with a thin layer (thickness ~ 10-250 nm) of Co/Cr and a radius of 10 nm. The force sensitivity of the cantilever is well within the range of calculated theoretical eddy current forces.

For the purpose of generating eddy currents in the sample, an air-core electromagnetic coil is designed with a radius of 6 mm and 100 turns of 36 gauge copper wire. The sample is placed on the coil and one face of the sample faces the circular end of the coil. The opposite face of the sample faces the cantilever with the magnetic film coated tip. The coil is excited by a sinusoidal radio frequency signal from a signal generator (HP 33120A) with appropriate frequency and amplitude. The strength of the eddy currents exponentially decreases as the distance increases from the coil into the sample. The circular eddy currents in the sample produce a magnetic field that is opposing the magnetic field of the coil. The combined electromagnetic force of oscillating magnetic field and the eddy currents in the conducting sample produces oscillations of the magnetic tip-cantilever. For the purpose of measuring eddy current forces, the cantilever-tip is positioned over the sample. The oscillation amplitude of the cantilever due to eddy current forces is detected by the four-quadrant photo-detector. The eddy current force is then determined by multiplying the amplitude with the spring constant of the cantilever. The amplitude of the oscillation of the cantilever is proportional to the conductivity of the sample material.

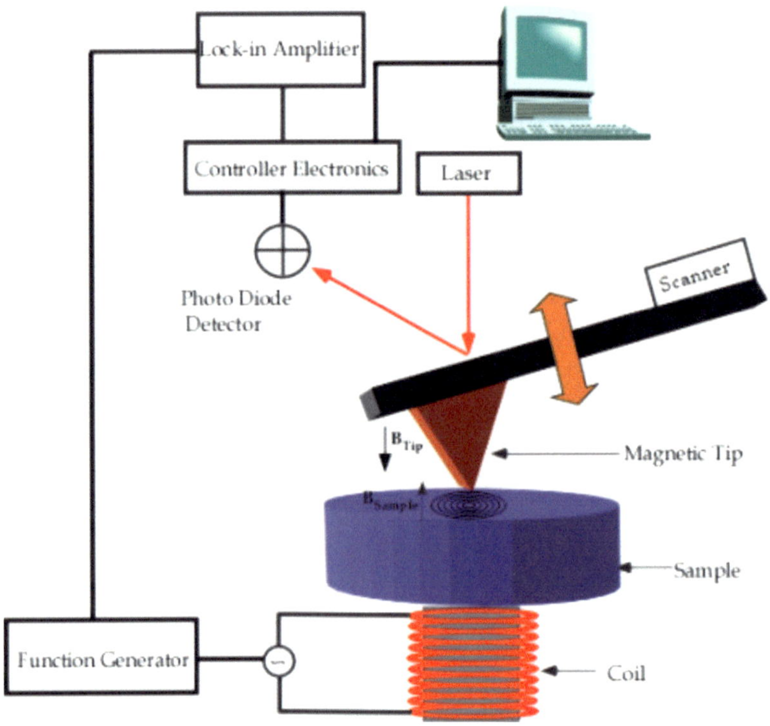

Fig. 2. A schematic diagram of the scanning eddy current force microscopy system

The electrical conductivity images were obtained using lift mode of the AFM. Lift mode allows the imaging of relatively weak but long-range interactions while minimizing the effects of topography. Measurements are taken in two passes across each scan line. In the first pass,the surface topography is obtained on one trace and retrace.The tip is then raised to the lift scan height and a second trace and retrace is obtained while maintaining constant separation between the tip and the surface topography. In the second pass, long range tip-sample interactions are measured. In the case of electrical conductivity imaging, the interactions are long-range magnetic forces between the magnetic tip and eddy currents in the sample. Therefore, electrical conductivity imaging is performed in non-contact fashion. The output of the photo-detector and the input signal to the coil are fed into a lock-in amplifier (SR 844). The lock-in amplifier measures the differential amplitude and the difference in the phase between the signal to the coil and the photo-detector signal. The difference in amplitude and the phase detected by the lock-in amplifier is proportional to the electrical conductivity of the sample under the magnetic tip. The output of the lock-in amplifier and the controller electronics of the AFM are used to generate surface topography and electrical conductivity images sample simultaneously.

3. Characterization of electrical properties

3.1 Experimental measurement of eddy current forces

Single crystal metallic samples of copper, cadmium, aluminum and polycrystalline platinum were chosen for the purpose of measuring eddy current forces. The electrical conductivity of these samples are respectively $5.961 \times 10^7\ (\Omega m)^{-1}$, $3.745 \times 10^7 (\Omega m)^{-1}$, $1.36 \times 10^7\ (\Omega m)^{-1,}$ and $0.94 \times 10^7 (\Omega m)^{-1}$.The eddy current force on each of the samples was measured in the following way. In the first step, an insulator was placed in the field of the coil excited with AC signal. The force on magnetic tip-cantilever due to coil's magnetic field was measured (F_{Ins}). In the second step, the insulator was replaced by the metallic sample and the force is measured (F_M). The difference between the two forces [F_{Ins}- F_M] is the eddy current force in the metallic sample and is directly dependent on the electrical conductivity of the metal. To determine the eddy current forces, the magnetic tip was positioned at a distance of 50 nm from the surface of the sample. The frequency of the excitation was chosen to be the resonant frequency of the cantilever while positioned over the sample.

Figure 3 compares the oscillation amplitudes of the AFM cantilever while positioned over different metallic samples. The frequency of the excitation was 82 kHz. It can be seen that the peak to peak amplitude is different for different metals. Platinum has the largest amplitude and copper has the least amplitude. The amplitudes of the cantilever over cadmium and aluminum are in between. In general, the amplitude of oscillation decreases with increasing electrical conductivity. The amplitude of oscillations, on the insulator is at least five times higher than that of the metals. Hence, it was not included in the figure for a direct comparison. The difference between the amplitude of oscillations between the insulator and the metallic samples is attributed to the generation of eddy currents in the metal. In an insulator, the magnetic field generated by the coil passes through the insulator without shielding. Hence, the entire magnetic field generated by the electromagnetic coil is sensed by the magnetic tip, producing large amplitude oscillations of the cantilever. On the other hand, in the presence of a metal, the oscillating electromagnetic field generates eddy currents in the metal shielding significant portion of the magnetic field that is sensed by the magnetic tip. The amplitude of oscillations of the cantilever on the metal is at least five times smaller than on insulator, because of the shielding effect.

The amplitude of the oscillations can be used to evaluate the eddy current force between the sample and the magnetic tip using Hooke's law. The spring constant of the cantilever is 0.1 N/m. However, since the cantilever was operated at the resonant frequency, the spring constant of the cantilever needs to be modified by the quality factor, Q of the cantilever to obtain dynamic spring constant. The dynamic spring constant was determined from the resonance curve for the cantilever using a method given in the literature (Sader, 1999). The eddy current force is calculated using the modified spring constant. In order to measure the eddy current force over a range of separation distance, the oscillation amplitude of the cantilever was measured at several fixed distances up to 550 nm. The effect of the separation distance on the eddy current force is shown in Fig. 4.

The eddy current force decreases exponentially over distance and the force is large when the separation distance is less than 100 nm. Above 100 nm separation, the force decreases rapidly and levels off after about 400 nm. In metals with higher conductivity, the exponential decrease of the force is much more pronounced while in lower conductivity metals, the decrease in the force as a function of distance is gradual. The solid lines in Fig.4 indicate the exponential fit to the data. The behavior seen in Fig.4 is similar to the force curves studied in MFM (Murphy & Spalding, 1999). The similarity is due to the fact that both MFM and eddy current force microscopy are functionally similar. The eddy current force distance curve is expected to follow inverse square law over the entire range of distance. Even in the force curves of MFM the inverse square behavior is not seen for all separation distances (Murphy & Spalding, 1999). The reason for this behavior is the contribution of other forces in the distance ranges. The same argument holds true in the case of eddy current forces also. The inverse square law is observed up to a separation distance of 300 nm. Beyond 500 nm, the eddy current forces are weak and the amplitude is in thermal noise range.

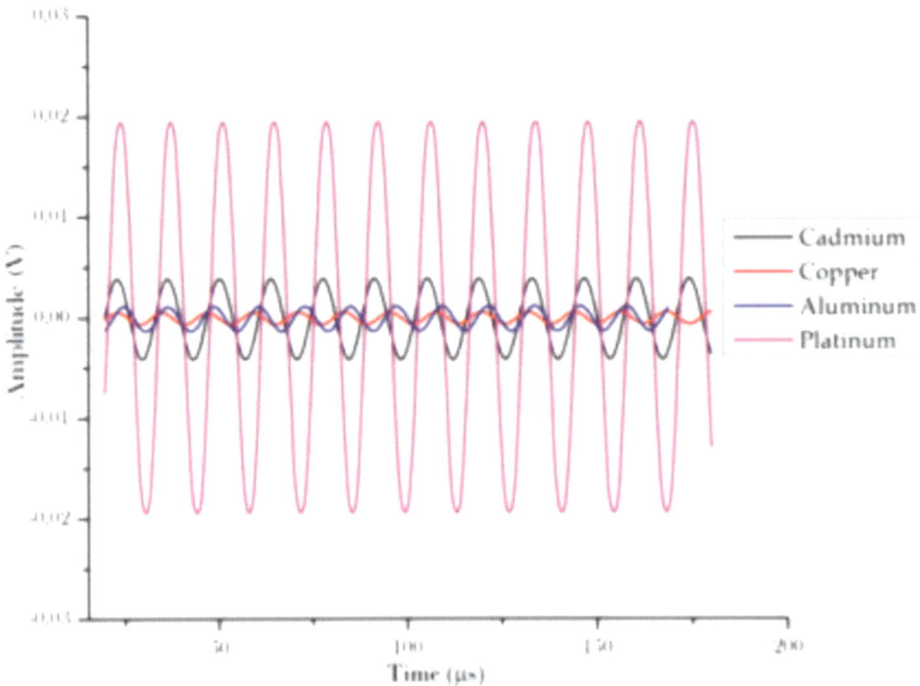

Fig. 3. Comparison of oscillation amplitudes of the AFM cantilever on different metallic samples at a separation distance of 50 nm and an excitation frequency of 82 kHz. The waveforms have been slightly shifted in time to show the waveforms separately.

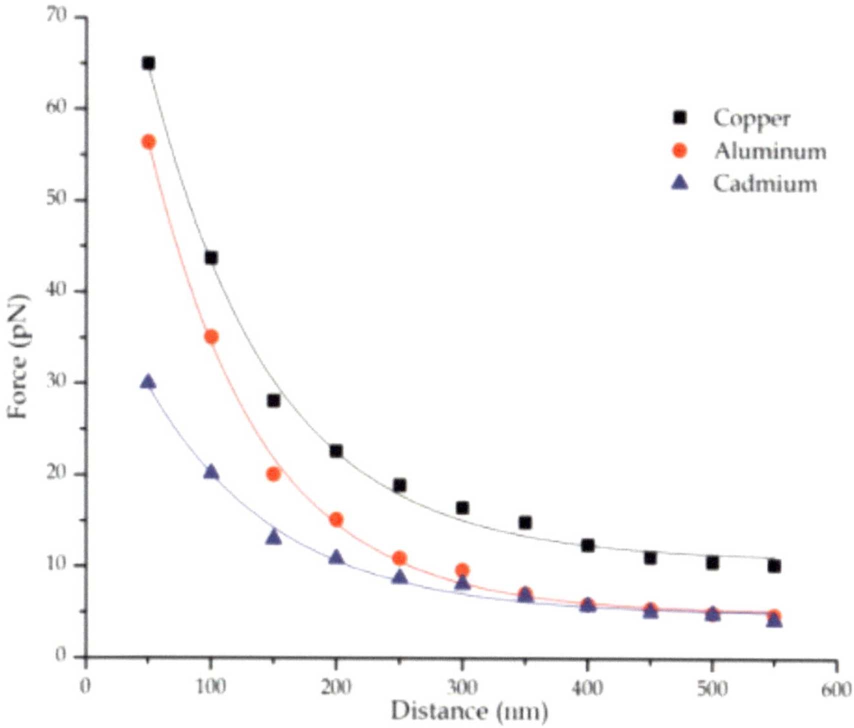

Fig. 4. Effect of separation distance between magnetic tip and sample surface on the eddy current force in different metals.

3.2 Imaging electrical conductivity variations in bulk conductors

It was shown that the force due to the eddy currents in a metal changes as a function of electrical conductivity. Therefore, by mapping the variations in the eddy current forces as the tip scans over the sample surface, one should be able to obtain an image of electrical conductivity. As the conductivity changes, the magnitude of the eddy current force changes and therefore, the image is a map of electrical conductivity variations. Before obtaining images, it is important to know the resonance spectra of the cantilever in order to achieve maximum sensitivity.A network analyzer (HP 8753D) was used to obtain the resonance characteristics of the cantilever coupled with the sample(Nalladega et al., 2008b). The resonance peaks of the cantilever while positioned over copper are shown in Fig. 5. The cantilever has resonance peaks at several frequencies, the dominant one being at 86 kHz with other peaks at 280 kHz, 508 kHz, and 580 kHz. While the images can be obtained at any of these frequencies, images obtained around 86 kHz will have maximum contrast in the images due to the maximum amplitude at this frequency. Similar experiment was also done for titanium and the resonance peaks in this case were observed at 92 kHz, 275 kHz, 510 kHz and 600 kHz. The resonance spectra of other metals (aluminum, cadmium) showed the peaks at similar frequencies. The differences in the resonant frequency can be attributed to many factors including thickness, conductivity, eddy current forces, penetration depth etc. (Siddoju et al., 2006). Therefore, the resonant characteristics of the cantilever should be characterized before obtaining an image.

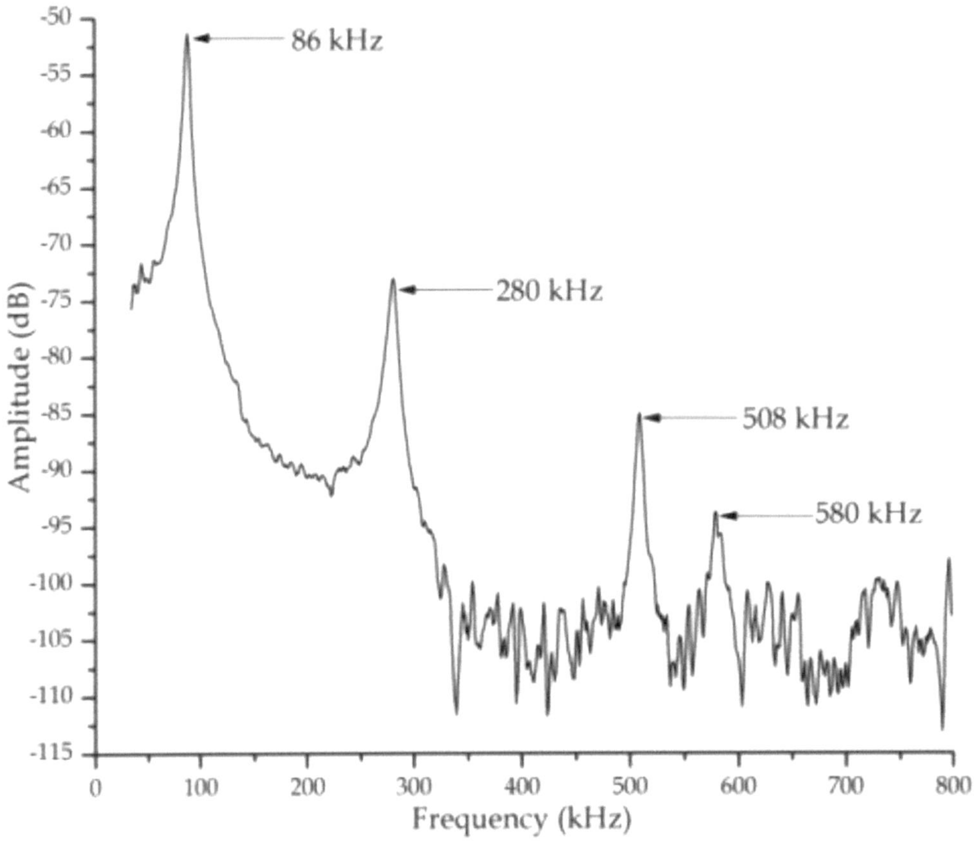

Fig. 5. Resonance spectra of the AFM cantilever positioned on copper showing multiple resonance peaks.

3.2.1 Carbon fiber reinforced composite

The eddy current imaging technique was first applied to image conductivity variations in a material with huge electrical conductivity variations. For this purpose, a carbon fiber composite with an average fiber diameter of 7 μm was chosen (Nalladega et al., 2007). The electrical resistivity of carbon and the polymer matrix is 0.006 Ω-cm and 1×10^{15}Ω-cm respectively. Figure 6 shows topography and electrical conductivity images of carbon fibers at an excitation frequency of 272 kHz and a lift height of 50 nm. The image on the left shows AFM topography and the image on the right is the eddy current image showing the electrical conductivity variations in the composite.

The contrast in the AFM image is due to variation of surface height and brighter regions indicate higher surface heights. Therefore, the carbon fibers appear bright in the image compared with the polymer matrix (Fig. 6a). Fig. 6b shows the eddy current image of the same region. In the eddy current image the fibers appear dark while the polymer matrix appears bright. The difference in the contrast is due to the differences in the electrical conductivity of fiber and polymer. The matrix is almost an insulator and the magnetic field generated by the coil passes through without shielding and hence less damping of the cantilever, producing

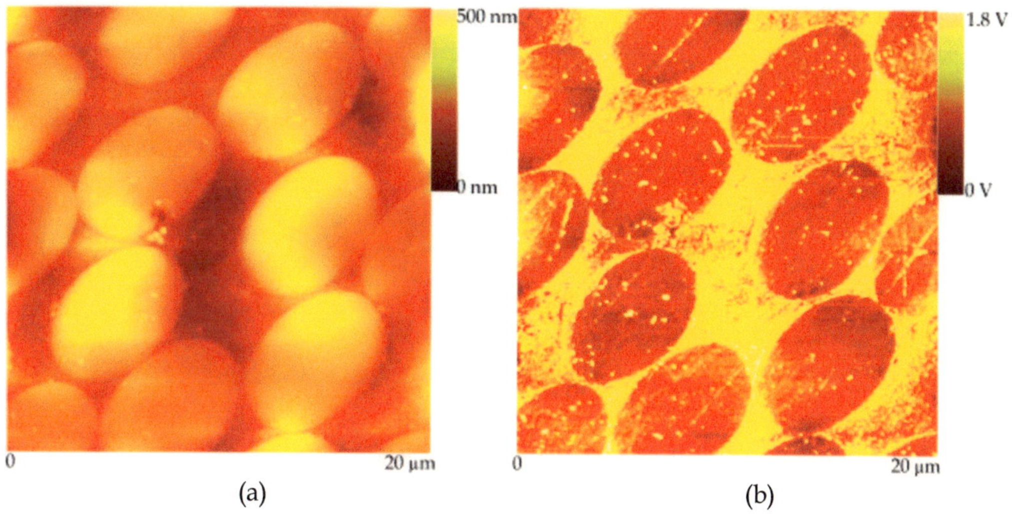

Fig. 6. (a) Surface topography and (b) eddy current images of carbon fibers reinforced in polymer matrix.

large amplitude of vibration of the tip-cantilever. Large amplitude produces significant output voltages from the photo-detector. When the tip is located on the carbon fiber, the magnetic field of eddy currents is shielded by the conductive fibers. Thus, vibration of the cantilever is dampened due to this shielding. Therefore,the magnetic tip-cantilever measures reduced amplitude compared with polymer matrix. Reduced amplitude of the cantilever produces lower output voltage from the photo-detector. Thus the carbon fiber appears darker than the polymer matrix in the eddy current image. Consequently, in eddy current images, darker contrast indicates higher conductivity regions and brighter contrast indicates less conductivity regions. The scale bar in the eddy current image represents the output voltage from the photodiode detector, which is proportional to the electrical conductivity.

A higher magnification conductivity image of the composite is shown in Fig.7. Along with the image a section analysis along the line shown in the image (Fig. 7b), is also presented. The profile above the center line represents the matrix and below the center line represents the fiber. The section analysis shows the variation of the amplitudes of the vibration of the cantilever as it scans the surface. The variation in amplitude is due to the variation in the magnetic field as the tip scans the surface. The strength of the magnetic field generated due to eddy currents is directly proportional to the electrical conductivity. Hence, the section analysis of the eddy current image shows the variations in relative conductivity. The section analysis shows that, at some regions there is a sharp transition in conductivity profiles at the boundary of matrix and fiber, while at other regions there is a gradual transition at the boundary, as shown by arrows in the image. In composite materials, it is very important for the proper bonding between the fiber and the matrix at the interface for maximum strength. Therefore, characterization of interface is important to estimate the strength of the composite. The image at the interface shows that high resolution imaging is possible using eddy current AFM. Thus, this technique is well suited for the characterization of the interface in composite materials based on the variations in electrical conductivity.

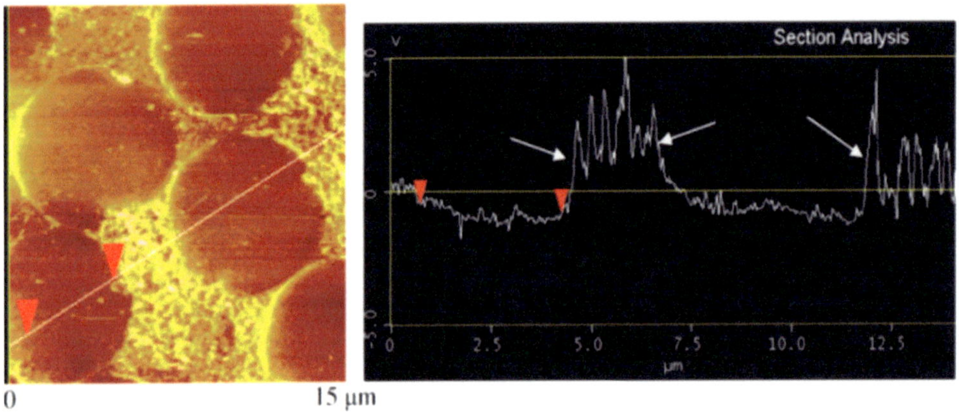

Fig. 7. A high resolution eddy current image of the carbon fibers and the section analysis along the line shown in the figure.

3.2.2 Dual phase titanium alloy

In electrically anisotropic metals, the electrical conductivity varies from grain to grain due to crystallographic orientation. An excellent example for the electrically anisotropic metallic alloy is a dual-phase titanium alloy, Ti-6Al-4V.Ti-6Al-4V alloys are used in high performance aerospace applications that require high toughness, good fatigue strength, and good corrosion resistance. Titanium exists in two crystal structures: hexagonal close packed (α-phase) and body centered cubic (β-phase). The two phases often exist together in α-β alloys. The hexagonal phase has anisotropic electrical conductivity. In pure titanium the electrical resistivity in the basal plane is 45.35 μΩ cm and 48 μΩ cm in a plane normal to basal plane (Meaden, 1965). Thus grains with different crystallographic orientation will have different electrical conductivity. The electrical conductivity of the two phases is expected to be close. In order to test the feasibility of imaging small changes in conductivity using the eddy current AFM, the technique was applied to a dual phase Ti-6Al-4V sample (Nalladega, et al., 2008b). The microstructure of the alloy consists of circular primary α grains with a grain size of 10-20 μm and fine lamellar $\alpha+\beta$ platelets.

Figure 8 shows surface topography and eddy current images on the Ti-6Al-4V sample. The image was obtained at an excitation frequency of 92 kHz at a lift scan height of 50 nm, between the tip and sample. From Fig. 4, it is evident that for higher sensitivity the separation distance should be less than 100 nm. Therefore, the eddy current images shown in this chapter are obtained at a lift height of less than 100 nm. The surface topography image shows α grains, the ($\alpha+\beta$) grains and α platelets inside ($\alpha+\beta$) grains with a maximum vertical height of 800 nm (Fig. 8a). The β between the α platelets cannot be observed clearly in the topography image. The eddy current force image (Fig. 8b) shows quite good contrast. The large α grains, the ($\alpha+\beta$) grains and the α platelets inside ($\alpha+\beta$) grains can be observed with significant contrast. Some of the large α grains that appear as a single grain in the AFM image appear to be consisting of smaller grains and α plates when viewed in the eddy current image. The difference in contrast in the eddy current imageis due to the difference in the electrical conductivities of α and β phases. Since the electrical conductivities of the two phases are different, the vibration amplitudes of the cantilever and hence the eddy current forces will be different in the two

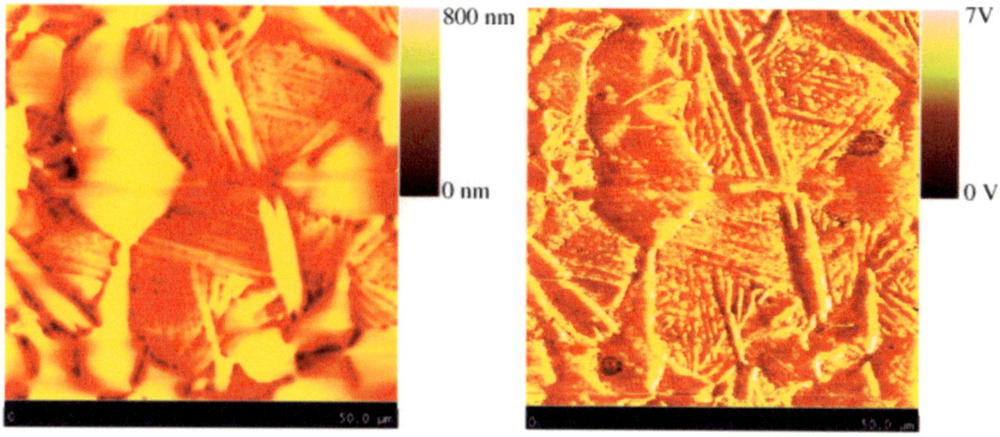

(a) (b)

Fig. 8. (a) Surface topography and (b) eddy current AFM images of dual phase titanium alloy, Ti-6Al-4V (Nalladega et al., 2008b). (Reprinted with permission from *American Institute of Physics*)

phases. Consequently, the contrast shows bright to dark variations depending on the changes in conductivity.The lamellar structure can be clearly resolved in the eddy current image. Contrast among large α grains can also be observed in the eddy current image. Thus, the conductivity image shows significantly more structure than the topography image. The eddy current image on the titanium shows that the technique is sensitive to the small changes in the electrical conductivity of the material. The conductivity of the β phase is different compared to α phase. The difference in the electrical conductivity between the two phases and the anisotropy in conductivity of the α phase enhances the contrast in eddy current images. These factors help in the observation of multiple grains, platelets in large α grains that appear to be single grain in AFM topography images. Although the surface roughness among the different phases and platelets are very small to show significant contrast in AFM images, the electrical conductivity is significantly different among different phases of the material and the anisotropy of electrical conductivity enhances the contrast in eddy current images.

A high magnification image of the platelets in the titanium alloy is presented in Fig. 9. A section analysis along the line shown in the images showing the variations in topography and conductivity respectively is also shown. In some regions it has needle like grains of alpha phase within a large grain. The surface topography image shows the needle like grains oriented in different directions (Fig.9a). The contrast is fuzzy and identifying individual platelets is difficult. On the other hand the eddy current image (Fig. 9b) shows the platelets with enhanced contrast at the boundary. The contrast at the grain boundary is quite strong because of significant change in the conductivity and due to modification of the eddy currents near the boundary. The conductivity is modified near the grain boundaries due to extra scattering of the electrons. The widths of the dark and bright phases shown in the line scan of the eddy current image are 560 nm and 520 nm respectively.

The eddy currents diffuse from the bottom of the sample through the thickness of the sample. Hence, the eddy current force is an average over the thickness of the sample. If the electrical conductivity is inhomogeneous in the thickness direction, the eddy current force is an average

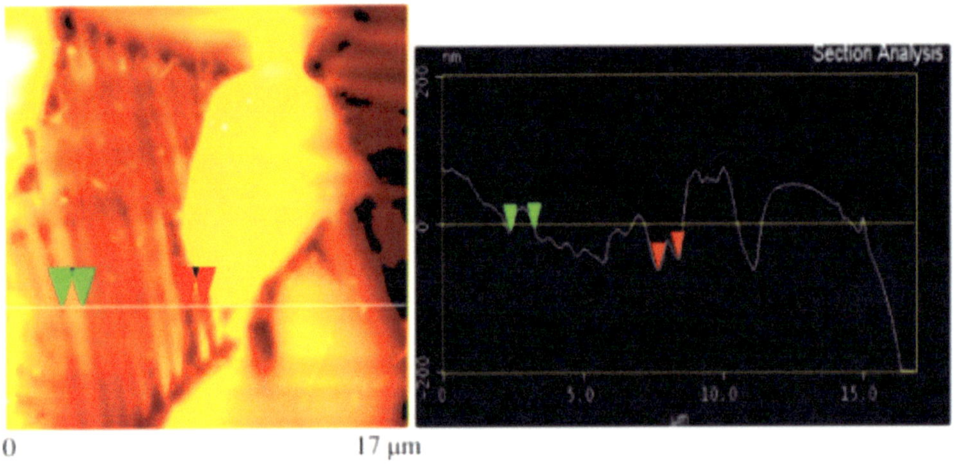

(a)

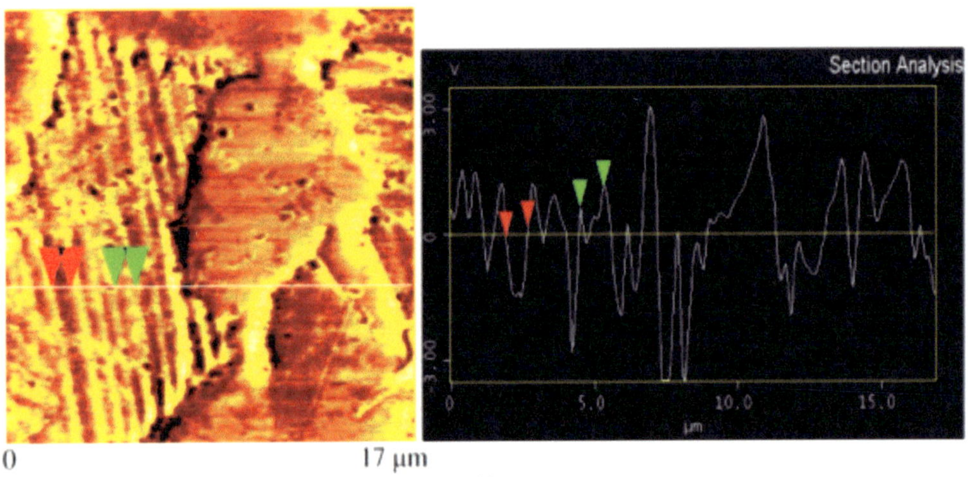

(b)

Fig. 9. (a) Surface topography and the corresponding line scan across the line shown in the topography image.(b) Eddy current image and the corresponding line scan across the line shown in the eddy current image of dual-phase titanium alloy.

over the sample thickness. While this appears to be a limitation of the technique, it may be useful in imaging subsurface defects or cracks that cause inhomogeneities in electrical conductivity. In fact this has been effectively used in eddy current NDE(Huang et al., 2006).

4. Spatial resolution of the eddy current AFM

A carbon nanofiber reinforced composite was used to determine the spatial resolution of the eddy current AFM (Nalladega et al., 2008b). The diameters of the fibers are in the range of 20–100 nm and the length is known to be in the range of fraction of microns to tens of microns. Figure 10 shows the topography and eddy current images of the carbon nanofibers, together with section analysis along a single fiber, as shown in the figure. The images were

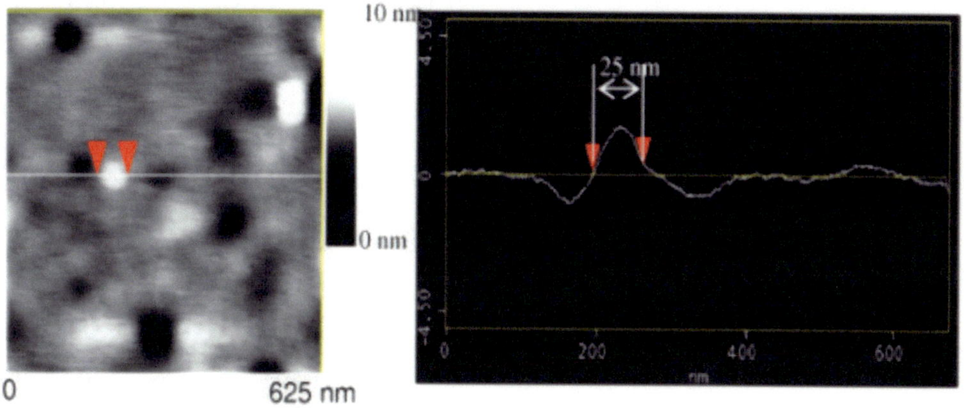

(a)

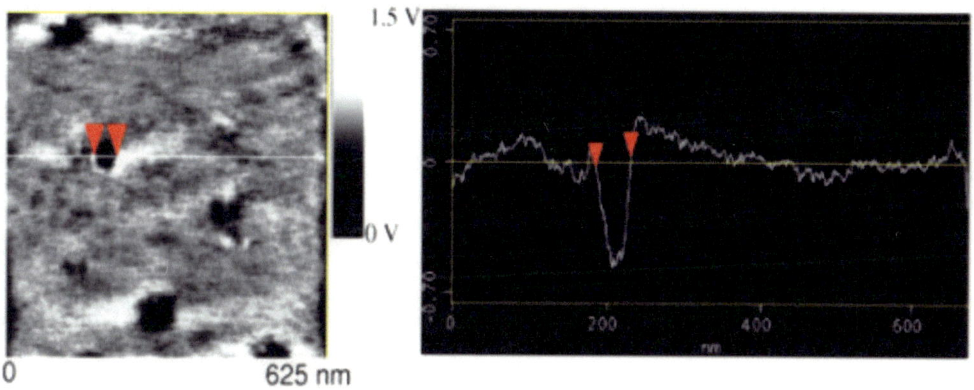

(b)

Fig. 10. (a) Surface topography and (b) eddy current images of carbon nanofibers (Nalladega et al., 2008b). (Reprinted with permission from *American Institute of Physics*).

obtained at an excitation frequency of 85 KHz and a lift height of 50 nm. A 25 nm diameter nanofiber can be resolved in eddy current images. In the experiment, a magnetic tip with a nominal diameter of 20 nm was used for imaging. The magnetic field sensed by the tip is larger than the diameter of the tip. The spatial resolution in AFM depends on the diameter of the probe and interaction forces between the tip and the sample (Banerjee et al., 1996; Hutter & Bechhoefer, 1993a). Long range tip-sample interaction forces degrade the resolution of the AFM (Goodman & Garcia, 1991; Hutter & Bechhoefer, 1993b). Since the magnetic forces due to the eddy currents are long range interaction forces and the magnetic tip interacts with a volume of magnetic field larger than the tip diameter, the resolution is slightly larger than the tip diameter.

5. Low-frequency electromagnetic properties of nanostructured materials

The reduction of the size of metals results in significant changes in the electromagnetic properties compared with the bulk metals. In particular, the electromagnetic response of metallic nanoparticles is the subject of many research efforts during the last decade (Kelly et al., 2003; Michaels et al., 1999; Tominaga et al., 2001). The metallic nanoparticles show plasmon resonance when subjected to electromagnetic fields. Plasmon resonance occurs due to coherent oscillation of the conduction band electrons induced by the incident EM field at opticalfrequencies. The surface plasmon resonance in metallic nanoparticles is usually characterized using near-field optical techniques such as scanning near-field optical microscopy (SNOM) (Courjon, 2003; Okamoto, 2006). The plasmon resonances in the metallic nanoparticles occur in the optical frequency range and they result in powerful localized sources of electric field. In these techniques, the *electric field* around the nanoparticle is imaged.

While the electromagnetic properties of nanoparticles have been extensively studied at optical frequencies, little information is available on the *low frequency* (less than 1 MHz) interaction of electromagnetic waves with metallic nanoparticles. It is known that in the presence of an external magnetic field, low frequency electromagnetic waves can propagate as helicons in metals (Maxfield, 1969; Petrashov, 1984). Helicons are circularly polarized electromagnetic waves that propagate with very low phase velocity. When the mean free path of electrons is sufficiently large and the frequency ω of the wave is low enough, the electrons affected by the Lorentz force would drift in the direction perpendicular to the plane formed by the uniform magnetic field and the electric field of the wave. Helicon resonance modes in metals can be observed whenever the helicon frequency approaches the cyclotron frequency, i.e. $\omega \rightarrow \omega_c$. The cyclotron frequency ω_c, is given by

$$\omega_c = \frac{eB_o}{m^* c} \tag{17}$$

where e is the electron charge, $\mathbf{B_0}$ is the static magnetic field, m^* is the effective mass of the electron and c is the speed of light. In addition, the local conditions, viz., $\omega\tau << 1$, and $kl<<1$ have to be satisfied. Here, ω is the excitation frequency, τ is the relaxation time of the electrons, $k = 2\pi/\lambda$ is the helicon wave vector and l is the electron mean free path. The local conditions are fulfilled for the frequencies $\omega < 10^8$ s^{-1}. The phase velocity of helicons is smaller than the speed of the light. At typical metallic densities, the plasma frequency, $\omega_p = 10^{16}$ sec^{-1}; for a frequency $\omega = 10^7$ sec^{-1} and $\omega_c = 10^{11}$ sec^{-1}, the phase velocity of a helicon is approximately 30 m/s. The magnetic field of helicon wave is much larger than the magnetic field of an ordinary electromagnetic wave with the same electric field.

Helicon wave propagation and its resonance have been observed in high purity metals only at very low temperatures (Bowers, 1961; Chambers, 1962; Houck, 1964; Taylor, 1963). At very low temperatures, the electron mean free path of most metals is in mm range. To observe helicon resonances, an electromagnetic coil is excited with frequency in MHz range. The sample and the coil are placed in a static magnetic field of 1-100 kG and cooled to liquid helium temperature. The impedance of the coil is measured as the magnetic field is varied. Helicon resonance modes in metals can be observed when the excitation frequency approaches the cyclotron frequency. In a typical metal placed in an external static magnetic field of 1 kG, the cyclotron frequency is about 10 MHz.

The electron mean free path of gold at room temperature is about 50 nm. However, the cyclotron resonance frequency at a magnetic field of 1kG is approximately 10^6-10^7 Hz. To observe helicon resonances at room temperature the electromagnetic frequencies have to be in the range of 10-100 MHz. Eq. (17) shows that helicon wave propagation at low frequencies in metallic nanoparticles can occur if either the effective mass is large or the static magnetic field is extremely high.

It was shown that the effective mass of electrons can be dramatically increased in artificial metallic lattice structures made of thin metallic wires (Pendry et al, 1996). When the metallic structure was subjected to an electromagnetic field, it has been shown that the effective mass of the electron increased by four orders of magnitude. By confining electrons to thin wires, an enhancement of their mass was achieved by 4 orders of magnitude. Consequently, the plasma frequency was reduced to GHz range from UV or optical frequency ranges. Following similar arguments, for a periodic array of metallic nanoparticles of gold or silver with diameter of 200 nm, and 1 μm spacing, it can be shown that the effective mass of the electron increases by two orders of magnitude and the effective electron density decreases by three orders of magnitude. The cyclotron resonance frequency was found to be few hundred kHz in a magnetic field of 1 kG, and the phase velocity of helicons was less than 100 mm/s. Thus it is possible to generate helicon waves and resonances in metallic structures at very low frequencies at room temperatures whenever these conditions are satisfied.

The eddy current AFM system was used to detect and image helicon resonances in a distribution of metallic nanoparticles (Nalladega et al., 2011). Nanoparticles of platinum were deposited on a glass substrate using a process known as Through Thin Film Ablation (TTFA)(Murray & Shin, 2008). The target consisted of 20 nm thin film of Pt that was sputter deposited onto a UV transparent fused silica. The target was then ablated through the silica support at an energy density of 0.5-1.0 J/cm^2. The nanoparticles are synthesized without agglomeration and have a uniform size distribution. Figure 11 shows a SEM micrograph of platinum nanoparticles synthesized using TTFA.

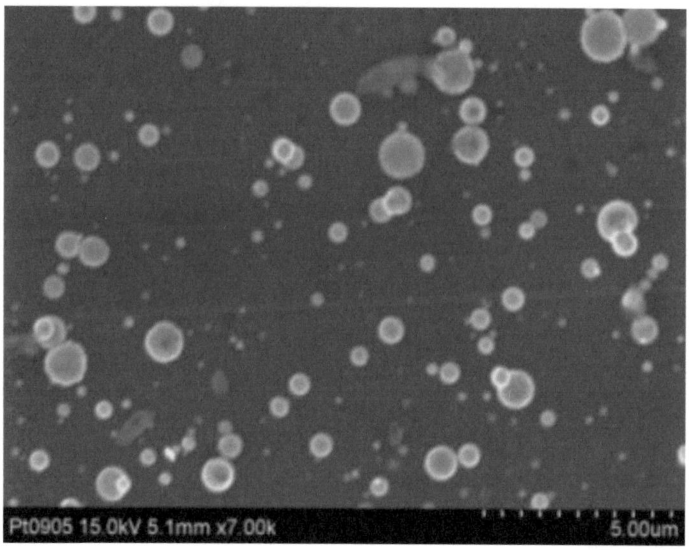

Fig. 11. Distribution of the platinum nanoparticles synthesized using through thin film ablation process.

Figure 12 shows surface topography and the eddy current images of individual platinum nanoparticles obtained at 90 kHz. Fig. 12a shows the surface topography and eddy current

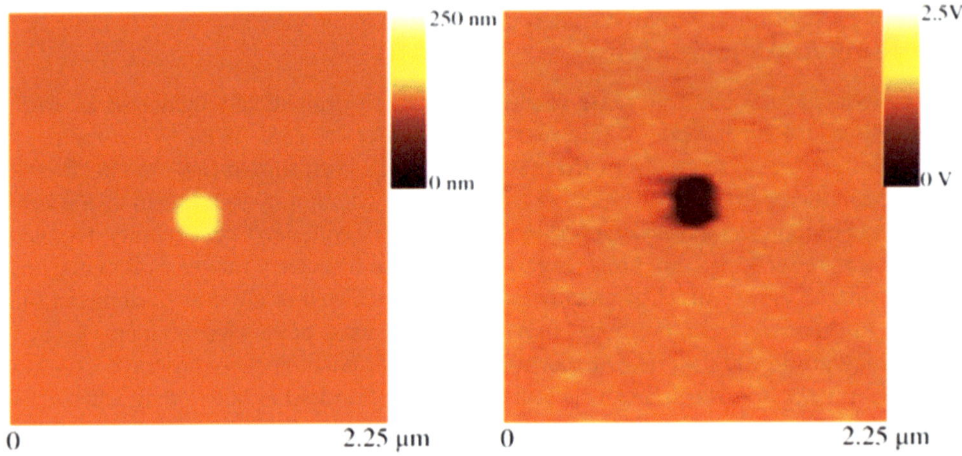

(a)

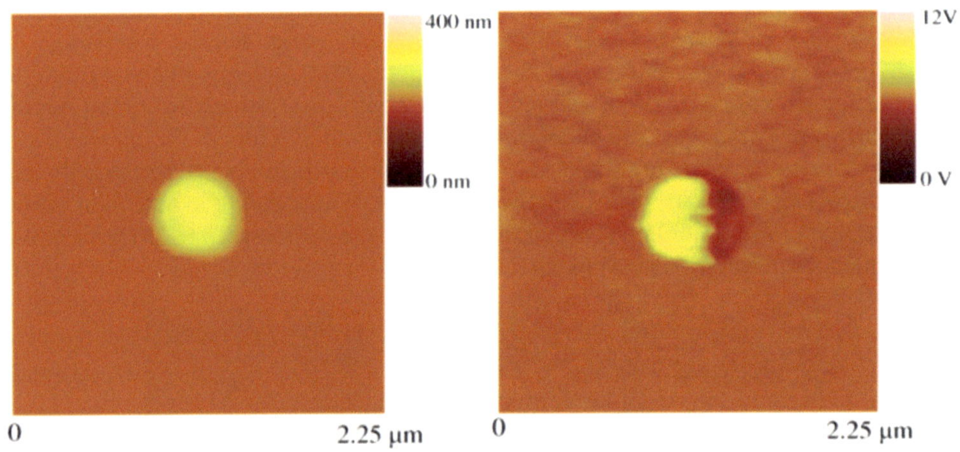

(b)

Fig. 12. Surface topographyand eddy current images of Pt nanoparticles (Nalladega et al., 2011). (Reprinted with permission from *American Institute of Physics*)

images of a 250 nm diameter nanoparticle. The eddy current image shows a darker contrast varying uniformly across the nanoparticle. Figure 12b shows another nanoparticle with 500 nm diameter at the same frequency. The surface topography image is similar to the 250 nm particle but the contrast in the eddy current image is very different. The image shows a particle with half bright and half dark regions. The contrast in eddy current AFM is based on the electrical conductivity. Thus, it appears there is a variation in electrical conductivity within the particle. However, it is extremely unlikely that a single particle can have two different conductivities especially if the particle is made of single material. Figure 13 shows the topography and eddy current images of platinum nanoparticles at a different region of the sample at 90 kHz. The images show three platinum nanoparticles of different size. Again, the eddy current image distinctly shows different contrast among the nanoparticles. Two of the nanoparticles are seen with a contrast similar to that of Fig.12b. A nanoparticle with 800 nm diameter and a dark region between two bright regions can also be seen in the image.

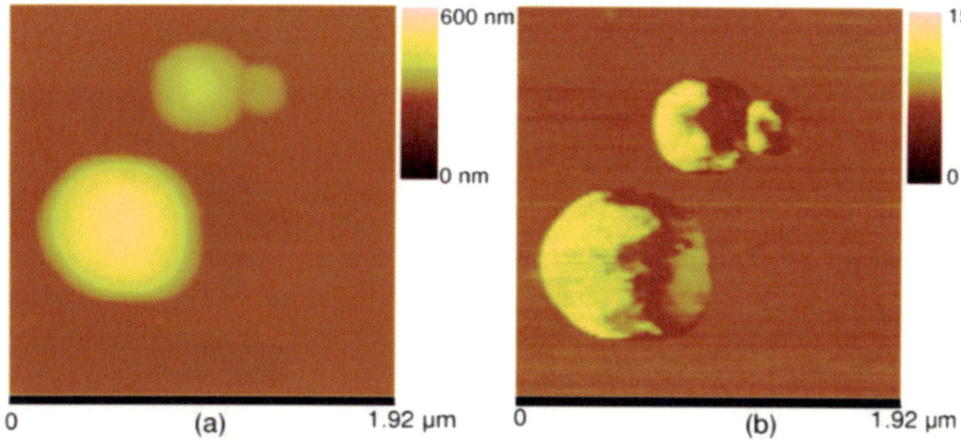

Fig. 13. (a) Surface topography and (b) eddy current images of platinum nanoparticles. The 800 nm Pt particle is split into three parts showing bright-dark-bright contrast within the particle (Nalladega et al., 2011). (Reprinted with permission from *American Institute of Physics*)

The eddy current images shown in Fig.12 and 13 are similar to the surface plasmon resonance images on spherical metallic nanoparticles reported in the literature (Kelly et al., 2003; Michaels et al., 1999; Tominaga et al., 2001;Courjon, 2003; Okamoto, 2006). Based on the similarity of the plasmon resonance images of nanoparticles with the images shown in Fig.12 and 13, it appears the images show the resonance behavior of the nanoparticles at low frequencies. While the optical techniques image the *electric field* around the nanoparticle, the eddy current AFM images the *magnetic field* around the nanoparticle. Therefore, we believe the source of the contrast seen in the above images is due to the resonances of low-frequency electromagnetic waves like helicons. This is further reinforced by the decrease in plasmon frequencies of artificial nanostructures (Pendry et al, 1996). Since a sharp magnetic probe is used instead of a pick-up coil to detect the helicons, it is now possible to *image* helicon resonance modes with nanometer resolution.

The magnetic field images of the nanoparticles can be compared with the magnetic lines of force of spherical particles subjected to electromagnetic radiation. The magnetic lines of force of spherical particles for different resonance modes are schematically shown in Fig.14. Higher order resonances have been observed in bigger particles in the sample showing regions of bright and dark band contrast. The shapes of the resonances have been observed to change with the diameter as well as the shape of the particle. The contrast can be explained based on accommodating tiny magnetic inside the particle with alternating the magnetic poles such that no two adjacent magnetic poles are alike.

Although the helicon resonance calculations assume periodic arrangement of square array, it was shown that for significant decrease of plasmon resonance frequency the actual lattice geometry may not be significant (Pendry et al., 1996). The experimental configuration developed in this work allows visualizing the helicon resonances in individual members of the structure with varying dimensions of the metallic nanoparticles and interparticle spacing.

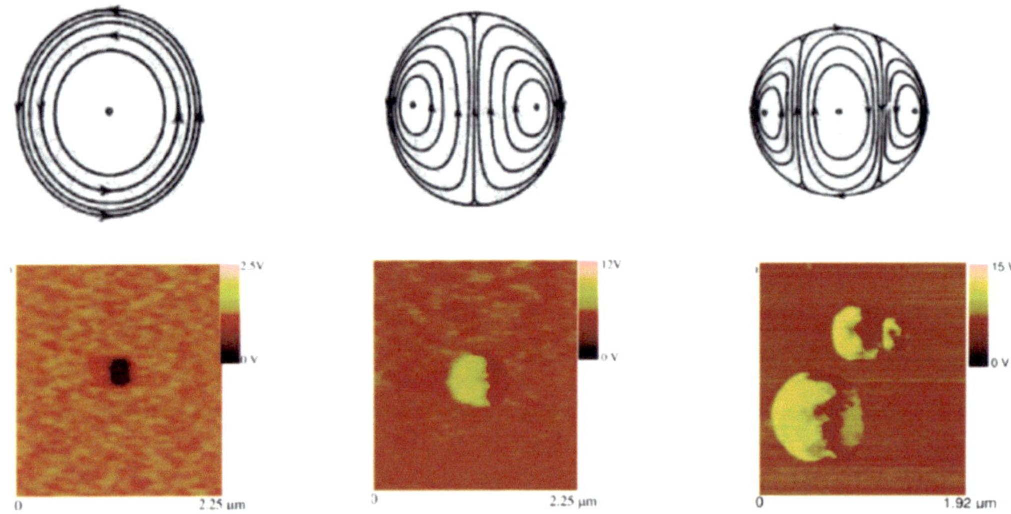

Fig. 14. Magnetic lines of force for the first, second, and third resonance modes in spherical particles and the correspondingmagnetic field images of Pt nanoparticles (Nalladega et al., 2011). (Reprinted with permission from *American Institute of Physics*)

6. Characterization of magnetic properties

In MFM, a ferromagnetic probe is attached the cantilever and scanned across a ferromagnetic sample to measure the magnetic interactions between the tip and the sample. For soft magnetic samples, the magnetization of the sample will be affected by the stray magnetic field of the tip. Therefore, the interpretation of the magnetic contrast in MFM images for such samples is difficult.This has led to the characterization of magnetic properties without a magnetic probe (Hoffmann et al., 1998). The technique uses a vibrating non-magnetic conducting tip to image the magnetic domains of a ferromagnetic sample. The oscillating conducting tip induces currents within the tip which is dependent on the magnetic properties of the sample.Experimental techniques to image magnetostrictive properties of magnetic materials using AFM have also been reported in the literature (Berger et al., 2000; Wittborn et al., 2000; Polushkin et al., 2003).

When a magnetic field is applied to a magnetic material, strains are generated in the material. This phenomenon is known as magnetostriction. When an oscillating magnetic field is applied to a ferromagnetic material, magnetostrictive strains are induced based on the frequency of the excitation. The magnetostrictive strain does not change the sign when the direction of the field is reversed. Therefore, the material produces strains at twice the excitation frequency. The strain due to magnetostriction is dependent on the magnetization of the domains and provides information about the magnetoelastic properties. Therefore, magnetostriction can be used to study the magnetic properties.The eddy current AFM system has been modified to image magnetic properties with a *nonmagnetic* probe (Nalladega et al, 2009). The approach is based on the measurement of strains due to magnetostriction in magnetic materials. Since the strains produced are a function of magnetization of individual domains, this technique can be used to image magnetic domains without the need of a magnetic probe. The technique is used to image differences in magnetostriction in amorphous and nanocrystalline magnetic melt spun ribbons. The contrast is explained based on the magnetostrictive interactions between a *nonmagnetic*probe and ferromagnetic material.

Amorphous and nanocrystalline magnetic ribbons of the nominal composition FeSiBNbCu alloy were used for this purpose. These materials are also known as FINEMET alloys (Yoshizawa et al., 1988; McHenry et al., 2003; Herzer, 1993). The nanocrystalline magnetic ribbon of FINEMET alloy is composed of nanosized crystallites embedded in a residual amorphous matrix. The nanocrystalline phase exhibit exceptionally soft magnetic properties which arise from a very small magnetostriction (Slawska-Waniewska & Lachowicz, 2003). An AC magnetic field with a frequency f=58 kHz was applied to the sample. Since magnetostriction is proportional to $2f$, the excitation frequency was chosen such that the $2f$ frequency is close to the resonant frequency of the cantilever in contact with the sample. A contact mode nonmagnetic AFM tip was scanned across the sample in contact mode. The magnetostrictive strains cause the cantilever to oscillate at $2f$ frequency. The amplitude of vibration of the cantilever at $2f$ frequency is detected by the lock-in amplifier to obtain topography and magnetostriction images simultaneously.

Figure 15 shows the topography and magnetostriction images on the amorphous sample. The surface topography image (Fig.15a) shows a uniform contrast with no visible features. However the magnetostrictive image (Fig. 15b) shows circular bright and dark regions within the scan area. The bright regions correspond to larger amplitudes of cantilever vibration and consequently higher strains. The dark regions indicate that the amplitudes of the cantilever vibration are small and thus correspond to relatively lower strains due to magnetostriction. Therefore, the image shows that there is a variation in the local magnetostriction. The amorphous magnetic ribbons exhibit higher saturation magnetostriction. This is confirmed by the brighter regions in the magnetostriction image. Since magnetostrictive strain is a function of the magnetization of each domain, the magnetostriction image can be used to identify magnetic domains. The bright and dark regions represent magnetic domains (Takata & Tomiyama, 2000), as shown by arrows in Fig. 15b.

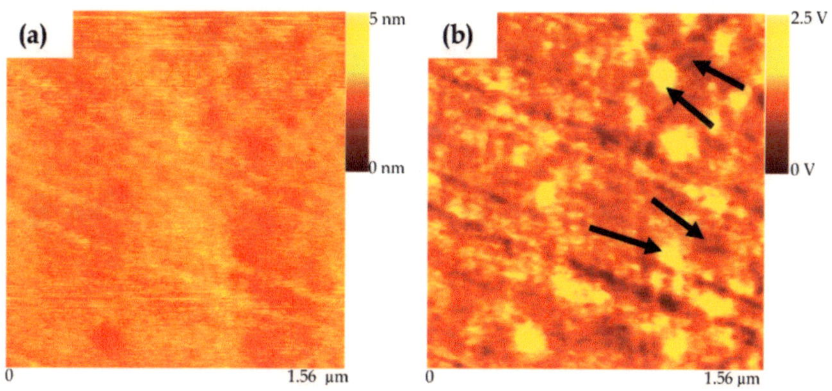

Fig. 15. (a) Surface topography and (b) magnetostriction image of amorphous magnetic ribbon sample. Some of the domains are indicated by arrows.

Figure 16 shows the surface topography and the magnetostriction images of the nanocrystalline sample. The images show the formation of crystals due to the annealing of amorphous sample with the size of the nanocrystals ranging from 50 nm-100 nm. The topography image (Fig. 16a) shows the nanocrystals embedded in amorphous matrix with a maximum vertical height of 200 nm. In the magnetostriction image (Fig.16b), most of the nanocrystalline phase appears with dark indicating the magnetostriction is smaller compared to that of amorphous phase. However, the presence of dark and bright regions locally can be observed. This indicates the presence of a small effective saturated magnetostriction. Figure 17 shows a higher magnification image of the nanocrystalline phase of the sample. Brighter regions within a domain can be observed in the magnetostriction image (Fig.17b). This indicates the saturation of magnetostriction in the domain. This results in a deformation, dependent on the magnetization direction of the domain (Takata & Tomiyama, 2000). At the domain wall the magnetic moments changes direction, resulting in the deformation of material. The magnetization of the sample is in-plane and the external magnetic field is perpendicular to the plane. Therefore each of the domains is subjected to rotational forces and consequently the amplitude of AFM cantilever is large near the domain walls. Based on this interpretation, the brighter region shown by the arrow (Fig. 17b) is a domain wall with a width of 40 nm.

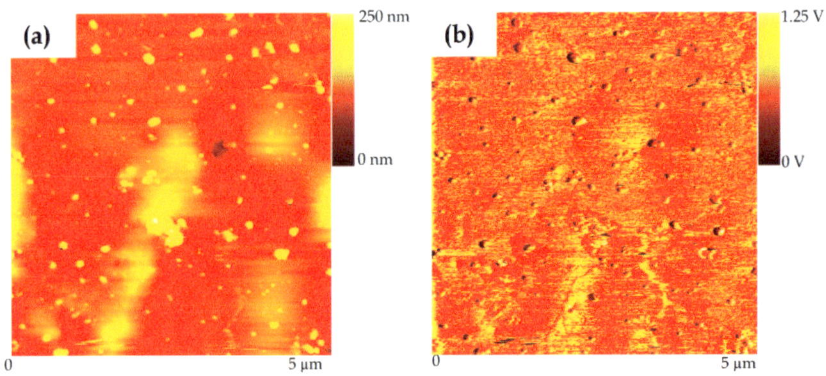

Fig. 16. (a) Surface topography and (b) magneto-elastic images of nanocrystalline magnetic ribbon sample. Nanocrystalline phase can be observed in the images.

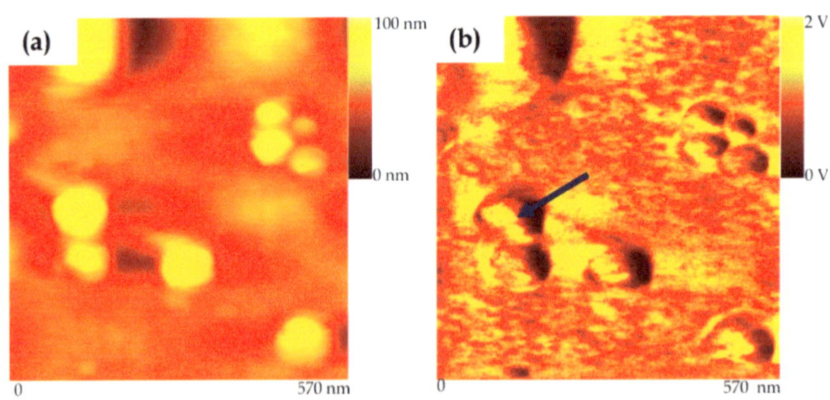

Fig. 17. A high magnification (a) surface topography and (b) magneto-elastic images of the nanocrystalline FINEMET alloy magnetic ribbon. A domain wall with a width of 40 nm is shown by arrow.

7. Conclusion

A new AFM based method to measure and image electrical properties of material with nanometer resolution and high sensitivity is presented. The technique is based on generating eddy currents in conductors and using a magnetic tip to image the magnetic field due to eddy currents. The magnetic fields generated are a function of the electrical conductivity of the sample. Thus the image is a representation of the electrical conductivity. A cantilever with a spring constant of 0.1 N/m was selected for the experiments for maximum sensitivity. Eddy currents in the samples were generated by a small electromagnetic coil. An external electronic instrumentation module was developed to generate images of both surface topography and electrical properties simultaneously at a given location on the sample. The electrical conductivity measurements are obtained in noncontact mode. Eddy current forces were experimentally measured in different metals. It was shown that the magnitude of the eddy current forces in the metal is related to its electrical conductivity. The effect of separation distance between the tip and the sample surface was also studied. It was shown that for maximum sensitivity the tip-sample separation should be less than 100 nm.

The technique was used to image carbon fibers reinforced in a polymer matrix. The contrast in the images is explained based on the huge electrical conductivity variation between the fibers and the matrix. Next, small variations in electrical conductivity variations in a dual-phase titanium alloy, Ti-6Al-4V were imaged. The difference in electrical conductivity from a and β phases of the alloy is small. Also, the difference in the electrical conductivity of the HCP phase is around 6%. The eddy current AFM was able to distinguish between two phases based on electrical conductivity variations. Contrast difference from grain to grain was also observed. The spatial resolution of the system was found to be approximately 25 nm.

In addition to the electrical characterization, the technique was also used to characterize magnetic and electromagnetic properties at the nanoscale with minor modifications. The low-frequency electromagnetic interaction with metallic nanoparticles was studied for the first time using the technique. It was demonstrated that it is possible to generate and image helicon wave resonances in metallic nanostructures with nanometer resolution. The same methodology was modified slightly to image magnetic properties using a *nonmagnetic* probe in contact mode. The approach was based on magnetostrictive strains in ferromagnetic samples subjected to alternating magnetic field. Magnetostriction images were obtained on amorphous and nanocrystalline magnetic ribbon samples. The images show that it is possible to image magnetic domains without a magnetic probe.

The eddy current AFM combines the localized nature and high sensitivity of eddy currents and high spatial resolution and the versatility of AFM. Since induced currents are used in this method, no bias voltage is required to be applied between the sample and the tip. This also means that conductive tips are not needed for conductivity imaging. The electrical conductivity images are obtained in non-contact mode.For higher sensitivity to electrical conductivity variations, flexible cantilevers are used. Since a coil is used to generate eddy currents, eddy current densities can be independently adjusted for different materials based on electrical conductivity of the sample. The vibration spectra of the cantilever over the sample surface showed that multiple resonance peaks are present. Thus the technique can be used to obtain conductivity images multiple frequencies. This facilitates the study of the dependence of frequency on the contrast observed in eddy current images. The same instrumentation can be used for characterization of multiple material properties with some modifications. Finally, since eddy currents are conventionally used in nondestructive evaluation of defects, the methodology can also be used to perform micro/nano NDE to detect nanoscale defects.

8. Acknowledgment

One of the authors (V.Nalladega) would like to acknowledge the financial support of Dayton Area Graduate Studies Institute in completion of the work. He also acknowledges the support and encouragement of Dr. Allan Crasto, Associate Director of University of Dayton Research Institute. The authors would also like to acknowledge the financial support from Wright-Patterson Air Force Base. The authors would like to thank Dr. Terry Murray (UDRI) for the platinum nanoparticles sample and Drs. Mike Gigliotti and P. R. Subramanian (GE Global Research) for providing the amorphous and nanocrystalline magnetic ribbon samples.

9. References

Banerjee, S.; Sanyal, M. K.& Datta, A. (1996). A Simulation Study of Multi-Atom Tips and Estimation of Resolution in Atomic Force Microscopy. *Applied Surface Science,*Vol.99, No.3, pp. 255-260

Berger, R.; Krause, F.; Dietzel, A.; Seo, J. W.; Fompeyrine, J. & Locquet, J.-P. (2000). Nanoscale Magnetostrictive Response in a Thin Film Owing to a Local Magnetic Field. *Appl. Phys. Lett.*, Vol.76, No.5, pp. 616-618

Betzig, E.; Trautman, J. K.; Harris, T. D.; Weiner, J. S.& Kostelak, R. L. (1991). Breaking the Diffraction Barrier: Optical Microscopy on a Nanometric Scale. *Science,*Vol.251, No.5000, pp. 1468-1470

Binnig, G.; Rohrer, H. & Gerber, Ch. (1986). Atomic Force Microscopy. *Phys. Rev. Lett.*, Vol.56, No.9, pp. 930-933

Bluhm, H.; Wadas, A.; Wiesendanger, R.; Roshko, A.; Aust, J. A.& Nam, D. (1997). Imaging of Domain-inverted Gratings in $LiNbO_3$ by Electrostatic Force Microscopy. *Appl. Phys. Lett.*, Vol. 71, No.1, pp. 146-148

Born, M.&Wolf, E. (1999). *Principles of Optics*, Cambridge University Press,Cambridge, UK

Bowers, R.; Legendy, C.& Rose, F. 1961. Oscillatory Galvanomagnetic Effect in Metallic Sodium. *Phys. Rev. Lett.*, Vol.7, No.9, pp. 339-41

Burnham, N.; Kulik, A. J.; Gremaud, G.; Gallo, P. J. & Oulevey, F. (1995). Scanning Local Acceleration Microscopy. *J. Vac. Sci. Technol.*, Vol. 14, No.2, pp. 749-799

Chambers, R. G. & Jones, B. K. (1962). Measurement of High-Field Hall Effect by an Inductive Method. *Proc. R. Soc. A, Vol.*270, No.27, pp. 417-434

Courjon, D. (2003). *Near-Field Microscopy and Near-Field Optics,* Imperial College Press, London, UK

Dinelli, F.; Assender, H. E.; Takeda, N.; Briggs, G. A.D. & Kolosov, O. (1999). Elastic Mapping of Heterogeneous Nanostructures with Ultrasonic Force Microscopy. *Surf. Interf. Anal.* Vol. 27, No.5-6, pp. 562-567

Franke, K.; Besold, J.; Haessler, W. & Seegebarth, C. (1994). Modification and Detection of Domains on Ferroelectric PZT Films by Scanning Force Microscopy. *Surf. Sci. Lett.*,Vol.302, No.1-2, pp. L283-L288

Gautier, B.; Fares, B.; Prudon, G. & Dupuy, J. (2004). Imaging by Atomic Force Microscopy of the Electrical Properties Difference of the Facets of Oxygen-Ion-Induced Ripple Topography in Silicon. *Appl. Surf. Sci.*, Vol.231-232, No.1, pp. 136-140

Goodman, F. O. & Garcia, N. (1991). Roles of the Attractive and Repulsive Forces in Atomic Force Microscopy. *Physical Review B*,Vol.43, No.6, pp. 4728-4731

Gruverman, A.; Auciello, O.; Hatano, J. & Tokumoto, H. (1996). Scanning Force Microscopy as a Tool for Nanoscale Study of Ferroelectric Domains. *Ferroelectrics*,Vol.184, No.1, pp. 11-20

Gu, Y. Q.; Ruan, X. L.; Han, L.; Zhu, D. Z.& Sun, X. Y. (2002). Imaging of Thermal Conductivity With Sub-Micrometer Resolution Using Scanning Thermal Microscopy. *International Journal of Thermophysics*,Vol.23, No.4, pp. 1115-1124

Hartmann, U. (1999). Magnetic Force Microscopy. *Annu. Rev. Mater. Sci.*, Vol. 29, pp. 53-87

Herzer, G. (1993). Nanocrystalline Soft Magnetic Materials. *Physica Scripta* Vol.T49A, pp. 307-314

Hirsekorn, S.; Rabe, U.; Boub, A.& Arnold, W. (1999). On The Contrast in Eddy Current Microscopy Using Atomic Force Microscopes. *Surface and Interface Analysis*,Vol.27, No.5-6, pp. 474-481

Hoffmann, B.; Houbertz, R.&Hartmann, U. (1998). Eddy Current Microscopy. *Appl. Phys. A*,Vol.66, Supplement No.1, pp. S409-S413

Houck, J. R. & Bowers, R. (1964). New Type of Flux Meter for the Measurement of High Magnetic Fields at Low Temperatures. *Rev. Sci. Instrum.*,Vol.35, No.9, pp. 1170-1172

Huang, P.; Zhang, G.; Wu, Z.; Cai, J. & Zhou, Z. (2006). Inspection of Defects in Conductive Multi-Layered Structures by an Eddy Current Scanning Technique: Simulation and Experiments. *NDT&E International*,Vol.39, No.7, pp. 578-584

Hutter, J. L. & Bechhoefer, J. (1993). Calibration of Atomic-Force Microscope Tips. *Rev. Sci. Instrum.*,Vol.64, No.7, pp.1868-1873

Hutter, J. L. & Bechhoefer, J. (1993). Manipulation of Van Der Waals Forces to Improve Image Resolution in Atomic Force Microscopy. *J. Appl. Phys.*,Vol.73, No.9, pp. 4123-4129

Karpen, W.; Becker, R. & Arnold, W. (1998). Characterization of Electric and Magnetic Material Properties with Eddy Current Measurements. *Nondestructive Testing Evaluation*,Vol.15, No.2, pp. 93-107

Kelly, K. L.; Corondo, E.; Zhao, L. L.& Schatz, G. C. (2003). The Optical Properties of Nanoparticles: The Influence of Size, Shape, and Dielectric Environment. *J. Phys. Chem.* B Vol.107, No.3, pp. 668-677

Kirby, M. W.& Lareau, J. P. (1991). Eddy Current Imaging of Aircraft Using Real Time Image Signal Processing,*Proceedings of the International Conference on Nondestructive Testing and Evaluation of Composite Structures*,Riga, Latvia, October 22-24, 1991

Krakowski, M. (1982). Eddy Current Losses in Thin Circular and Rectangular Plates. *Archiv für Elektrotechnik*,Vol.64, No.6, pp. 307-311

Landau, L. D. & Lifshitz, E. M. (1960). *Electrodynamics of Continuous Media*,Pergamon,Oxford, UK

Libby, H.L. (1971). *Introduction to Electrodynamic Nondestructive Test Methods*,Wiley-Interscience,New York, USA

Matey, J. R. & Blanc, J. (1985). Scanning Capacitance Microscopy. *J. Appl. Phys.*,Vol.57, No.5, pp. 1437-1444

Maxfield, B. W. (1969). Helicon Waves in Solids. *Am. J. Phys.*,Vol.37, No.3, pp. 241-269.

McHenry, M. E.;Johnson, F.; Okumura, H.; Ohkubo, T. et al. (2003). The Kinetics of Nanocrystallization and Microstructural Observations in FINEMET, NANOPERM and HITPERM Nanocomposite Magnetic Materials. *Scripta Materialia*,Vol.48, No.7, pp. 881-887

Meaden, G. T. (1965). *Electrical Resistance of Metals*, Plenum, New York, USA

Michaels, A. M.; Nirmal, M.& Brus, L .E. (1999). Surface Enhanced Raman Spectroscopy of Individual Rhodamine 6G Molecules on Large Ag Nanocrystals. *J. Am. Chem. Soc.*,Vol.121, No.43, pp. 9932-9939

Murphy, W. L. & Spalding, G. C. (1999). Range of Interactions: An Experiment in Atomic and Magnetic Force Microscopy. *Am. J. Phys.*, Vol.67, No.10, pp. 905-908

Murray, P. T.& Shin, E. (2008). Formation of Silver Nanoparticles by Through Thin Film Ablation. *Materials Letters*,Vol.62, No.28, pp. 4336-4338

Nalladega, V.; Sathish, S.; Klosterman, D.; Jata, K. V. & Blodgett, M. P. (2007). Atomic Force Microscopy Based Eddy Current Imaging and Characterization of Composite and Nanocomposite Materials, *Proceedings of SAMPE Technical Conference*, Baltimore, Maryland, USA, June 3-7, 2007

Nalladega, V.; Sathish, S. & Brar, A. S. (2008a). Characterization of Defects in Flexible Circuits with Ultrasonic Atomic Force Microscopy. *Microelectronics Reliability*,Vol.48, No.10, pp. 1683-1688

Nalladega, V.; Sathish, S.; Jata, K. V. &Blodgett, M.P. (2008b). Development of Eddy Current Microscopy for High Resolution Electrical Conductivity Imaging Using Atomic Force Microscopy. *Rev. Sci. Instrum.*, Vol.79, No.7, pp. 073705-073705-11

Nalladega, V.; Sathish, S.; Gigliotti, M. F. X.; Subramanian, P. R. & Iorio, L. (2009). Characterization of Magnetoelastic Properties at Nanoscale Using Atomic Force Microscopy. *Proceedings of 14th International Workshop on Electromagnetic Nondestructive Evaluation*, ISBN 978-1-60750-553-2, Dayton, Ohio, USA, July, 2009

Nalladega, V.; Sathish, S.; Murray, T.; Shin, E.; Jata, K. V. & Blodgett, M. P. (2011). Experimental Investigation of Interaction of Very Low Frequency Electromagnetic Waves with Metallic Nanostructure. *J. Appl. Phys.*, Vol.109, No.11, (June 2011), pp. 114907-114907-9

Nonnenmacher, M.; O'Boyle, M. P. & Wickramasinghe, H. K. (1991). Kelvin Probe Force Microscopy. *Appl. Phys. Lett.*,Vol.58, No.25, pp. 2921-2923

Nyffenegger, R. M.; Penner, R. M. & Schierle, R. (1997). Electrostatic Force Microscopy of Silver Nanocrystals with Nanometer-Scale Resolution. *Appl. Phys. Lett.*,Vol.71, No.13, pp. 1878-1880

Oh, J. & Nemanich, R. J. (2002). Current-voltage And Imaging of $TiSi_2$ Islands on Si(001) Surfaces Using Conductive-Tip Atomic Force Microscopy. *J. Appl. Phys.*,Vol.92, No.6, pp. 3326-3331

Okamoto, H. & Imura, K. (2006). Near-field Imaging of Optical Field Wavefunctions in Metal Nanoparticles. *J. Mater. Chem.*, Vol.16, No.40, pp. 3920-3928

Olbrich, A.; Ebersberger, B. & Boit, C. (1998). Conducting Atomic Force Microscopy for Nanoscale Electrical Characterization of Thin SiO_2. *Appl. Phys. Lett.*,Vol.73, No.21, pp. 3114-3116

Petrashov, V. T. (1984). An Experimental Study of Helicon Resonance in Metals. *Rep. Prog. Phys.*,Vol.47, No.1, pp. 47-110

Pendry, J. B.; Holden, A. J.; Stewart, W. J.& Youngs, I. (1996). Extremely Low Frequency Plasmons in Metallic Mesostructures. *Phys. Rev. Lett.*,Vol.76, No.25, pp. 4773-4776

Poltz, J. (1983). On Eddy Currents in Thin Plates. *Archiv für Elektrotechnik*, Vol.66, No.4, pp. 225-229

Polushkin, N. I.; Rao, K. V.; Wittborn, J.; Alexeev, A. M. & Popkov, A. F. (2003). Visualization of Small Magnetic Entities by Nonmagnetic Probes of Atomic Force Microscope. *J. Magn. Magn. Mater.* Vol.258-259, No.1, pp. 29-31

Rabe, U. & Arnold, W. (1994). Acoustic Microscopy by Atomic Force Microscopy. *Appl. Phys. Lett.*, Vol. 64, No.12, pp. 1493-1495

Ruskell, T. G.; Workman, R. K.; Chen, D.; Sarid, D.; Dahl, S. & Gilbert, S. (1996). High Resolution Fowler-Nordheim Field Emission Maps of Thin Silicon Oxide Layers. *Appl. Phys. Lett.*,Vol.68, No.1, pp. 93-95

Sader, J. E. (1995). Parallel Beam Approximation for V-Shaped Atomic Force Cantilevers. *Rev. Sci. Instrum.*,Vol.66, No.9, pp. 4583-4587

Siddoju, A.; Sathish, S.; Ko, R. & Blodgett, M. P. (2006). Electrical Circuit Model of an Eddy Current System for Computing Multiple Parameters. *AIP Conference Proceedings Review of Progress in Quantitative Nondestructive Evaluation*, ISBN 978-0-7354-0399-4, Portland, Oregon, USA, July 30-August 4, 2006

Ślawska-Waniewska, A. & Lachowicz, H. K. (2003). Magnetostriction in Soft Magnetic Nanocrystalline Materials. *Scripta Materialia*,Vol.48, No.7, pp. 889-894

Stern, J. E.; Terris, B. D.; Mamin, H. J.& Rugar, D. (1988). Deposition and Imaging of Localized Charge on Insulator Surfaces Using a Force Microscope. *Appl. Phys. Lett.*,Vol.53, No.26, pp. 2717-2719

Takata, K. & Tomiyama, F. (2000). Strain Imaging of a Magnetic Material. *Jpn. J. Appl. Phys.*,Vol.39, No.5B, pp. 3090-3092

Taylor, M. T.; Merrill, J. R.& Bowers, R. (1963). Low-frequency Magnetoplasma Resonance in Metals. *Phys. Rev.*,Vol.129, No.6, pp. 2525-2529

Tominaga, J.; Mihalcea, C.; Büchel, D.; Fukuda, H. et al. (2001). Local Plasmon Photonic Transistor. *Appl. Phys. Lett.*,Vol.78, No.17, pp. 2417-2419

Wadas, A. & Hug, H. J. (1992). Models For The Stray Field From Magnetic Tips Used in Magnetic Force Microscopy. *J. Appl. Phys.*,Vol.72, No.1, pp. 203-206

Weaver, J. M. R. & Abraham, D. W. (1991). High Resolution Atomic Force Microscopy Potentiometry. *J. Vac. Sci. Technol. B*, Vol.9, No.3, pp. 1559-1561

Wittborn, J.; Rao, K. V.; Nogues, J. & Schuller. I. K. (2000). Magnetic Domain and Domain-wall Imaging of Submicron Co Dots by Probing the Magnetostrictive Response Using Atomic Force Microscopy. *Appl. Phys. Lett.*, Vol.76, No.20, pp. 2931-2933

Williams, C.C. (1999). Two-dimensional Dopant Profiling by Scanning Capacitance Microscopy. *Annu. Rev. Mater. Sci.*, Vol.29, pp. 471-504

Yamanaka, K.; Ogiso, H. & Kolosov, O. (1994).Ultrasonic Force Microscopy for Nanometer Resolution Subsurface Imaging. *Appl. Phys. Lett.*, Vol. 64, No.2, pp. 178-180.

Yoshizawa, Y.; Oguma, S. & Yamauchi, K. (1988). New Fe-based Soft Magnetic Alloys Composed of Ultrafine Grain Structure. *J. Appl. Phys.*,Vol.64, No.10, pp. 6044-6046

2

Tuning Fork Scanning Probe Microscopes – Applications for the Nano-Analysis of the Material Surface and Local Physico-Mechanical Properties

Vo Thanh Tung[1,3], S.A. Chizhik[2], Tran Xuan Hoai[3],
Nguyen Trong Tinh[3] and V.V. Chikunov[2]
[1]*Hue University of Sciences*
[2]*A.V. Luikov Heat and Mass Transfer Institute of National Academy of Sciences of Belarus*
[3]*Institute of Applied Physics and Scientific Instrument of Vietnamese Academy of Science and Technology*
[1,3]*Vietnam*
[2]*Belarus*

1. Introduction

The atomic force microscopes (AFMs) or scanning force microscopes (SFMs) are a very high resolution type of scanning probe microscope (SPM), with demonstrated resolutions of a fraction of a nanometer. The AFMs are one of the foremost tools for imaging, measuring and manipulating matters at the nanoscale level. AFMs operate by measuring the atomic interaction (attractive or repulsive) forces between a tip and a sample [1]. Today, most AFMs employ an optical lever sensor - an expensive system that can achieve pico-meter resolution. The optical lever (Fig.1 (a)) operates by reflecting a laser beam off the cantilever. The angular deflection of the cantilever causes a large twofold of the laser beam. The reflected laser beam strikes a position sensing photodetector consisting of two side-by-side photodiodes. The difference between the two photodiode signals indicates the position of the laser spot on the detector and thus the angular deflection of the cantilever. Because the cantilever-to-detector distance generally measures thousands of times the length of the cantilever, the optical lever greatly magnifies motions of the tip [2]. The advantages of optical lever are possibility to measure absolute force and commercial available but there are many disadvantages: external optical measuring system needed, very low stiffness, fragile and high cost of AFM.

Recently, the quartz tuning fork sensors (QTFs) are being developed and it is well established that quartz tuning forks can be used as sensors for acoustic [3] and force microscopy [4]. Quartz tuning forks were originally introduced into the field of scanning probe microscopy (SPM) by Gunther *et al.* [3], and later by Karrai and Grober [5]. Giessibl *et al.* [6] employed them for atomic resolution AFM imaging. Tuning forks were used as sensors at low temperatures and in high magnetic fields by Rychen *et al.* [7]. The forks have several advantages: high amplitude and phase sensitivities, high mechanical quality factor

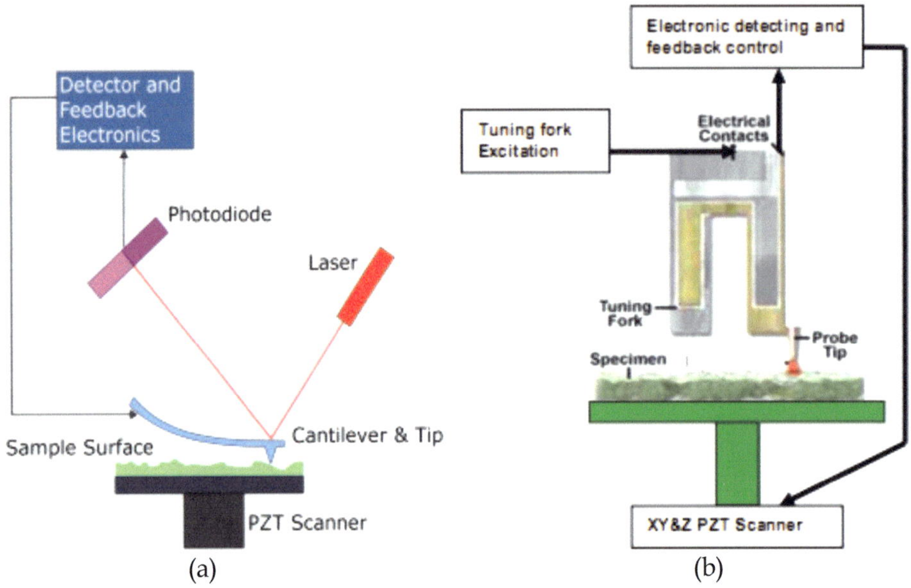

Fig. 1. (a) The traditional AFM (b) AFM with Tuning fork sensor

Q, large spring constant which allows the detection of pico-newton forces and the acquisition of true atomic resolution images. Their advantage is that the measurement of their oscillation amplitude uses the piezoelectric effect native to quartz crystal, yielding an electrical signal correlated to the applied forces (amplitude, phase and resonance frequency are correlated to the applied force) and making them small, robust and simple to operate compared to optical force measurement schemes. The feedback control can be constructed in a fully digital-electronic manner, so that the implementation of the feedback control is greatly simplified. The quartz tuning forks have been successfully demonstrated under various conditions [3, 5, 7-9].

K. Karrai and I. Tiemann [10] reported the results of the measurements with a tuning fork that was able to detect the friction force in the pN range. In their measurements, the tuning fork was excited by mechanical coupling with an external piezo oscillator, so that the Q-factor of the system could not be controlled. Furthermore K. Karrai and I. Tiemann reported difficulty in obtaining images with tuning fork sensors in air and liquid environments due to fluctuations of the resonant frequency.

Recently, electronic devices have been developed that allow the Q factor of an oscillating tuning fork in an AFM to be varied in a controlled manner [11]. It was possible to lower the Q factor of the probe, reducing the scanning time of the microscope accordingly. In these cases, not only did the images have a better signal-to-noise ratio, but they were also obtained at a faster scan speed. More details on the relation between the Q factor of the probe and the microscope sensitivity can be found in previous publications [12, 13, 14].

In this chapter, we demonstrate atomic force microscopes in the ambient conditions using the Q control for quartz tuning fork. We will describe the use of the tuning fork as a force sensor and use it in some applications of scanning probe microscopy (Fig.1 (b)). The advantages and disadvantages of a device with Q controlled will be discussed also.

2. Quartz tuning fork

A tuning fork is a two-pronged, metallic fork with the tine formed from a U-shaped bar of elastic material (usually steel). A tuning fork resonates at a specific constant pitch when set vibrating by e.g. striking it against a surface or with an object, and after waiting a moment to allow some high overtones to die out. The pitch generated by a particular tuning fork depends on the length of the two prongs, with two nodes near the bend of the U [15-17].

Figure 2 (a) shows a picture of the tuning fork appears as a metallic cylinder 8 mm in height, by 3 mm in diameter, holding a two-terminal electronic component. The packaging of the tuning fork can easily be opened by using tweezers to clamp the cylinder until the bottom of the cylinder breaks. A more reproducible way to open the packaging is to use a model-making saw to cut the metallic cylinder, keeping the bottom insulator as a holder to prevent the contact pins from breaking (Fig. 2 (b)).

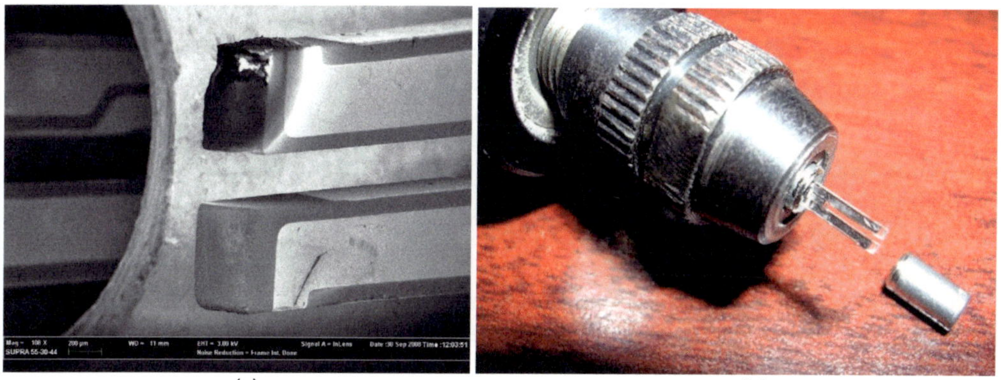

(a) (b)

Fig. 2. (a) SEM image of tuning fork displaying the layout of the electrodes. (b) A tuning fork just removed from its packing, and the metallic enclosure that would otherwise keep it under vacuum.

Quartz tuning forks are primarily designed for frequency controls and time-based applications. Furthermore, applications of quartz tuning fork resonators seem to be an attractive alternative to the conventional mass measurement techniques since the tuning fork resonators combine the high Q-factor in the ambience of a quartz resonator and the flexural oscillation mode of a cantilever [18]. A number of tuning fork designs were developed that exploits the mechanical resonance such as flexure, extensional, torsion and shear modes. The sensitivity of these mode frequencies to external perturbations such as mass loading, force, pressure, and temperature quartz oscillators are suitable for sensor technologies [8, 19-21].

A tuning fork of a commercially available type fabricated for "quartz" clocks was used (type 74-530-04 of ELFA Company with a standard resonance frequency 32757Hz, and a theoretical quality factor Q=15000). The fork was modeled in a standard way as a series R-L-C circuit. The R-L-C model provides a convenient electrical analogue to the mechanical properties of the tuning fork. (Its mass m, stiffness or spring constant k, and damping due to internal and external dissipative forces are represented by L, C, and R respectively.) This model is usually further improved by the inclusion of a parallel shunt capacitance Co corresponding to the package capacitance. The admittance was measured as a function of frequency using a signal synthesizer and lock-in amplifier. The theoretical spring constant is obtained from the formula $k = \frac{E}{4} W \left(\frac{T}{L} \right)^3$ [22] where $E = 7,87.10^{10}\, N/_{m^2}$ is the Young modulus of quartz. The length (L), thickness (T) and width (W) of the tuning fork used are 6.01, 0.35 and 0.61 mm, respectively. Using these parameters, we obtain k ≈ 7kN/m, which agrees reasonably well with our experimental result.

3. Tips and probes

The tips attached to the prong of the tuning fork were silicon tips and tungsten tips. The tip was placed on the tuning fork using an optical microscope equipped with a micro-positioning stage. The mass of the silicon tip was so small that the reduction in the quality factor as well as the resonance-frequency shift of the fork was very small (both changes were less than 1%). For the tungsten tip, it was fabricated by electrochemical etching achieving a radius of about 30-70nm. When the tip was glued directly to a prong of a tuning fork, the quality factor decreased from 14000 to 7000-9000.

A method of connecting a tip (silicon tips or tungsten tips) to the tuning fork is gluing with two component epoxy. Epoxy glue was chosen because of its strong adhesive properties and the drying rate was not too fast. The added mass by the glue was able negligibly small. Figure 3a shows the result of this process with a tungsten tip across attached to one prong of the tuning fork. Figure 3b shows the result of a very sharp silicon tip only a few ten of micrometers attached to one prong of the tuning fork. Figure 4 represents the resonance curves of amplitude in ambience for a tuning fork before, and after it is mounted with a tip on one prong, as described earlier.

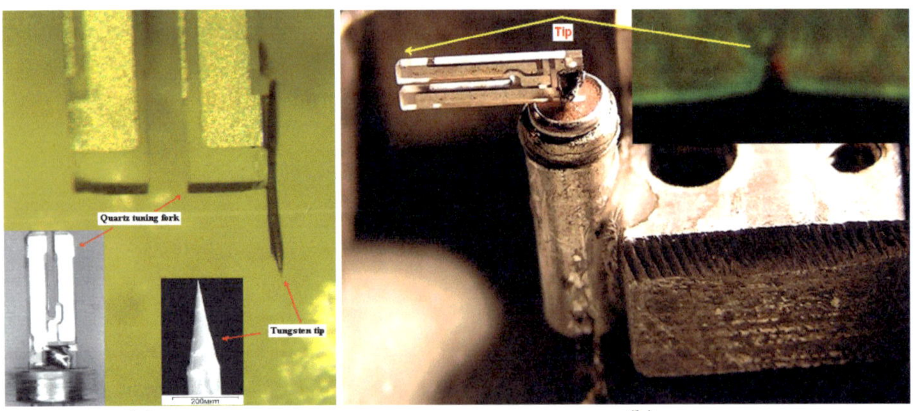

(a) (b)

Fig. 3. Image of tungsten tip (a) and silicon tips (b) attached to tuning fork face surface

4. Principle operation of quartz tuning fork base on atomic force microscopy (Fork-AFM)

A scanning probe microscope, based on the above described quartz tuning fork, is shown in Fig. 5 (a). The mechanical part of the tuning fork that connecting to the atomic force

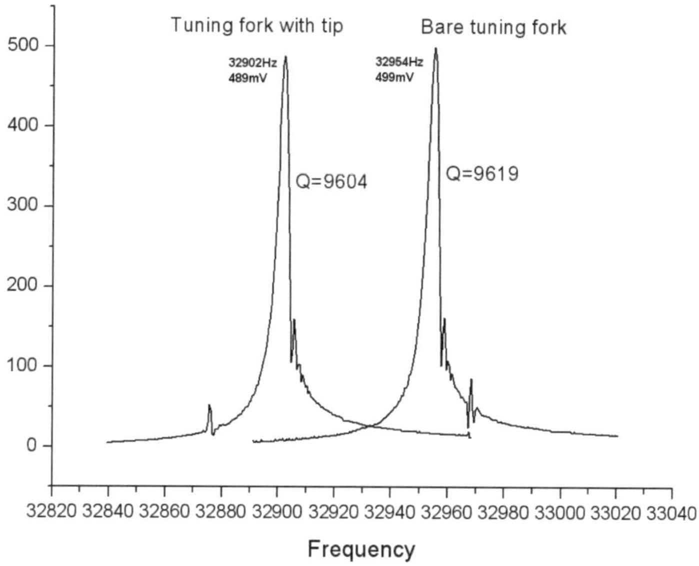

Fig. 4. Spectrum of bare tuning fork and tuning fork with commercial AFM cantilever tip

microscopy is shown in Fig. 5 (b). This mechanical part consists of two main units: a holder (1) and a base plate (2). The holder is designed as the unit housing most of the mechanical components of the shear force microscope. Two fine-pitch screws (4) are fixed to the holder in the standard arrangement for probe-sample coarse and fine approaching. The tuning fork sensor (5) as the heart of the system is attached to the holder with cyanoacrylate glue, and connected to the AFM through the cable (8). The holder with tuning fork is placed on the base plate (2) and secured using the outside metal box (6). The metal box could be moved in the sample state (7).

There are two basic methods of dynamic operation: intermittent contact mode and shear-force (or lateral mode) operation. Figure 5 (c) displays the basic principle of Fork-AFM in ambient conditions using a quartz tuning fork in these modes. As shown in Fig. 5 (c-lower part), the tip is mounted perpendicular to the tuning-fork prong so that the tip oscillates perpendicularly to the sample surface. This is the intermittent contact mode. In the lateral force sensor mode (Fig. 5 (c-upper part)), the tip is mounted parallel to the tuning fork prong and oscillates nearly parallel to the surface of a sample.

A constant sine wave voltage is applied to the one of the connectors of the tuning fork to drive the fork sensor. The other connector of the tuning fork is connected to a reference signal generator of a lock-in amplifier. An I-V converter is used to convert the net current to a voltage (Fig. 6). The current to voltage (I –V) gain of the circuit is calibrated from dc to 100 kHz. Under the resonant condition, the tuning fork arm has the biggest displacement that corresponds to the maximum or peak output voltage amplitude. The upper current-to-voltage converter will sense both the piezoelectric currents from the tuning fork oscillation and additional stray currents. The lower op-amp in Fig. 6 is not always necessary. However, it is present in order to cancel currents from stray capacitance (Co and other capacitance from wires). The lower op-amp *(I- V)* converter allows the cancellation of stray currents by adjusting the variable capacitor away from the resonance.

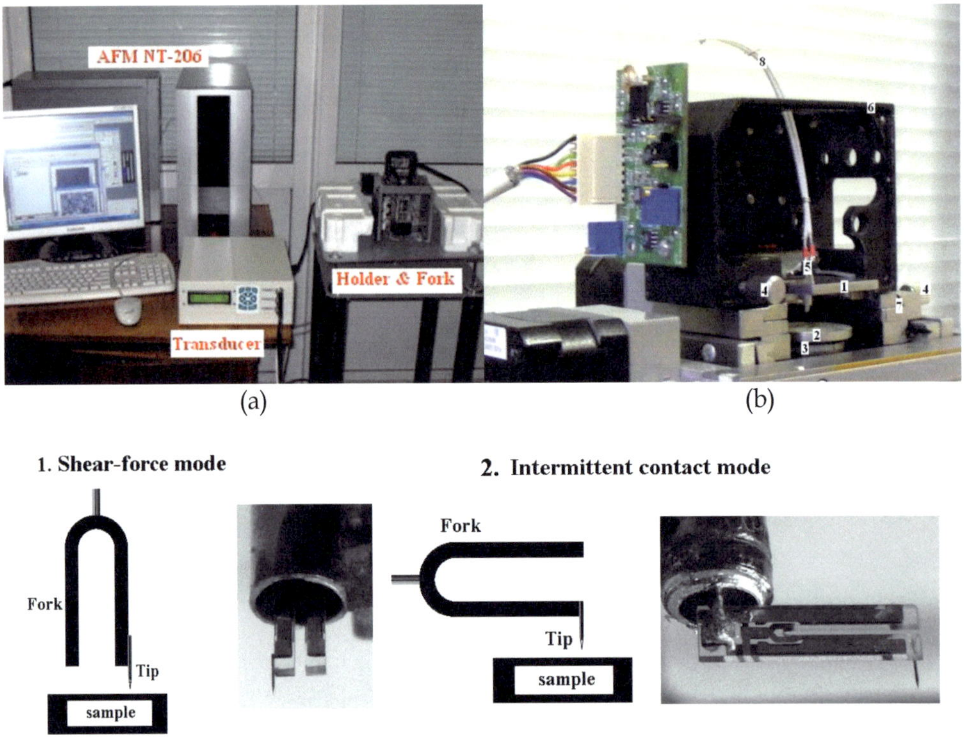

(c)

Fig. 5. (a) Photography of Atomic force microscopy using a quartz tuning fork (Fork-AFM) (b) Header of AFM using a quartz tuning fork (Fork-AFM) (1- holder, 2- base plate, 3- piezoscanner, 4- screws, 5- tuning fork sensor, 6- metal box, 7- sample state, 8- cable). (c) Principle of two operation modes: shear-force and intermittent contact mode.

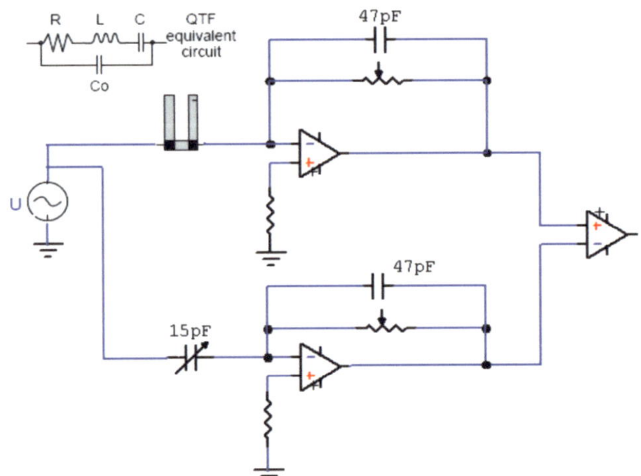

Fig. 6. Essential features of tuning fork sensing system and the equivalent circuit. Current through the tuning fork is converted to a voltage by an Op-Amp (OA). The lock-in reference signal is derived directly from the driving source and is not shown.

When the probe is approaching near the sample surface, the oscillation of the sensor is dampened due to probe-sample force interactions, resulting in a decrease in the output signal of the lock-in amplifier. The decreased signal is compared with a set-point of the feedback circuit and the resulting difference is fed back to the scanner via the high voltage amplifier in order to control the probe-sample distance during scanning. During the scan, the phase and amplitude of the quartz tuning fork oscillations are also recorded.

5. Tuning the quality factor (Q-Control)

The quality factor Q is widely used when discussing oscillators, because this property is useful for predicting the stability of the resulting frequency around the resonance [14, 23 - 26]. Furthermore, we can infer that the quality factor can be increased by injecting energy into the tuning fork during each cycle. Similarly, the quality factor can be decreased by removing energy during each cycle. These two cases can be accomplished by adding a sine wave at the resonance frequency with the appropriate phase. However, in practice, a quartz tuning fork works at a low enough frequency to allow classical operational amplifier based circuits to be used for illustrating each step of quality factor tuning.

Figure 7 illustrates a possible implementation of the circuit including an amplifier, a phase shifter, a bandpass filter, and an adder. The feedback gain defines the amount of energy fed back to the resonator during each period; the phase shift determines whether this energy is injected in phase with the resonance (quality factor increase) or in phase opposition (quality factor decrease). The resonance frequency shift was associated with a feedback loop phase that is not exactly equal to -90°. The phase shift was set manually, using a variable resistor and an oscilloscope in XY mode, until a circle was drawn by an excitation signal and by the phase-shifted signal, allowing for a small error in the setting. Figure 8 (a, b) display a measurement of the decreasing of the quality factor based on a discrete component implementation of the circuit in Fig. 7.

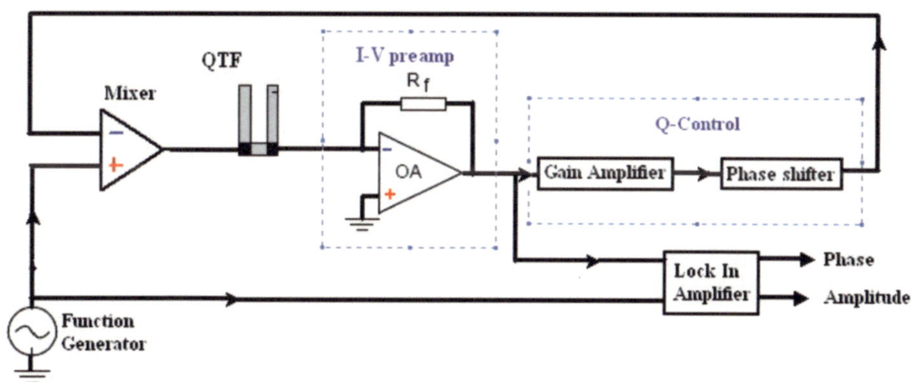

Fig. 7. Principle diagram of Q-control feedback circuit and the block diagram of I-V preamplifier and the Q-control system.

As can be seen, the tuning fork response is phase shifted and amplified before being added to the drive. The effective damping is altered by feeding back some of the response, with the phase of this feedback signal, relative to the fork velocity, determining whether the damping

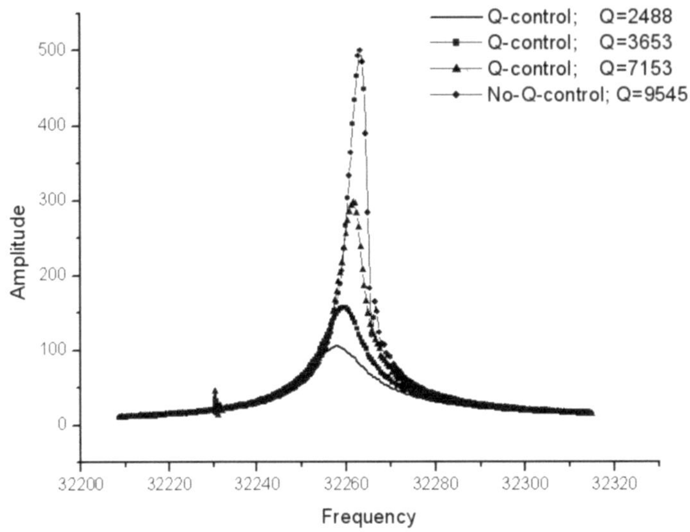

(a)

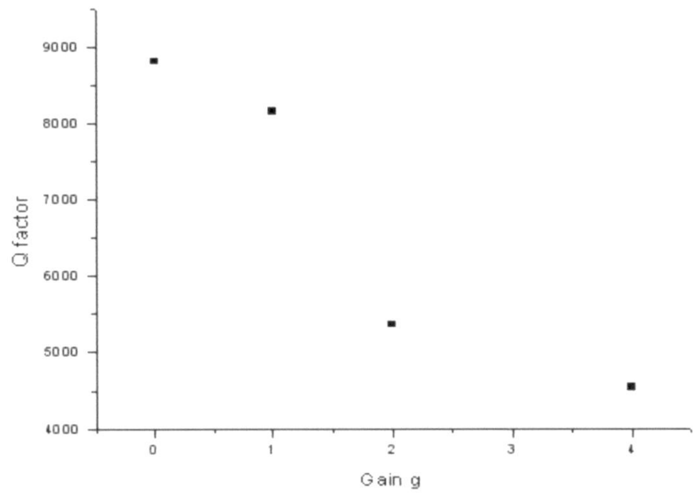

(b)

Fig. 8. Result of using Q-control, the amplitude response as a function of the driving frequency with different feedback g under condition of the phase shift -90 degree

is effectively increased or decreased. By choosing an appropriate value of the gain g, the damping of the cantilever motion is partially compensated, and hence the degree of alteration of the damping (related to Q) is changed by the gain setting.

The effect of gain g on the amplitude curves (at phase -90 degree) and constant driving force is shown in Fig. 8 (a). While the Q seems largely unchanged with the smaller gains ($g<1$), the increasing of g ($g>1$) strongly decreases the quality factor Q, oscillation amplitude, shifts the resonance frequency to lower values, and broadens the amplitude curves. This behavior is consistent with the Q-control signal adding to the damping force, not canceling it. The amplitude at resonance can be decreased by two times of magnitude with respect to the initial system, while the resonance frequency changes by a small factor (0.2%) and, as a result, g value increases from 1 to 4). Figure 8 (b) considers the effective Q as a function versus gain g, respectively. These results show that the increasing of g from 0 to 4 produces a decrease of Q of about two times of magnitude (from 8000 to 4000).

6. Fork-AFM imaging

To test the achievable resolution on our new system, a sample calibrating grating TGZ-02 was measured in two modes. The image size was about 14x14 μm^2 (256x256 pixels), the scan direction was from left to right, obtained with a scan rate of 0.3-0.5 Hz per line. The set point for the engaged tip vibration amplitude was fixed to about 90% of the maximum amplitude. Figures 9 (a, b) show topographical images obtained in the shear and the tapping modes, respectively. The abrupt changes in these images were attributed to instabilities such as signal drifts or tip contaminations. In the results, the effects of the radius of the tungsten tip are clearly visible. Furthermore the quality of the image shows that the tuning fork with these tip-scanners is capable of handling large variations in the height of the sample.

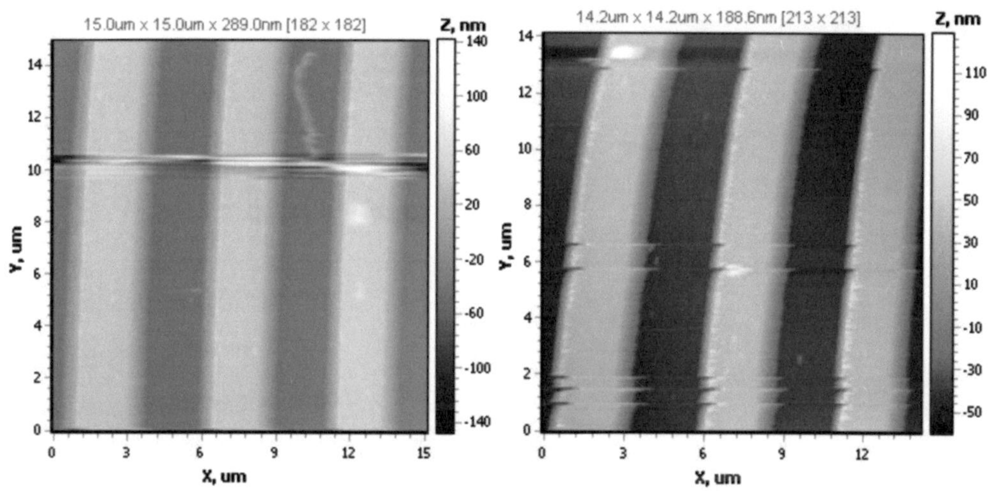

(a) (b)

Fig. 9. Topographies obtained in (a) shear mode, and (b) tapping mode for sample calibration grating TGZ02. The dimensions of the images are about 14x14(μmxμm).

To demonstrate that the Fork-AFMs using tungsten tips were successfully used in imaging samples in two modes: tapping and shear modes, we showed related experimental results in Fig. 10. A "rigid" sample: alumina Al2O3 and a "non-rigid" sample PVP were measured. The scanning speed of two mode operations is 0.3Hz per line, the current through the tuning fork is 3nA (the prong vibration amplitude is around 3.8nm) for shear force mode and 2nA for tapping mode. The entire image was obtained in about 15 minutes for a resolution of 256x256 pixels. The set point of the feedback circuit was set at 90% maximum amplitude on resonance.

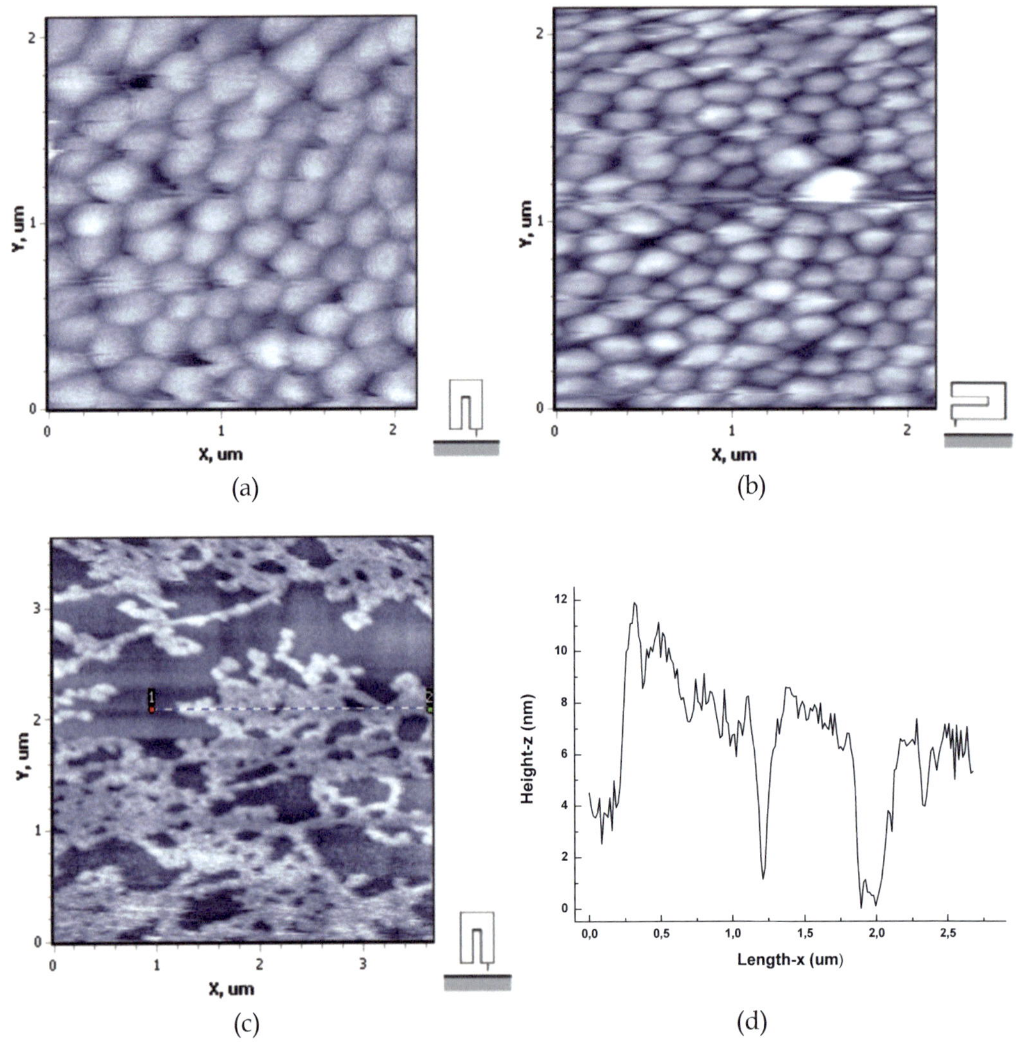

Fig. 10. Topography images of (a, b) alumina Al2O3, (c) "non-rigid" sample PVP in the shear-force and intermittent contact modes using Q-control feedback (Q<2500). The dimensions of images is noted in the figure. A line profile from 1 to 2 in (c) is show in figure (d)

By comparing images obtained in both shear mode (Fig. 10 (a)) and tapping mode (Fig. 10 (b)), we found that good images could be achieved in Fork-AFM using tungsten tips. Furthermore, a remarkable advantage of the presented technique that the length of the tungsten tips attached to the tuning fork (0.7mm) was much longer than that of commercial tips (a few micrometers). Therefore the distance between the side electrodes of the turning fork and the sample surface was much larger.

However, when using tungsten tips, the extra mass and mechanical properties of the metallic wire and the glue used to attach it to the prong can radically alter the resonance frequency (shifted by about ~ 1000Hz), and the amplitude of the oscillation and especially, the quality factor Q are significantly reduced (by more 20%). In addition to affect the amplitude and the quality factor, the symmetry property of the oscillation of quartz tuning forks is broken because of its balance. It can affect the stability and measurement results of tuning forks with atomic force microscopes. For making balance, we usually counterbalance by a suitable mass on the other prong of a tuning fork but the amplitude and resonant frequency will suffer a rather large decrease. Moreover, the radius of tungsten tips is rather large, about 30-70nm. It is not easy to consistently make small, sharp tips with reproducible properties. Thus when the oscillating average amplitude of a tip was high, especially in shear force mode, the poor contrast images were received (Fig. 10 (a), (c)).

Figures 11 (a - d) show topographical images obtained in the tapping and shear modes, respectively, of alumina Al2O3 and PVP samples in the air at room temperature when using the cantilever silicon tips. Figures 11 (a, c) show the characteristic of shear-mode images on these materials. In these images, we do not observe the long scratches caused by the vibration of tip such as tungsten tips. The appearance of the abrupt change in the shear-mode image in Fig. 11 (a, c) could be attributed to instabilities, such as signal drifts, tip or surface sample contaminations. Figures 11 (b, d) show images of the surface of the two samples in tapping mode. Similar to the case of using the tungsten tips, the resolution of images in the tapping-mode is better than the shear-force mode. This time, the topography images can be seen as clearly as in Fig. 11(a - d). Moreover, figures 11(e, f) show the averaged line profile for PVP in two modes. From this line profile, we observed the optimum signal in the tapping mode. Additionally, the average height of the features indicated in the image is about 10 nm, and the resolution is achieved to 50nm.

It is interesting to note that we could obtain the high contrast images in shear force and tapping mode operation when using the cantilever silicon tips. Using silicon tip is rather sensitive to the signal interaction because: (i) the area contact belongs to the sharp of the tips, for the cantilever tip AFMs, because the radius of tips is rather small (<10nm), when vibrating, the amplitude tip is not large, especially with the low applied control voltage, vibration is not significant, thus the region contact between tip and sample may achieve the atom interaction; (ii) due to the short tip length (10–20 μm), the distance between the side of the tuning fork body and the sample surface is very small, and consequently this distance modulation due to the oscillation of the tip may induce undesirable noise, (iii) the mass of the tip is not significant, thus it does not influence the properties of fork, especially the symmetry. As a result, we found that a much better signal control could be achieved, and we could obtain a high resolution images with the different samples with the cantilever silicon tip.

We succeeded in getting images with discernible bit line in both cases. Furthermore by comparing force images obtained in both shear mode (Fig. 11 (a, c)) and tapping mode (Fig. 11 (b, d)) with one type of tips, we found that a much better signal could be achieved in tapping mode operation. To explaining for this result, we proposed some assumptions for explanations in the following way: in the shear-force mode, because the tip oscillates parallel to the surface of sample, the area contact between tip and sample is about 30-40nm. Therefore, in this process, the instability such as signal drifts or tip contaminations may happen and influence the resulted scanning. In the intermittent contact mode, the area contact between tip and sample is much smaller than shear force mode (about 10-15nm). As a result, the region contact between tip and sample may achieve the atom interaction preventing sample damages, and we could obtain the images with high contrast resolutions. The results showed that this Fork-AFM system with a good, sharp tip can obtain high-resolution images of samples in ambient conditions.

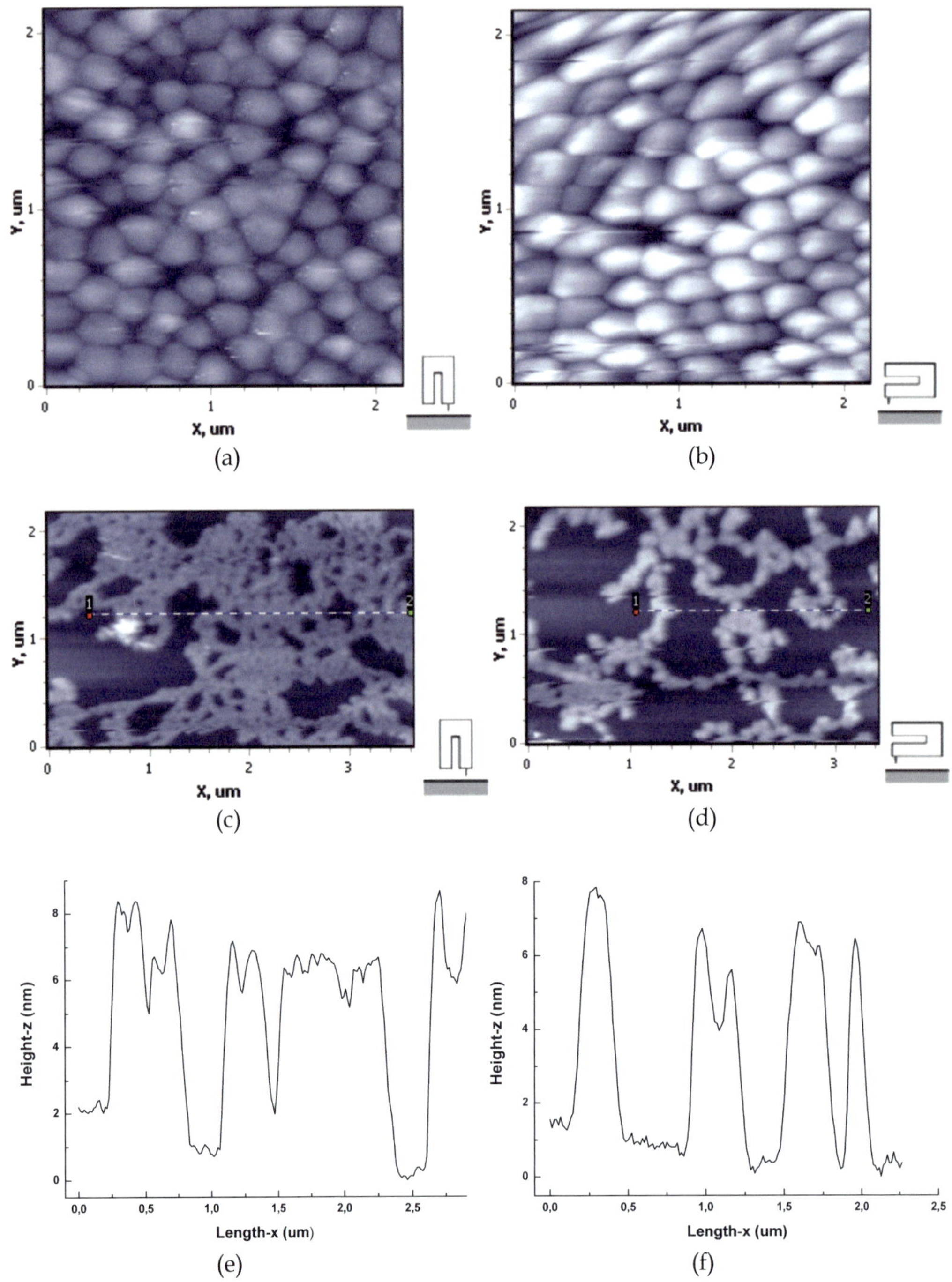

Fig. 11. Topography images of (a, b) alumina Al2O3, (c) "non-rigid" sample PVP in the shear-foce and intermittent contact mode using Q-control feedback (Q>9500). The dimension of images is noted in the figure. A line profile from 1 to 2 in (c, d) is show in figure (e, f)

We also developed a Fork-AFM system for investigations of human blood cell morphology, namely erythrocytes with high-resolution imaging. Erythrocytes samples were prepared by different ways. The first method was standard for clinical laboratory in which a drop of fresh human blood (10-20μl) was applied on a first glass surface and smeared by a second glass. The thickness of the smear decreases along a direction of smearing. In the second method, some drops of fresh human capillary blood were fixed in a 2% aqueous glutaraldehyde solution. The cover slips were rinsed with a 2% aqueous glutaraldehyde solution and washed with phosphate buffered saline (PBS). The freshly extracted blood was then diluted in PBS and this solution was then added again for 1 minute to rigidify the cell. Then the cells were washed with PBS. A preparation was dried up on air at room temperature for several hours.

Figures 12 (a - d) showed topographical images and line profile obtained in the shear force and intermittent contact modes, respectively, of erythrocytes samples in the air at room temperature. The scanning speed of two mode operations was 0.3Hz per line, the current through the tuning fork was 3nA (the prong vibration amplitude was around 3.8nm) for the shear force mode and 2nA for the intermittent contact mode. The entire image was obtained in about 15 minutes for a resolution of 256x256 pixels. The set point of the feedback circuit was set at 90% maximum amplitude on resonance.

The topography of the erythrocytes samples was obtained using shear force detection in air in Fig. 12(a). Figure 12 (b) shows the averaged line profile. From this line profile, the maximum height of the feature indicated in the image was about 1.8 μm. Furthermore, we observed the abrupt change in the shear-mode image. To explaining for the abrupt change, we bring out some assumption for explanation in the following way: in this mode, because the tip oscillates parallel to the surface of sample, the area contact between tip and sample is about 30-40nm. Furthermore, the differential height of sample is rather large. Therefore, in this process, the instability such as signal drift or tip contamination maybe appear and influence the results scanning.

Figure 12 (c) shows the topography of erythrocytes in intermittent contact mode. The image in Fig. 12 (c) can be seen as more obviously than the result in Fig. 12 (a). From this line profile (Fig. 12 (d)), the maximum height in the image is about 2.3 μm. Clearly, in this mode, the area contact between tip and sample is much smaller than shear force mode (about 10nm). As a result, the region contact between tip and sample may achieve the atom interaction, thus it prevents sample damage, and we could obtain the images with high contrast resolution.

7. Fork-AFM in analysis of property surface [29]

In this section we describe the measurement of the tip-sample interactions of a tuning fork with a tungsten tip in the shear-force mode operation on two classes of samples: i) samples with the "non-rigid surface" (soft) (polyethylene and fiber plastic) and ii) samples with the "rigid surface" (hard) (alumina Al_2O_3). In our study, we constructed a controllable Q-factor system, so that we could investigate the tip-sample interaction through the variation of the pre-set quality-factor (meaning the Q-factor at the free oscillation, when the tip is far from the sample surface, which we call hereafter Q_∞ for short). The results were interpreted with simple models that considered the damping forces between the tip and the surface polymer

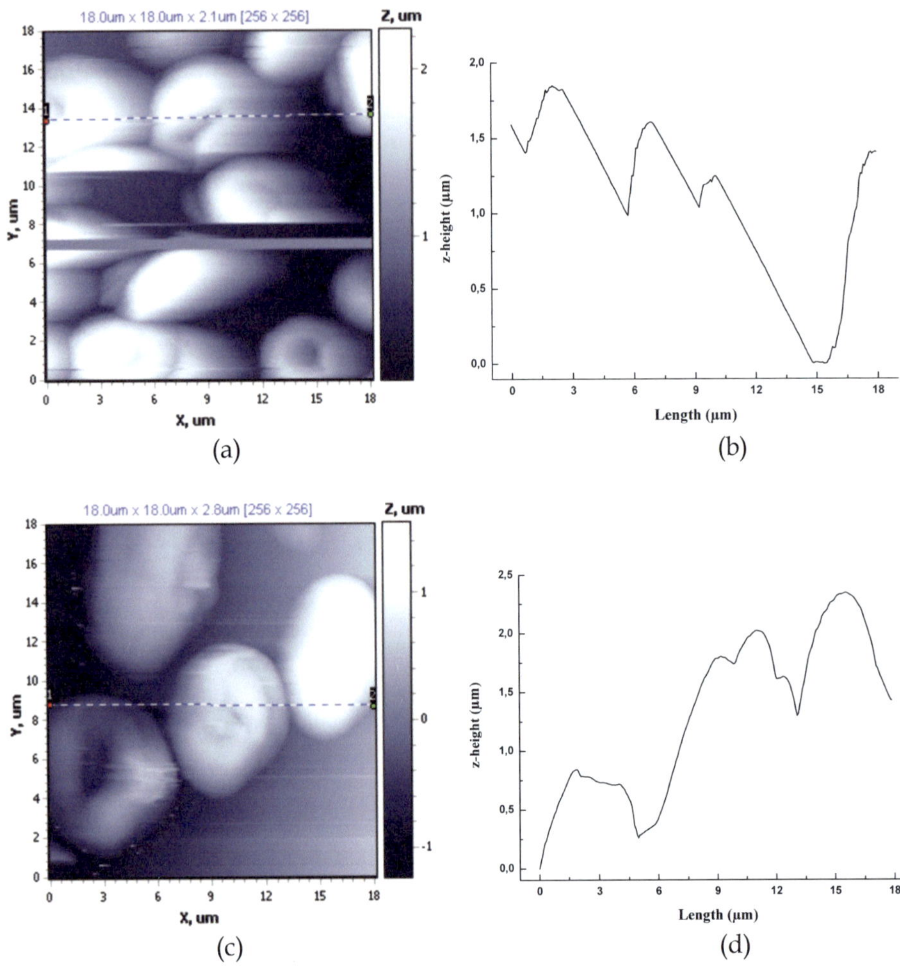

Fig. 12. (a, c) Shear force and intermittent contact mode topographical images of erythrocytes. A line profile from 1 to 2 in (a, c) is show in figure (b, d)

and the elastic forces of the polymer around the tip. The influence of the Q-factor on the interaction components of tip-sample was investigated. The results showed that the capabilities of the combination of a tuning fork and an AFM can be extended to quantitatively analyze the properties of the surface of nano-materials with high precision.

7.1 Dynamic force spectroscopy and theory of friction and elastic forces with controllable Q-factor

The amplitude-frequency-distance characteristic was measured during approach only or during retraction only to avoid backlash effects. In this measurement, we set the beginning, the final positions and the operation step of the probe. At the same time, we set up the driving frequency, ordinarily around the resonant frequency. When operating, we obtained the change of the amplitude-frequency when the tip was approaching to the surface sample very slowly. Figure 13 shows 2-D images taken in order to investigate the distance

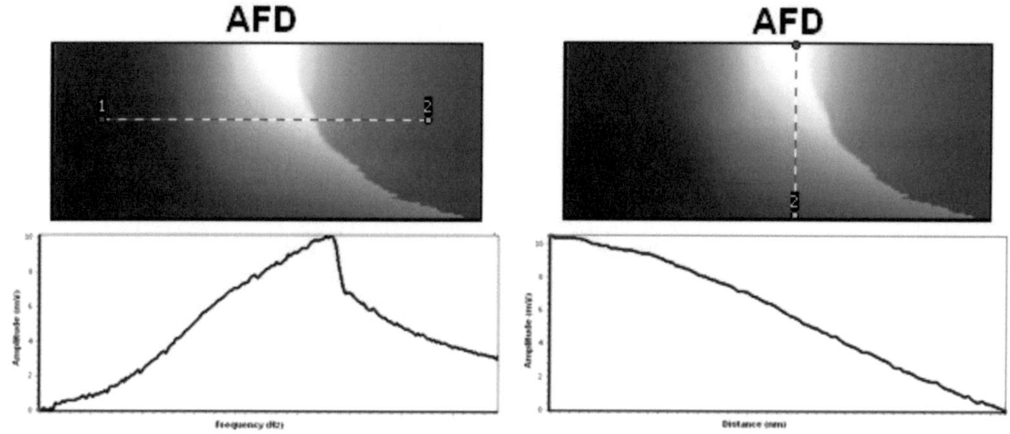

(a) (b)

Fig. 13. The 2D images of characteristic of spectroscopy amplitude-frequency-distance that operating in the dynamic shear-force. The 2-D grey scale images show the amplitude of the tuning fork vs. frequency (horizontal) and distance (vertical). In (a) the lower graph shows the amplitude as a function of frequency at fixed distance along the line 1-2 in the grey scale image; In (b) the amplitude is shown as function of distance along the line 1-2 at the resonant frequency.

dependence of these interactions on polyethylene polymer. In these images, the amplitude oscillation of the tip, that corresponded to vertical axis and defined by light-scale, was shown as a function of frequency (horizontal axis) and distance between tip and sample (vertical axis).

The tip position, d, is reported as a distance from an arbitrary reference point far from the surface, where Q-infinity is preset. An increase in d corresponds to a decrease in tip/sample separation. At each distance the amplitude is recorded as a function of frequency. All the data presented is as-measured, with no smoothing algorithms applied.

The frequency-dependent oscillation amplitude u of the arm of the tuning fork using a Q-control system obeys Newton's equation of motion given by [25, 27]:

$$\frac{\partial^2 u(t)}{\partial t^2} + \left(\gamma + \gamma_\infty + \frac{g}{\omega} \right) \frac{\partial u(t)}{\partial t} + \frac{K+k}{M} u(t) = \frac{F_D}{M} \tag{1}$$

The effective mass M of the arm, which is proportional to the mass distributed along the tine, is defined as: $M = \frac{K+k}{(2\pi f)^2} = 0.966x10^{-6} kg$. The gain g is the adjustable coefficient of Q-control system. The force F_D drives the fork harmonically at a frequency f. The damping of the motion is included in the damping rate term $\gamma + \gamma_\infty$, where gamma_inf corresponds to resonance parameters in a free oscillation. The damping rate intrinsic to the fork, where the tip disengaged from the sample, is γ. The tip-sample interaction is included in equation (1) through terms k and γ, respectively. We can then determine the tip sample-interaction-induced damping rate γ in relation to γ_∞ as

$$\gamma = \gamma_\infty \left(\frac{f_\infty u_\infty}{fu} - 1 \right) + \frac{g}{\omega} \left(\frac{u_\infty}{u} - 1 \right) \tag{2}$$

where f_∞ and f are the resonant frequency of quartz at free oscillation and at a height position, of which an interaction between tip and sample, respectively.

The damping force $F_f = M\gamma \dot{u}$ due to the tip-sample viscous damping is a drag force opposing the oscillatory motion of the tip. Using equations (1) and (2), F_f is given by:

$$F_f = \frac{i}{\sqrt{3}} \frac{K u_\infty}{Q_\infty} \left[\left(1 - \frac{fu}{f_\infty u_\infty} \right) + \left(\frac{Q_\infty}{Q_{eff}} - 1 \right) \left(1 - \frac{u}{u_\infty} \right) \right] \tag{3}$$

Here, we introduce a new coefficient Q_{eff}, that satisfies: $\frac{1}{Q_{eff}} = \frac{1}{Q_\infty} + \frac{\sqrt{3} g M}{K}$.

The elastic force F_e due to the reactive tip-sample interaction, which is a restoring force along the oscillation motion of the tip, can be expressed as $F_e = ku$. The theoretical spring constant K can be obtained from $K = \frac{E}{4} w \left(\frac{t}{l} \right)^3$ where $E = 7{,}87x10^{10}\, N/m^2$ is Young's modulus of quartz. For our tuning fork - width $w \approx 0.38 mm$, thickness $t \approx 0.6 mm$, and length $l \approx 5.00 mm$ - we obtain $K = 12\ kN/m$, which agrees reasonably well with our experimental results [22]. The local tip-sample interaction equivalent spring constant k can be obtained by using the relation for K in the equation: $k = \left[\left(\frac{f}{f_\infty} \right)^2 - 1 \right] K$, which gives F_e as

$$F_e = \left(\frac{f^2}{f_\infty^2} - 1 \right) K u \tag{4}$$

To interpret expressions (3) and (4), it is necessary to define the relationship between the measured output voltage V_{out} and the oscillation amplitude of one arm of the tuning fork. The output voltage is sensitive to the asymmetric mode of the tuning fork, therefore, $V_{out} = c(u_1 - u_2)$, where c is a constant and u_1, u_2 are the amplitude of motion of the two arms of the fork. When driving the fork with an external voltage, as in our experiment, only the asymmetric mode is excited, yielding $u_1 = -u_2$. Thus, we define: $V_{out} = 2cu_1 = u_1 / \alpha$, where $\alpha = 253 pm / mV$ as calculated in [22].

Using equations (2), (3) and (5) and with the combination of above relationship between the measured output voltage and the oscillation amplitude of one arm of the tuning fork, we have:

$$F_f = \frac{i}{\sqrt{3}} \frac{K V_\infty}{Q_\infty} \alpha \left[\left(1 - \frac{f V_{out}}{f_\infty V_\infty} \right) + \frac{\sqrt{3} g M}{K} Q_\infty \left(1 - \frac{V_{out}}{V_\infty} \right) \right] \tag{5}$$

$$F_e = \frac{K}{f_\infty^2} \alpha \left(f^2 - f_\infty^2 \right) V_{out} \tag{6}$$

with V_{∞} is the voltage corresponding to the free vibration of tuning fork. Based on these equations and the experimental data, we measured the amplitude and resonant frequency of prong of tuning fork when approaching the surface of polyethylene, and from there we determined the interaction force with surface the sample.

Figure 14(a) shows the results of the dependency between oscillation amplitude and oscillation frequency on the distance between tungsten tip and polyethylene surface where an increasing distance (d) corresponds to a decreasing tip/sample separation. We used a tuning fork with a tungsten tip having $Q_{\infty} = 1860$, resonant frequency 30584Hz. The applied voltage to the fork is 2.32V. As the tip approaches the surface, the amplitude *vs.* distance curve shows a monotonic decrease in amplitude. On the contrast, the resonant frequency f seems to increase monotonically though it is noisier. It can be seen that close to the surface of sample, the frequency reveals a more pronounced change as compared to the amplitude signal. Therefore, using modulated frequency in the near-contact region might give a better signal for distance control.

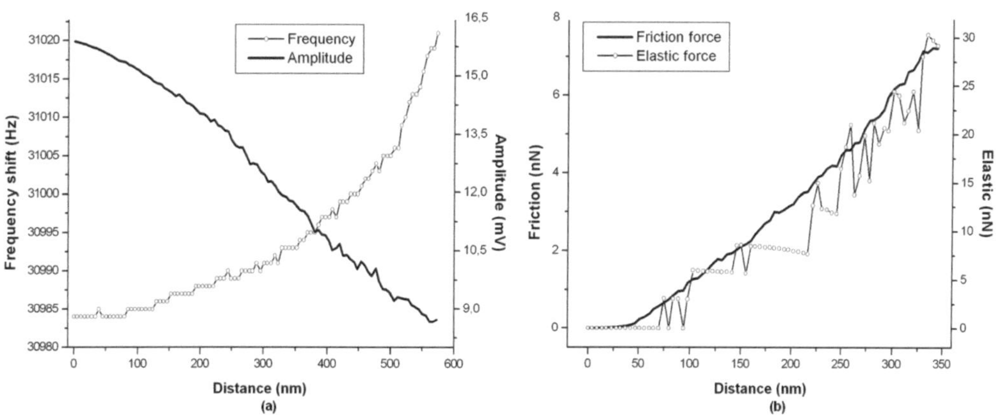

Fig. 14. The curves demonstrating the dependence of a) amplitude and frequency, b) force interactions (damping and elastic force) as a function of the tip-sample distance when the tip approaches to the surface of polyethylene. The tuning fork gluing tungsten tip with quality factor Q=1860, resonant frequency f=30584Hz, the applied voltage to the fork is 2.32V.

From the measurement results of amplitude and frequency, using equations (6) and (7) the damping and elastic forces are calculated and shown in Fig. 14 (b). The damping force changed when the separation tip-sample decreased. Similar to the damping force, the elastic interaction changed negligibly in the pre-setting contact of tip and sample. However, close to the surface, a different form of the elastic force, a "sawtooth" and "unstable" profile is observed. Therefore we could split into two interaction regions: (1) the vibrating tip is posited far from the sample; here there is no interaction between tip and sample, (2) when the tip is close to the surface of sample (the contact region), the damping force increases rapidly and monotonically. Our results of the dependence of amplitude, frequency shift, and force interactions (elastic and damping forces) in Fig. 14 (a, b) are similar to the works by Khaled Karrai et al. [27].

7.2 Influence of pre-set quality factor on parameters of tuning forks

There have been several experiments by different groups before (see, e.g., [25, 26, 28]) where the influence of a Q-control system on spectroscopy curves and on results of scanning has been investigated. But a quantitative interpretation of those experimental results is still lacking. So, for more clearly understanding, we surveyed the effect of the change of Q_∞ on the change of amplitude oscillation and quality factor of tuning fork. For a better comparison, we specifically choose one tuning fork, using the same applied voltage and measuring under the same conditions of operation, with no change of the area of the sample approached by the tip.

Firstly, we surveyed the influence of the pre-set quality-factor Q_∞ on the variation of Q factor of tuning fork during the tip approaching to the surface on these materials. Figures 15 (a, b) show the curves of quality factor versus distance for the approach on the polyethylene and alumina. We observe that the quality factor Q decreases as distance between tip-sample is reduced. For alumina the decrease of Q is rapid and rather abrupt. For polyethylene the change in Q is slow and not sudden. However, here we do not observe the influence of Q_∞ on these curves. It is seen that the variation of quality factor Q only depends on the properties of surface of materials, but not upon Q_∞. Therefore, it would be interesting to use these results for examining the general structures of the surface sample.

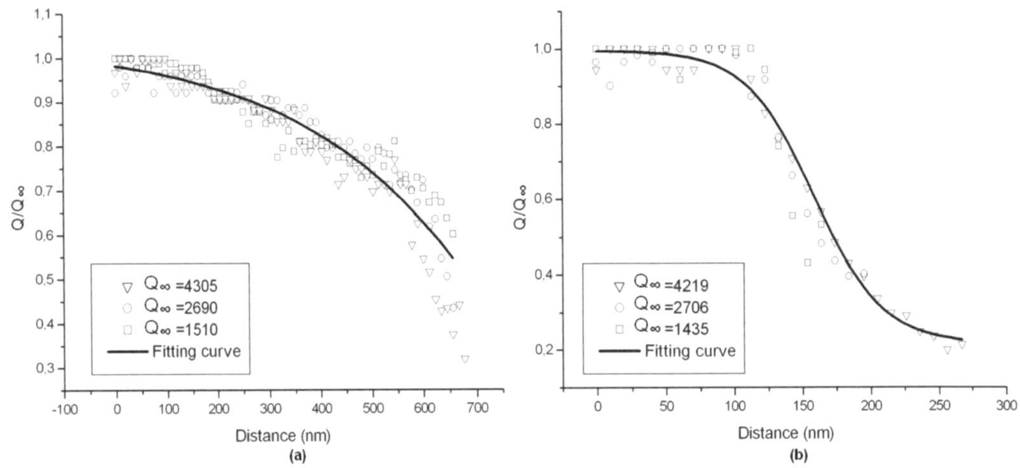

Fig. 15. The curves of the variation of quality factor versus distance between tip-sample that using the different pre-set quality factor Q_∞ on (a) fibre plastic, (b) alumina. The dark curves are the fitting curves of the experimental data.

In order to illustrate the application of the variation of quality factor Q with sample, we performed the above experiments on three materials: polyethylene, fiber plastic and alumina Al_2O_3 (in Fig. 16). We observed an obvious difference between these materials. While the change of the slope Q versus the distance between tip-sample for polyethylene and fiber is rather shallow, the result for alumina Al_2O_3 is a steep slope. Therefore we are confident that we could expand this measure to identify the surface property of the other materials: "rigid" or "non-rigid" materials.

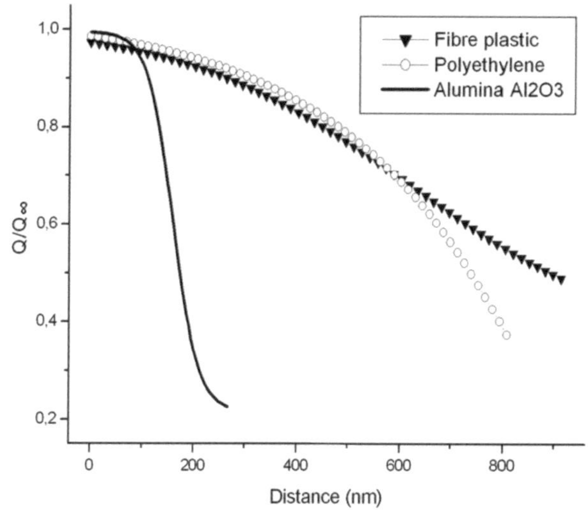

Fig. 16. The fitting of the measured data demonstrating the dependence of the change of quality factor on distance of tip-sample when the tip approaches to the surface on three samples: polyethylene, fiber plastic and alumina Al2O3.

In order to understand the observed material dependence of the variation of Q during the tip approaching the surface, we make some assumptions for two classes of materials: for the alumina sample, the property of surface is rather homogeneous, and has a rigid structure. In the shear-force mode, the tip oscillates horizontally. On the tip abruptly enters the influence region of the atomic forces. Therefore the variation of the interaction curves of alumina sample is rather abrupt. In contrast, the surface of polyethylene or fiber plastic has no tight structure and is not smooth. The influence of the atomic forces is not uniform and there is not a clear bound of the interaction region. So when approaching closer to the "non-rigid" sample, the tip is not affected suddenly, but rather smoothly. This observation may be very useful for surveying the properties of surface for the different types of samples.

As mentioned above, the variation of the quality factor Q during the tip approach is not affected by the change of Q_∞. On the other hand, the change of the amplitude oscillation is strongly affected by the change of Q_∞. To understand this effect, we measured the change of the amplitude oscillation of the tip as it approached closer to the sample using different pre-set Q_∞ on three samples: polyethylene, fiber plastic, alumina, as shown in Fig. 17 (a - c). For alumina, the larger Q_∞ resulted in the shallower slope of the curve of the change amplitude versus distance between tip-sample. However the trends differed for polyethylene and fibre plastic. Furthermore, we realized that for "non-rigid" materials, at positions closer to the sample, the amplitude oscillation of the tip increased when Q_∞ decreased. We did not observe this behavior on alumina. These results explained that we could be able to obtain the information about images as well as the properties of the surface of "non-rigid" samples more advantageously when reduced Q_∞. It is suitable for the observations that we could achieve the best scan rate and obtained the real time imaging and high resolution with low quality factor Q_∞ [25].

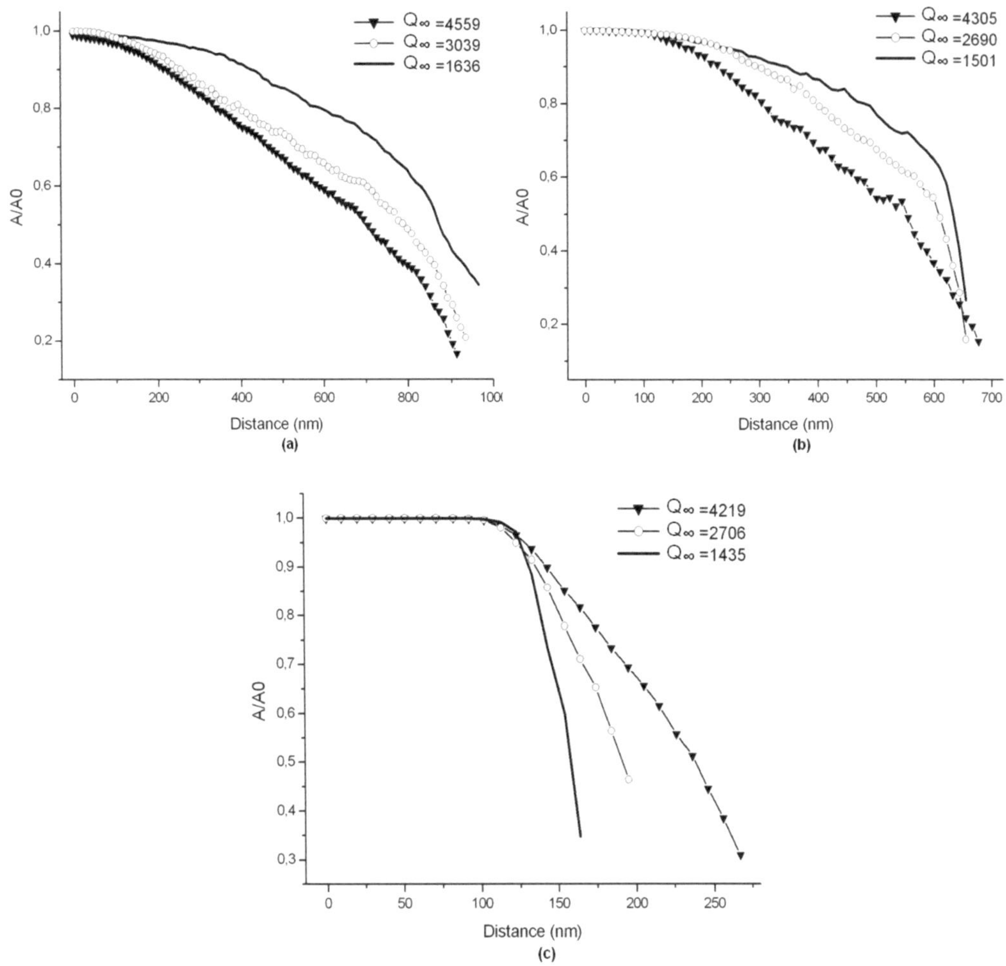

Fig. 17. The amplitude dependence on tip-sample distance for three pre-set quality factor Q_∞ on (a) polyethylene, (b) fiber plastic, (c) alumina Al2O3.

7.3 Influence of pre-set quality factor on the force interactions

Figures 18 (a - d) showed the resulting force interaction curves with the different pre-set quality factors on the "soft" material: polyethylene, and "rigid" material: alumina (Al_2O_3). A clear effect of the variation of factor Q_∞ on the damping force is observed for "soft" materials (polyethylene or fiber plastic). At the same distances between the tip and sample, corresponding to lower preset factors Q_∞, the damping force is larger (Fig. 18 (a)). These results demonstrated that further reduction of damping force can be achieved by decreasing the fork's quality factor. We predicted that with the "soft" materials in the semi-contact and contact regimes we could change the quality factor and therefore increased the effective tip-sample interaction, leading to higher stabilization in shear-force operation. In contrast, for alumina, the major difference was observed in the results of the curves of the damping force. In Fig. 18 (c), the damping force for alumina depended on the change of the pre-set Q_∞.

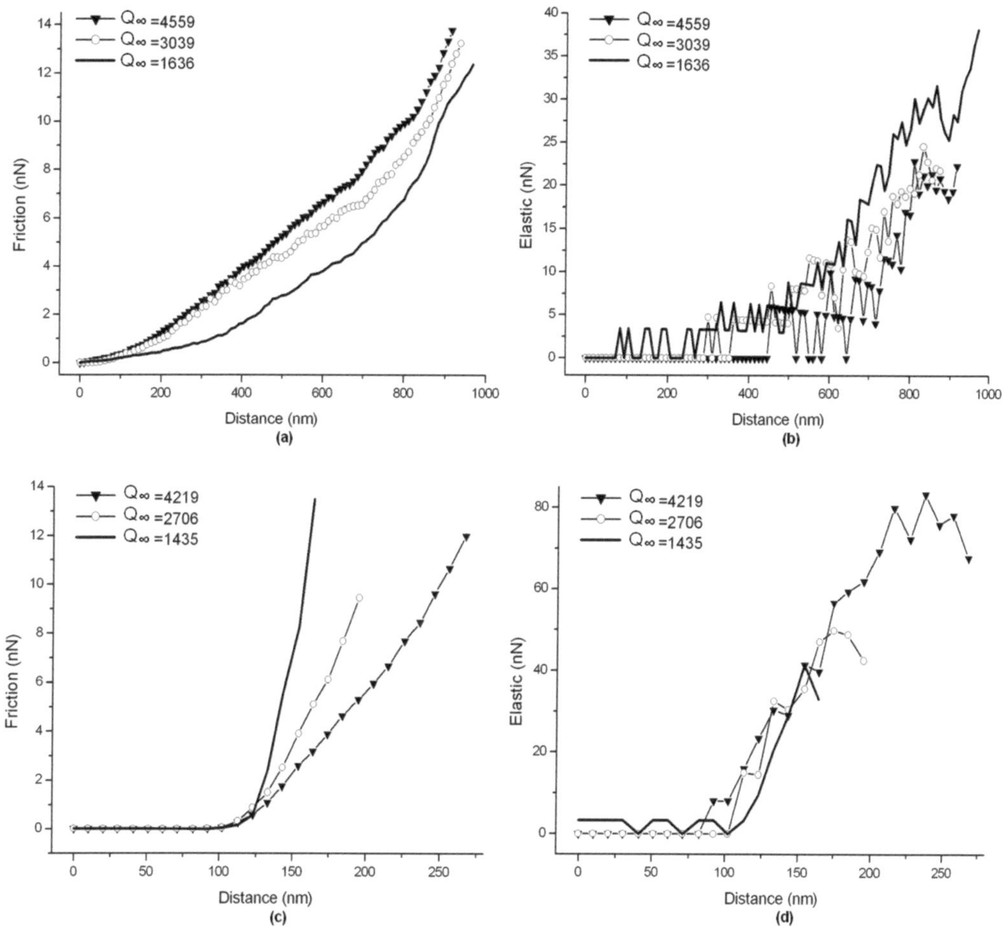

Fig. 18. The force spectroscopy curves demonstrating the influence of the change of the free quality factor on force interactions (damping and elastic force) when the tip approaches to the surface of (a, b) polyethylene sample; (c, d) alumina Al2O3.

However, the larger Q_∞, the smaller the variation of the slope curves of the damping force. We do not compare the increase of the damping force at the positions closer to samples corresponding to the three different pre-set quality factors Q_∞.

8. Discussions and conclusions

We have discussed the quartz tuning fork, a two-terminal electronic component whose use is essential in applications that require an accurate time reference. We have also demonstrated atomic force microscopies using quartz tuning with and without Q-control fork in air conditions in two operation modes: shear force and intermittent contact modes. The scanning time can be reduced so that for the images of smaller size one can acquire high resolution images almost in real time, which may allow the recording of high resolution videos. Reproducible topographic images have been obtained on hard and soft samples. Furthermore, it suggested that Fork-AFMs can become a very useful and reliable tool in the study of biomolecules. The obtained results showed that the capabilities of the combination of AFM and tuning fork could enable quantitatively analyses of the properties of surface with high precision and resolution.

Furthermore, we have measured the dynamic spectroscopy of amplitude-frequency-distance and developed a model to determine the force interactions (damping and elastic forces) using tuning fork gluing tungsten tips in shear-force mode. A survey of the influences of the pre-set quality factor (by using the Q-control) on the interaction components between the tip and sample was presented. They also opened up new ways to explain the advantages of using the low pre-set quality factor for tuning forks to get better information on the surface of "non-rigid" materials. In the future, this will provide a framework to extend applications of the tuning fork based on AFM in qualitative analysis in nanotribology, and to provide insight into determinations of parameters and properties of surface of soft material, such as selection of structure surface of materials, viscosity coefficient and shear modulus.

9. Acknowledgments

The authors would like to thank Professor J. Maps from University of Minnesota Duluth, USA, for helpful discussions.

The work was carried out in the frame of the Belarusian Project 1.9 of SSTP "Scientific Equipment".

10. References

[1] Binnig, G.; Quate, C. F. & Gerber, C. (1986). Atomic force microscope, *Phys. Rev. Lett.*, Vol. 56, pp. 930-933

[2] Tran Xuan Hoai; Tran H. Hung; Au D. Tuan; Phan Canh; Phung T. Thuc & Vo T. Tung (2004). Design a DSP controlling atomic force microscope working with Linux operating system, *Proceeding of ICT.rda'04, Hanoi*, pp. 31-39, Vietnam, September 18-20, 2004

[3] Guethner, P.; Fischer, U. & Dransfeld, K. (1989). Scanning near-field acoustic microscopy, *Appl. Phys. B. Photophys. Laser Chem.*, Vol. B 48, pp. 89-92

[4] Giessibl, F.J. (1998). High-speed force sensor for force microscopy and profilometry utilizing a quartz tuning fork, *Appl. Phys. Lett.*, Vol. 73, pp. 3956-3958

[5] Karrai, K. & Grober, R.D. (1995). Piezoelectric tip-sample distance control for near field optical microscopes, *Appl. Phys.Lett.*, Vol 66, pp. 1842-1844

[6] Giessibl, F.J. (1989). High-speed force sensor for force microscopy and profilometry utilizing a quartz tuning fork, *Appl. Phys. Lett.*, Vol. 73, pp. 3956-3958

[7] Rychen, J.; Ihn, T.; Studerus, P.; Herrmann, A. & Ensslin, K. (1999). A low-temperature dynamic mode scanning force microscope operating in high magnetic fields, *Rev. Sci. Instrum.*, Vol. 70, pp. 2765-2768

[8] Zang, J. and O'Shea,S. (2003). Tuning forks as micromechanical mass sensitive sensors for bio- or liquid detection, *Sensor and Actuator*, Vol. 94, pp. 65-72

[9] Rensen, W. H. J. (1995). Imaging soft samples in liquid with tuning fork based shear force microscopy, *Appl. Phys.Lett.*, Vol. 66, pp. 1842-1844

[10] Karrai, K. and Tiemann, I. (2000). Interfacial shear force microscopy, *Phys. Rev. B*, Vol. 62, pp. 13174-13181

[11] Tamayo, J.; Humphris, A.D.L. & Miles, M.J. (2000). Piconewton Regime Dynamic Force Microscopy in Liquid, *Appl. Phys. Lett.*, Vol. 77, pp. 582

[12] Antognozzi, M.; Binger, D.R.; Humphris, A.D.L.; James, P.J. & Miles, M.J. (2001). Modeling of cylindrically tapered cantilevers for transverse dynamic force microscopy (tdfm), *Ultramicroscopy*, Vol. 86, pp. 223

[13] Bhushan, B. (2006). *Springer Handbook of Nanotechnology* 2nd, Springer, Heidelberg

[14] Lei, F.H.; Nicolas, J.-L. & Troyon, M. (2003). Shear force detection by using bimorph cantilever with the enhanced Q-factor, *J. Appl. Phys.*, Vol. 93, pp. 2236–2243

[15] M. P. Forrer, (1969). A flexure-mode quartz for an electronic wrist-watch, *Proceedings 23rd ASFC Ann. Symp. on Frequency Control*, pp. 157–162

[16] Yoda, H.; Ikeda, H. & Yamabe, Y. (1972). Low power crystal oscillator for electronic wrist watch, *Proceedings 26th ASFC Ann. Symp. on Frequency Control*, pp. 140–147,

[17] Ong, P. P. (2002). Little known facts about the common tuning fork, *Phys. Educ.*, Vol. 37, pp. 540–542

[18] Matsiev, L.F. (1999). Application of flexural mechanical resonators to simultaneous measurements of liquid density and viscosity, Ultrasonic symposium, Vol. 1, pp. 457–460

[19] Chuang, S.S. (1981). Quartz Tuning Fork Crystal Using Overtone Flexure Modes, 35th Annual frequency control symposium, pp. 130–143

[20] Itoh, H.; Aoshima, Y. & Egawa, T. (2001). Model of a quartz crystal tuning fork using torsion spring at the joint of the arm and the base, *Proceedings* of the 2001 IEEE International Issue Date, pp. 592–596

[21] Blaauwgeers, R.; Blazkova, M.& Človečko, M. (2007). Quartz tuning fork: thermometer, pressure - and viscometer for Helium liquids, *Low temperature physics*, Vol. 146, pp. 537–562

[22] Vo Thanh Tung, Chizhik, S.A.; Chikunov, V.V.; T.V. Nguyen & X.H. Tran (2006). Influence of additional mass on quartz tuning fork in dynamic operation mode, *Proceeding of 7th Int. BelSPM-7*, Belarus, pp. 236–240

[23] Lei, F.H.; Angiboust, J.F.; Qiao, W. (2004). Shear force near-field optical microscope based on Q-controlled bimorph sensor for biological imaging in liquid, *Journal of Microscopy*, Vol. 216, pp. 229–233,

[24] Vo Thanh Tung (2008). Influence of Q-control on Shear-force Detection with Quartz Tuning Fork in Atomic Force Microscopy, *Молодежь в науке – 2007: прил. к журн. «Весці Нацыянальнай акадэміі навук Беларусі»*. – Минск: Белорус. наука, Vol. 3, pp. 85–89

[25] Vo Thanh Tung & Chizhik, S.A. (2007). Quartz tuning fork atomic force microscopy using quality-factor control, *Proceding. of Int. Physics, Chemistry and Application of Nanostructures: Reviews and Short Notes to Nanomeeting-2007 Minsk, Belarus, (World Scientific Pub Co Inc.)*, pp. 535–538

[26] Friedt, J.-M. & Carry, É. (2007). Introduction to the quartz tuning fork, *Am. J. Phys.*, Vol. 75, pp. 415–422,

[27] Karrai, K. & Tiemann, I. (2000). Interfacial shear force microscopy, *Phys. Rev. B*, Vol. 62, pp.13174-13181

[28] Callaghan, F. D.; Yu, X. & Mellow, C. J. (2005). Variable temperature magnetic force microscopy with piezoelectric quartz tuning forks as probes optimized using Q-control, *App. Phys. Lett.*, Vol. 87, pp. 214106

[29] Vo Thanh Tung; Chizhik,S. A. & Tran Xuan Hoai (2009). Parameters of tip-sample interactions in shear mode using a quartz tuning fork AFM with controllable Q-factor, Journal of Engineering Physics and Thermophysics, Vol. 82, No. 1, pp. 140-148

Section 2

Surface Morphology

3

Statistical Analysis in Homopolymeric Surfaces

Eralci M. Therézio[1], Maria L. Vega[2],
Roberto M. Faria[3] and Alexandre Marletta[1]
[1]*Universidade Federal de Uberlândia, Instituto de Física*
[2]*Universidade Federal do Piauí, Departamento de Física*
[3]*Universidade de São Paulo, Instituto de Física de São Carlos*
Brazil

1. Introduction

Surface topology investigation of luminescent polymeric films is a promising research area due to the interest in their technological application, mainly to large-area electroluminescent displays (Holzera et al., 1999). Surface topology of organic flexible materials is generally controlled by a chemical and physical process, the polymer coatings are generally heterogeneous, and the most significant changes occur in the nanometer scale (Gobato et al., 2002). The Atomic Force Microscopy (AFM) technique, in special, has made significant experimental contributions to three-dimensional imaging with high-resolution in a sub-nanometer scale, without specific sample preparation (Bar et al., 1998; Binnig et al., 1986; Gua et al., 2001). In soft biological or polymeric materials, the permanent contact of the tip produces irreversible deformation, modifying the surface topology (Magonov & Whangbo, 1996; Weisenhorn et al., 1992). To minimize these deformations, the AFM tapping mode was introduced as a non-destructive technique (Marti et al., 1999). In general, AFM systems include commercial software to control the probe and capture process, and analyze surface images such as height or phase. The subjective analysis of AFM images that sometimes depends on the high quality of the experimental data is commonly found in the literature. For example, quantitative analysis, uniformity, fractality, interface, nanostructure, composition, and others cannot be accomplished. The difficulty in performing the quantitative characterization of the surface increases during homopolymer films surfaces investigations, due to macro-molecule interpenetration, thus forming a very complex system (Marletta et al., 2010).

In this chapter, we focus on the study of surface homopolymer films, introducing first and second order statistics analysis. Poly(*p*-phenylene vinylene) films were investigated since they are easily processed by the casting technique. Casting PPV films were processed by a conventional precursor polymer route and thermal annealed at 230°C, under vacuum (10-3 mbar) (Marletta, 2000). AFM images were performed by the commercial *NanoScope® IIIa Multimode ™ of Digital instruments*, Santa Bárbara, CA, in a tapping mode. The first order statistical analysis was used to quantify the surface's topology, by calculating surface height distribution and roughness distribution moments mean squares, Skewness, and Kurtosis. Additional analysis was performed by a Scanning Probe Image Analysis (SPIA) (Costa et al., 2003) customized program to obtain the number of peaks, using the maximum regional concept and distances between selected peaks. The second order statistical analysis was used to calculate the 1D height auto-covariance function of determined surface nanostructures. In addition, we have discussed the possibility to correlate the scattering of lights and the surface´s roughness. Finally, surface changes due to photo-bleach effects, oxidation phenomena, caused mainly by the oxidative processes, were investigated.

2. Statistical analysis of topologic surfaces

2.1 First-order statistical analysis

A random rough surface can be described by defining its height function, $h = h(\vec{r})$. The height h on a surface position $\vec{r} = (x, y)$ is obtained relatively to the mean height ($\bar{h}$), and the simplest approach to describe the surface is through height distribution *p(h)*. It gives the probability to find the height h between h and $h+dh$ at any point $(\vec{r})$ on the surface, and is positive and normalized as (Thomas, 1999; Zhao et al., 2001):

$$\int_{-\infty}^{+\infty} p(h)dh = 1 \tag{1}$$

It is important to emphasize that the purely random process, i.e. the random surface height distribution, is generally described by a Gaussian function. However, different statistical processes may be related to the treatment or growth mechanism of the surfaces. It is important to stress that the height distribution $h = h(\vec{r})$ was obtained by the SPIA program (Costa et al., 2003), where $\bar{h} = 0$. The first step to check whether the surface will be flat and/or rough, is through roughness measurement. The most used AFM technique is the roughness mean square (σ_{RMS}) (Palasantzas, 1993; Simpson et al., 1999), and the profile of the height distribution *p(h)* is quantified by the value of the central moment, which is defined as (Zhao et al., 2001):

$$\sigma_n = \int_{-\infty}^{+\infty} \left(h - \bar{h}\right)^n p(h)dh. \tag{2}$$

The 1st-order moment is the height average: $\sigma_1 = \bar{h}$; the 2nd-order moment of the variable h is the root-mean-square (RMS) roughness (σ_{RMS}), which is given by:

$$\sigma_{RMS}^2 = \sigma_2 = \int_{-\infty}^{+\infty} \left(h - \bar{h}\right)^2 p(h)dh. \tag{3}$$

σ_{RMS} describes the fluctuations around the mean value $\bar{h}$, as the surface height approaches to random behavior, σ_{RMS} tends towards the width of the Gaussian distribution.

$$p(h) = \frac{1}{\sqrt{2\pi\Delta^2}} \exp\left(-\frac{(h-\bar{h})^2}{2\Delta^2}\right) \tag{4}$$

Figure 1 shows the typical height distribution for normal surfaces and a surface covered with bumps and pits. However, it is important to emphasize that different rough surfaces can have the same $p(h)$ and σ_{RMS}, but different height fluctuation frequency.

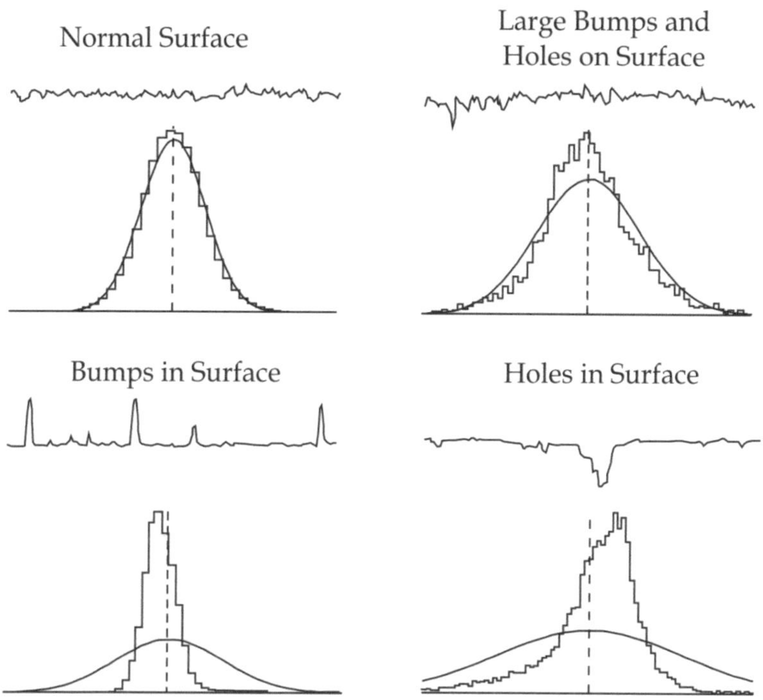

Fig. 1. Typical height distribution functions and equivalent Gaussians for a normal surface and a surface covered with bumps and pits (Bennett & Mattsson, 1999).

The higher-order moments give us more information about the surface height distribution, which is useful for analyzing surfaces in more detail. The Skewness (SK) is the 3rd-order moment given by:

$$\sigma_{SK} = \frac{\sigma_3}{\sigma_{RMS}^3} = \frac{1}{\sigma_{RMS}^3} \int_{-\infty}^{+\infty} \left(h - \bar{h}\right)^3 p(h) dh . \quad (5)$$

And σ_{SK} is sensitive to the asymmetry of the distribution. The Gaussian distribution has Skewness equal to zero, since it presents equally distributed peaks and valleys. Height distribution with negative Skewnees ($\sigma_{SK} < 0$) is due to a larger number of valleys and that with positive Skewnees ($\sigma_{SK} > 0$) is attributed to the larger number of peaks. By and large, this parameter gives an indication of the existence of a deep valley or sharp peaks. The 4th-order moment defines the Kurtosis (KU), given by

$$\sigma_{KU} = \frac{\sigma_4}{\sigma_{RMS}^4} = \frac{1}{\sigma_{RMS}^4} \int_{-\infty}^{+\infty} \left(h - \bar{h}\right)^4 p(h) dh . \quad (6)$$

σ_{KU} is the measurement of the height distribution sharpness, and it describes the randomness of the surface related to that perfectly random surface (Gaussian distribution), where $\sigma_{KU} = 3$. For $\sigma_{KU} > 3$, the distribution is platykurtic (mild peak) and for $\sigma_{KU} < 3$ the distribution is leptokurtic (sharp peak). The parameters σ_{RMS} (eq. 3), σ_{SK} (eq. 5), and σ_{KU} (eq. 6) are dependent on the on the measurements conditions such as: a) scanning size, b) lateral and vertical resolution and, c) sampling density (Bennett & Mattsson, 1999; Thomas, 1999; Zhao et al, 2001).

2.2 Second-order statistical analysis

First-order statistics also gives information about the surface height at the individual position $\vec{r}$, that is, it does not reflect the correlation between the two different points $\vec{r}_1$ and $\vec{r}_2$. To take into account such specific situation in *homopolymers*, we introduced the second-order statistics, by calculating the height auto-covariance function $G(\vec{r}_1,\vec{r}_2)$, defined as (Zhao et al., 2001):

$$G(\vec{r}_1,\vec{r}_2) = \int_{-\infty}^{+\infty}\int_{-\infty}^{+\infty} h_1 h_2 p(h_1,h_2;\vec{r}_1,\vec{r}_2)dh_1 dh_2 \ . \tag{7}$$

Equation 7 gives the probability to find the height h_2 at $\vec{r}_2$ provided that we have the height h_1 at $\vec{r}_1$. For the homogeneous and isotropic rough surface, almost a normal surface, we consider that the $G(\vec{r}_1,\vec{r}_2)$ depends only on the distance between $\vec{r}_1$ and $\vec{r}_2$

$$G(\vec{r}_1,\vec{r}_2) = G(\rho)\,, \tag{8}$$

where $\rho = |\vec{r}_1 - \vec{r}_2|$.

is the quantity and ρ is the translation coordinate, sometimes called the lag or the slip. For $\rho = 0$, $G(\rho)$ is equal to the variance of the surface height ($G(0) = \sigma_{RMS}^2$). For a random rough surface, the height auto-covariance function ($G(\rho)$) decreases to zero when the lateral distance increases ($\rho \rightarrow \infty$). In the first case, the shape of the function $G(\rho)$ depends on the type of random surface and the distance over which two points become uncorrelated. The lateral correlation length ξ of an auto-correlation function defines the representative lateral dimension of the rough surface or the radius where two points cannot be considered correlated any more. The intensity of the correlation is defined as the radius where the function decays to $1/e$ of its zero-value:

$$G(\xi) = \frac{G(0)}{e} = \frac{\sigma_{RMS}^2}{e} \ . \tag{9}$$

Useful height auto-covariance functions are exponential (Eq. 10a), Gaussian (Eq. 10b), and self-affine (Eq. 10c) (Bennett & Mattsson, 1999; Palasantzas, 1993; Thomas, 1999). These functions are quite useful to describe surface topology. For example, $G(\rho)$ displays harmonic oscillation (sine or cosine) correlated with the surface periodic structure in a nanoscale dimension.

$$G(\rho) = G(0)\exp\left(-\frac{\rho}{\xi_{\exp}}\right), \tag{10a}$$

$$G(\rho) = G(0)\exp\left(-\frac{\rho^2}{\xi_G}\right), \tag{10b}$$

$$G(\rho) = G(0)\exp\left[-\left(\frac{\rho}{\xi_a}\right)^{2\alpha}\right]. \tag{10c}$$

In surface scattering theory, the position of the outgoing light is not necessarily the same as that of the incoming position. In this chapter, we consider that the light emitted by the PPV is scattered on the surface film/air and the light analyzed is obtained outside the material. It is possible to determine this interface's scattering intensity quantitatively from the correlation function of the rough surface profile. Using the inverse Fourier transform of the function $G(\vec{r}_1, \vec{r}_2)$ and the height auto-covariance function (Eq. 7), we obtain the following expression (Assender et al., 2002):

$$\langle I_s(\xi,\eta)\rangle = CTE * R * \iint_A G(\vec{r}_1, \vec{r}_2) \times \exp\left\{-\frac{2\pi i}{\lambda f}(\xi r_1 + \eta r_1)\right\} dr_1 dr_2 \tag{11}$$

where ξ and η are local variables, $R = \exp\left[-\left(\frac{4\pi\sigma_{RMS}}{\lambda}\right)^2\right]$ is the reflectance of the surface, σ_{RMS} is the root-mean-square roughness (eq. 3), λ is the scattered wavelength, $CTE = \frac{E_0^2 4k_Z^2}{2A^2 Z_W}$ is the constant, E_0 is the amplitude of the plane wave with air resistance $Z_W = \sqrt{\frac{\mu\mu_0}{\varepsilon\varepsilon_o}} \approx 377\Omega$, $\mu(\mu_0)$, k_z is the wavelength number, and A is the area of the laser excitation. Considering the homogeneous and isotropic rough surface (eq. 8), i.e., normal surface, we can rewrite the equation 10, by considering the maximum scattering intensity at $\xi = 0$ and $\eta = 0$ for the light outcome in the interface polymer/air and the normal surface approximation (eq. 8) as:

$$\langle I_{MAX}\rangle = \langle I(\xi = 0, \eta = 0)\rangle \propto \frac{R}{\sigma_{RMS}^2} \iint_A G(|\vec{r}_1 - \vec{r}_2|) dr_1 dr_2 \tag{12}$$

Where *G(ρ)* is the height auto-covariance function.

3. Casting PPV films

In this section, we will present the topological and optical study of casting PPV films in regards to thickness, stretching, and photo-blanching effects.

3.1 Topological analysis

By using 1st-order statistical analysis of surface homopolymer films, it is possible to investigate thickness effects on surface topology. It is important to emphasize that the phase image does not contribute, in the present study, with any additional information. Casting PPV films were processed from a conventional precursor polymer route, thermal annealed at 230°C, under vacuum (10^{-3} mbar), and thickness of 0.4, 0.9, 2.2, and 3.2 μm.

Figure 2 displays AFM height images, in a 10x10 nm^2 area for casting PPV films thickness of 0.4 μm (Fig. 2a) and 3.2 μm (Fig. 2b). A simple visual inspection shows a lot of peaks with height of about 50.0 nm for the thick film. However, the topology of a thin film is more homogeneous. So, what is the height distribution profile? Do the highest peaks represent a significant area on the film? What are the changes on the surface topology? Well, to answer these questions, we quantified the images using the 1st-order statistical analysis.

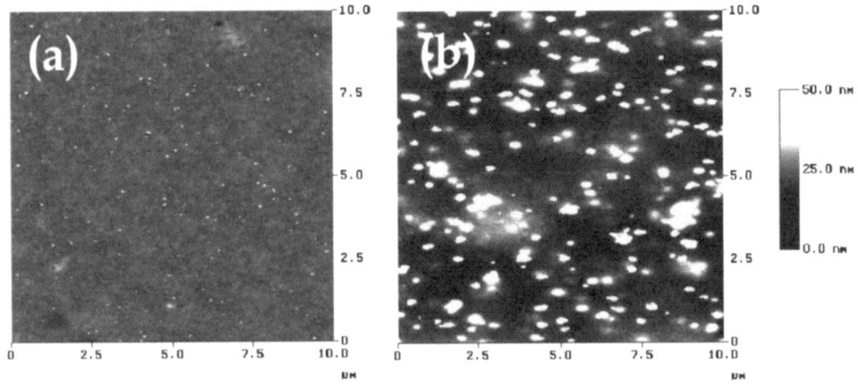

Fig. 2. AFM height image for *casting* PPV films with different thickness: (a) 0.4μm and (b) 3.2 μm.

Figure 3a and 3b shows the height histogram for 0.4 μm and 3.2 μm films, respectively, in which we can observe considerable dispersion of heights for the thickest film. One has thinner films, with less than 4.5% of peaks' heights (between 25 and 100 nm). As for thick films, the percentage increases considerable to 67.6%, in the same range. Latter data is not consistent with the visual observation of the AFM image! It is important to stress that the histogram in Figure 3 is a direct count of the heights of the AFM image in Figure 2. That type of histogram does not allow us to infer about the height distribution *p(h)* and to compare it with the types in Figure 1.

In order to improve the quantitative analysis of the height distribution, figure 4 plots the *p(h)* (Eq. 1) height distribution considering $\sigma_1 = \bar{h} = 0$ for thickness films of 0.4 μm (Fig. 2a) and 3.2 μm (Fig. 2b). It is interesting to observe the profile of both histograms where the thinnest film (Fig. 2a) is close to a normal surface and the thicker film is close to bumps on the surface (Fig. 1). The simple histogram analysis in Figure 2 is not able to demonstrate the behavior conclusion. To quantify the profile of *p(h)* (Figure 4), we calculated the values of σ_{RMS} (Eq. 3), σ_{SK} (Eq. 4), and σ_{KU} (Eq. 5). It is important to emphasize that these parameters depend on the AFM experiment, i.e., scanning size and lateral and vertical resolution (Bennett & Mattsson, 1999; Thomas, 1999; Zhao et al., 2001). Table 1 shows the evolution of the parameters for each sample. By increasing film thickness, the σ_{RMS} rises considerably, according to the AFM image in Figure 2b. The σ_{SK} parameter does not change significantly and presents relative low value (~3-5), closer to a normal surface for all samples. σ_{KU} decreases significantly when sample thickness increases, indicating the presence of sharp peaks on the surface. Finally, we can consider the values for σ_{SK}, σ_{KU}, and σ_{RMS} in Table 1 as closer to the Gaussian distribution ($\sigma_{RMS} = \Delta$, $\sigma_{SK} = 0$, and $\sigma_{KU} = 3$), for all samples.

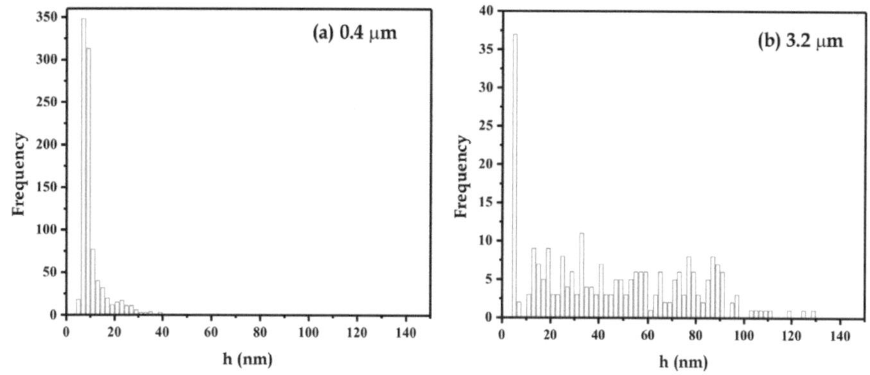

Fig. 3. Height histogram of AFM images in Figure 2, with thickness of (a) 0.4μm and (b) 3.2 μm.

Thickness (μm)	0.4	0.9	2.2	3.2
σ_{RMS} (nm)	1.5 ± 0.1	7 ± 4	10 ± 1	14 ± 2
σ_{SK}	5 ± 1	3.6 ± 0.3	3.1 ± 0.1	3 ± 1
σ_{KU}	54 ± 5	22 ± 3	13 ± 3	13 ± 4

Table 1. First-statistical parameters: roughness (σ_{RMS}), Skewness (σ_{SK}) and Kurtosis (σ_{KU}) in function of film thickness of 0.4, 0.9, 2.2, and 3.2 μm.

Fig. 4 (continuous line) shows data adjustment for *p(h)*, using the equation. The mean square root between the histogram and the fit, using the Gauss function, is about 97% for all samples, according to the approach mentioned above. Figures 4a and 4b show the contribution of high height values (|h|>25nm) for films with thickness of 0.4 and 3, 0% and 9% respectively. These findings corroborate with the discussion above mainly for thicker films, where the number of peaks above 25 nm is smaller than the calculation using the height histogram in Fig. 3. Finally, we can conclude that surfaces are essentially topological structured at low height values (< 25nm) and the thickness of PPV casting films, after being processed, does not change the polymeric surface significantly.

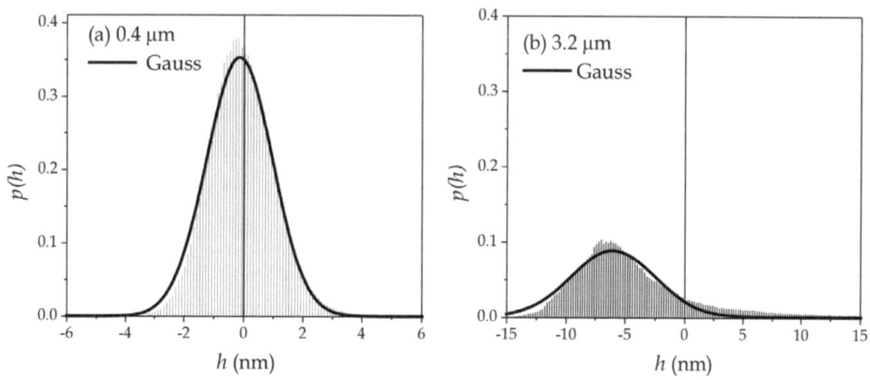

Fig. 4. Height distribution *p(h)* for film thickness 0.4 μm (a) and 3.2 μm (b). The continuous line represents the adjustments using Gaussian function (Eq. 4).

We have also analyzed the distance between the first adjacent selected peaks, 50, 100, 150, 200, and 250, using the methodology developed by Costa et. Al., 2003. In summary, the count is performed by slicing the AFM image along its thickness in a step $\Delta Z=(h_{max}-h_{min})/n$, where n is the number of slices, adding new peaks after each slice. The mean distance between first-neighbor introduces a number of combinations equal to $n!/(2!\,(n-2)!)$. It shows, mainly, how uniform the peak distribution is on the film surface. The mean distance <*d*> was calculated as:

$$\langle d \rangle = \frac{\sum_j d_j \cdot f_j}{\sum_j f_j} \tag{13}$$

where d_i is the distance between two selected peaks with frequency f_i. Figure 5 shows the mean distance for the nearest selected peaks for all samples, considering ΔZ = 0.1nm. The value of <*d*> (Eq. 13) and the mean height $\sigma_1 = \bar{h}$ (Eq. 2) are also listed in Table 2.

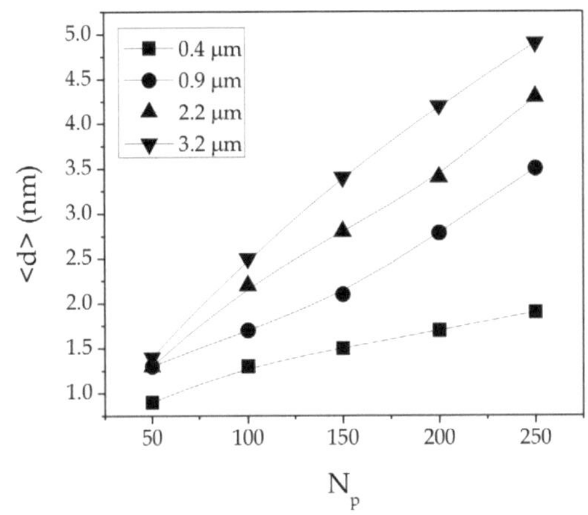

Fig. 5. Average distance <*d*> (Eq. 9) in function of the first-neighbor selected peaks, 50, 100, 1500, 200, and 250, for sample thickness of 0.4, 0.9, 2.2, and 3.2 μm.

Thickness (μm)	$\bar{h}$ (nm)	$\langle d \rangle$ (nm) - *First neighbor*				
		50	*100*	*150*	*200*	*250*
0.4	12.6	0.9	1.3	1.5	1.7	1.9
0.9	13.7	1.2	1.7	2.1	2.5	2.9
2.2	21.2	1.3	2.2	2.8	3.4	4.1
3.2	28.5	1.4	2.5	3.4	4.2	4.9

Table 2. Mean height $\sigma_1 = \bar{h}$ (Eq. 2) and average distance <*d*> (Eq. 15) of adjacent selected peaks in function of the number of the first neighbor: 50, 100, 150, 200, and 250 for all casting PPV films.

The distance between adjacent peaks was obtained by choosing 5 positions on AFM images, randomly. A monotonic increase of the mean distance between first-neighbors (selected peaks) where the number of peaks rises from 50 to 250 can be observed. This last result is consistent with the above observation, where casting PPV films do not change the surface topology significantly due to thickness. This last result is confirmed by observing the peaks with height equal to 50 nm (white pixels) in the AFM image in Fig 2. The mean height parameter is equal to 12.6 and 28.6 nm, where the thickness is equal to 0.4 and 3.2 μm, respectively. In fact, the parameter allows us to confirm the presence of a larger number of high height peaks in thick PPV films (see Fig 4). We can infer that the surface is more heterogeneous when the thickness increases.

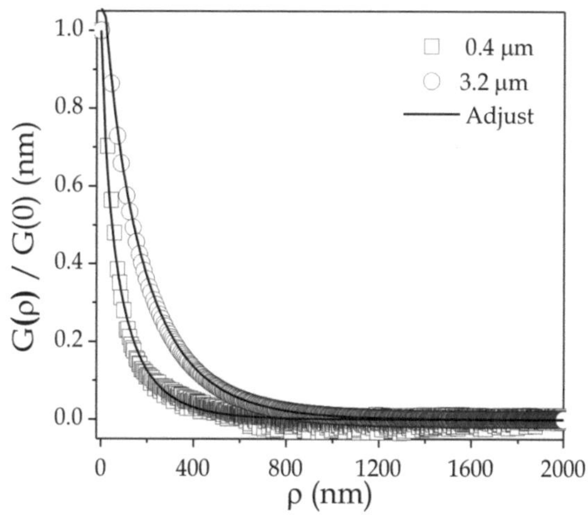

Fig. 6. Height auto-covariance curves for the AFM image (Fig. 1) for casting PPV films thickness of 0.4 μm (open cube) and 3.2 μm (open circle). The continuous line represents the adjustment using a self-affine equitation (Eq. 10c).

To confirm the previous finding that the statistical analysis of first and second order, where the thickness does not change significantly the surface topology of casting PPV films, we estimated the maximum of the film light scattering $<I_{MAX}>$ (Eq. 10). The result of the calculation for all samples is presented in Table 3. A non-significant variation of the scatter intensity was obtained, which corroborated with the experimental measurement of photoluminescence (PL), excitation 457 nm of the argon ion laser, and normalized at the thickness (NPL) (see Fig 7). Both NPL spectra present the same line shape and approximately the same intensity, confirming the data presented in Table 3.

Thickness (μm)	0.4	0.9	2.2	3.2
$G(0)$	2.3 ± 0.1	59 ± 4	142 ± 6	203 ± 10
$G(0)/e$	0.9 ± 0.1	22 ± 2	42 ± 3	75 ± 5
ξ_a (nm)	72 ± 5	152 ± 7	158 ± 7	163 ± 8
α	0.37 ± 0.02	0.41 ± 0.03	0.43 ± 0.03	0.45 ± 0.03
$\langle I_{MAX} \rangle$	0.028	0.027	0.026	0.025

Table 3. Best adjustment parameters for casting PPV films thickness of 0.4, 0.9, 2.2, and 3.2 μm, using the self-affine function and the maximum scattering intensity $<I_{MAX}>$ (Eq. 12).

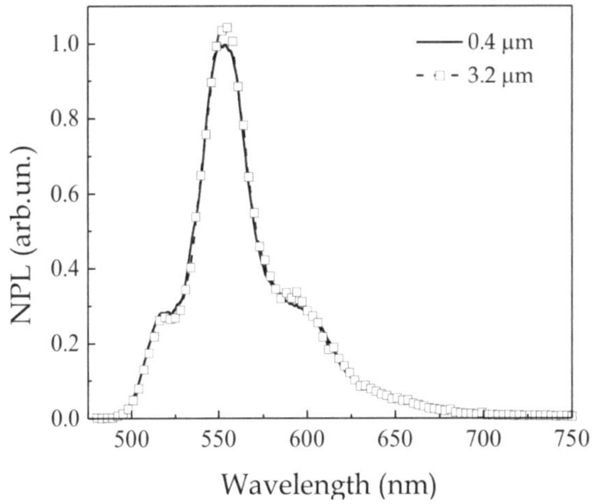

Fig. 7. Normalized photoluminescence spectra (NPL) for casting PPV films with thickness of 0.4 μm (line) and 3.2 μm (square-line). Spectra were normalized by the thickness.

3.2 Stretched casting PPV film

Polymer light emission diodes are generally processed as a thin film by the spin-coating technique on a transparent metal. As intrinsic characteristics of the spin-coating technique, the films are isotropic and the emission does not show linear polarization, as well. This is particularly interesting for the information industry (Cimrová et al., 1996). Polymers films are mechanically stretched easily, and the macromolecules are aligned, inducing anisotropy along the stretching direction, and the optical properties, in particular luminescence, are strongly affected by molecular anisotropy (Alliprandini et al., 2009; Therézio et al., 2011a, 2011b). This finding is corroborates with the fact that the emission polarization has a direct relationship with the orientation of the molecular transition dipole (electric) moment. Moreover, in most of the works done on emission polymers properties, the photo-physical process of energy transfer between conjugated polymers are poorly understudied due to the complex morphology of these materials. In the present section, we investigate stretch effects on surface topology of casting PPV films deposited on poly(vinylidene fluoride) (PVDF) tapes.

Fig. 8 displays AFM height images in a 10x10 nm^2 area for casting PVDF/PPV films stretched at 0% (ST0% - Fig. 8a), 50% (ST50% - Fig. 8b), and 100% (ST100% - Fig. 8c) in their length (axis y). Examples of height profiles for each AFM image are presented in Fig 8.A non-intentional alignment of the PPV casting film on a PVDF tape is observed (see Fig. 8a); the PVDF tape in their fabrication is stretched, and the surface ripple is expected, self-organizing the PPV chains due to the polymer/polymer interface. Where stretching was applied to ST50% and ST100% films, the alignment is more evident. The height profile in Fig. 8b and 8c shows a periodic oscillation perpendicular to the stretch direction. Next, the root-mean-square roughness (σ_{RMS}) increases as the perceptual of stretching increases; it is equal to 15, 21, and 30 nm for a stretching percentage equal to 0%, 50%, and 100%, respectively.

Figure 9 presents the height distribution (Eq. 1) for all AFM images in Fig. 8. For ST100% sample, the function $p(h)$ reduces considerably the width at half height, and it is closer to a Gaussian distribution (Eq. 4), χ^2 ~10^{-7}. Table 4 lists the parameters obtained using equation 4.

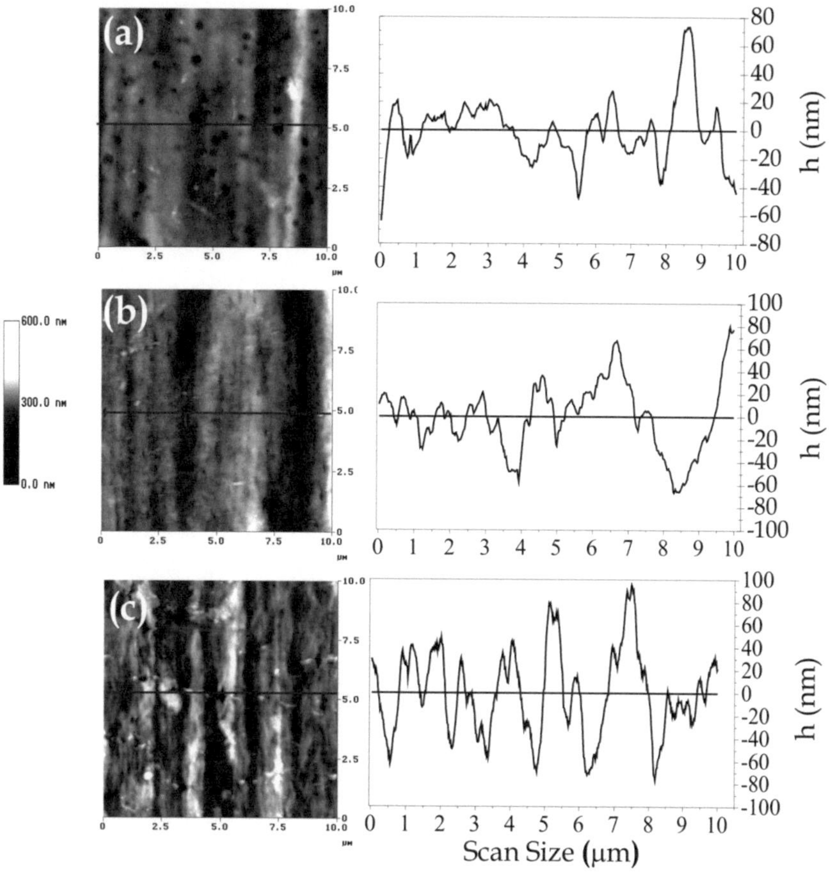

Fig. 8. AFM images (left) and height profile (right) for casting PVDF/PPV films stretched (a) 0%, (b) 50%, and (c) 100%.

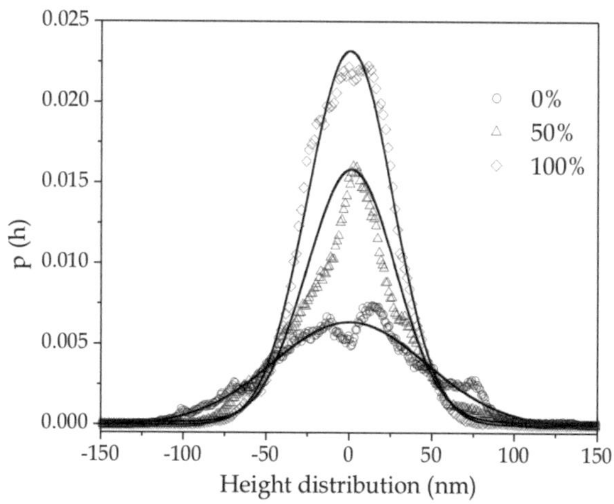

Fig. 9. Height distribution *p(h)* for ST0%, ST50% and ST100% films. The continuous line represents adjustments, using the Gaussian function (Eq. 4).

Sample	Stretching (%)	$\bar{h}$ (nm)	Δ(nm)
ST0%	0	0.22±0.04	92.1±0.9
ST50%	50	1.15±0.03	60.4±0.6
ST100%	100	1.74±0.01	51.5±0.2

Table 4. Best fit parameters of the height distribution p(h) for ST0%, ST50% and ST100% films in Figure 9.

By increasing the stretching perceptual, the first moment, $\sigma_1 = \bar{h}$, displays values near the normal surface, where $\sigma_1 \rightarrow 0$. This is corroborated with the half height (Δ) decrease shown in Table 4, for function *p(h)*.

The second statistical order was applied to the AFM images in Fig. 8, considering the approach of expression 7. Results are presented in Fig. 10. The white pixels in Fig. 10 represent heights with higher correlation. It is interesting to observe the presence of anisotropy in the direction of the stretching direction.

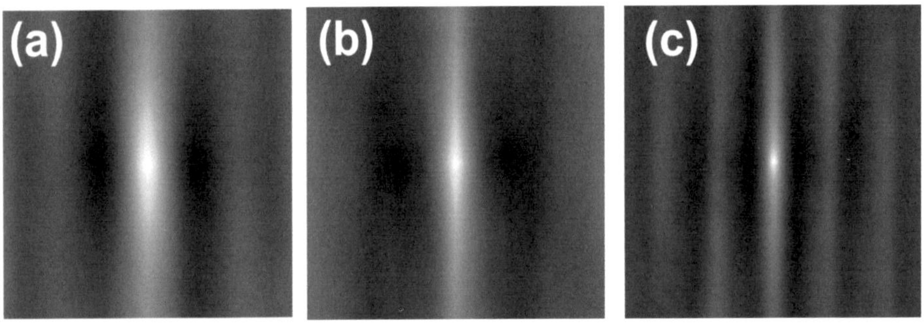

Fig. 10. Height auto-covariance image of casting PVDF/PPV films stretched at (a) 0%, (b) 50%, and (c) 100% for the AFM images in Fig. 8.

We can define the height-difference correlation function *g(ρ)* as(Bennett & Mattsson, 1999; Marletta et al., 2010), by using the height auto-covariance function *G(ρ)* (Eq. 7),:

$$g(\rho) = 1 - \frac{G(\rho)}{\sigma_{RMS}^2} . \tag{14}$$

Figure 11 displays the component of height-difference correlation for all samples. Fig.11a shows the component in perpendicular direction g(x) and Fig. 11b in parallel direction g(y) of the stretch direction (axis y). The main result is the periodic oscillation of the function g(x) for ST50% and ST100% films. Moreover, when the stretching increases (100%), the period of the oscillation decreases, in agreement with the images in Figure 10. This analysis allows for the verification of the anisotropy effects introduced due to mechanical stretching of PVDF/PPV film. For the component g(y), the correlation of the heights follows the behavior observed for casting PPV films, section 3.1, and a large length of non-correlated heights is observed. To quantify the correlation length, Table 5 presents the lateral correlation length parameters ξ (Marletta, 2010). The correlation length for non-stretched sample (ST0%) is different in both analyzed directions. In principle, casting films should not present that

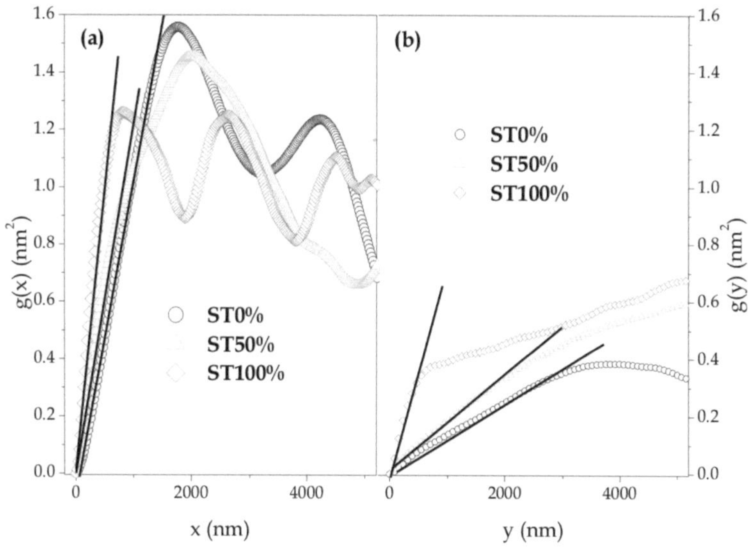

Fig. 11. (a) Perpendicular - *g(x)* and (b) parallel - *g(y)* components of height-difference correlation function for AFM image (Fig. 8) of casting PVDF/PPV films stretched at 0%, 50%, and 100%. Stretch direction is the axis y.

Sample	Stretching (%)	Direction	ξ_a (nm)	α
ST0%	0	x	623±5	0.97±0.02
ST50%	50	x	452±6	0.81±0.03
ST100%	100	x	260±2	0.99±0.3
ST0%	0	y	418±5	0.49±0.01
ST50%	50	y	273±3	0.47±0.02
ST100%	100	y	270±3	0.34±0.01

Table 5. Lateral correlation (for casting PVDF/PPV films of ST0%, ST50%, and ST100%.

difference. However, the AFM image (Fig. 8a) shows an initial anisotropy in a PPV film on a PVDF tape. By increasing the stretching percentage, that parameter decreases considerably, showing the strong dependence between both analyzed directions.

Finally, to confirm the anisotropy of the PPV moieties along the stretching direction, Fig. 12 presents the polarized absorbance (Fig. 12a) and the polarized emission (Fig. 12b). The enhancement in the absorbance and emission in the parallel direction, proving the presence of anisotropy, is evident.

3.3 Diffraction grating

In this section, we report and analyze the physical effects of the photo-oxidation reaction on PPV cast films, using the AFM technique. The experimental setup for the grating recording is shown in Fig. 13. A linearly polarized Ar ion laser beam, operating at 488 nm, is used to induce grating. This laser beam passes through a half-wave plate, to control its polarization,

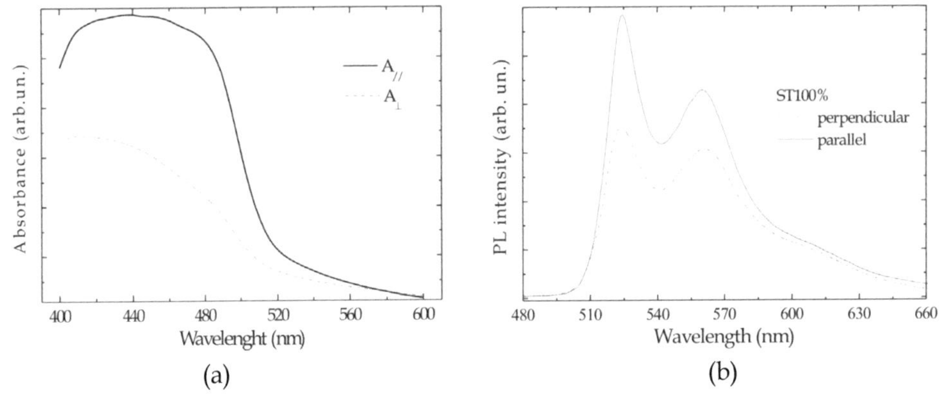

Fig. 12. (a) Polarized absorbance spectra for casting PVDF/PPV film stretchered at 100% (ST100%) parallel ($A_{//}$ - black line) and perpendicular ($A_{\perp}$ - red line) to the stretching direction. (b) Polarized photoluminescence spectra for casting PVDF/PPV film stretched at 100% (ST100%) parallel (black line) and perpendicular (red line) to the stretching direction.

and is expanded and collimated before shining the sample. Half of the collimated beam impinges directly the sample, while the other portion is reflected onto the sample from a aluminum coated mirror. The grating was recorded with s and p polarization. The intensity of the recording beam after the collimation system was 200 mW/cm^2,and the recording time was about 2 hours. The incident angle of the recording beams was selected at 10°, resulting in a grating spacing of about 2 μm.

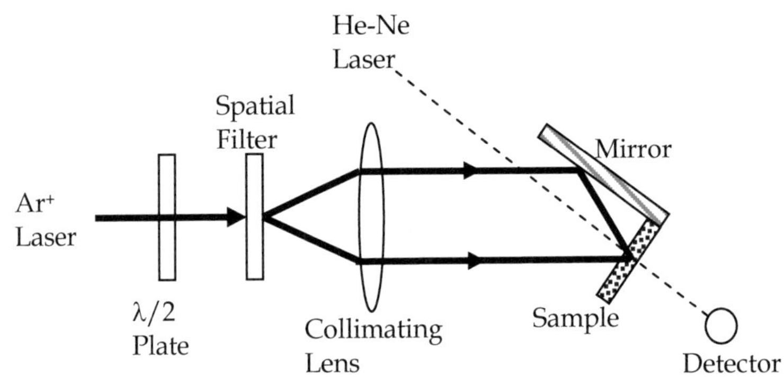

Fig. 13. Experimental setup of the holography diffraction grade recording.

Macroscopic modifications in the PPV surface were verified by using optic microscopy images. Fig. 14 shows the non-irradiated (NI) and irradiated (I) regions, clearly distinguishable in the AFM image, which was obtained by using a CCD camera coupled to the microscope, using the illumination in the transition setup. The majority of the light was transmitted in the irradiate region (I), in agreement with the absorption spectra (Gobato et al., 2002). In this macroscopic scale, the image shows qualitatively low non-homogeneity in both surface regions, with morphological defects (peaks) distributed randomly on the surface. Such morphological defects in the NI region were produced during the thermal conversion process.

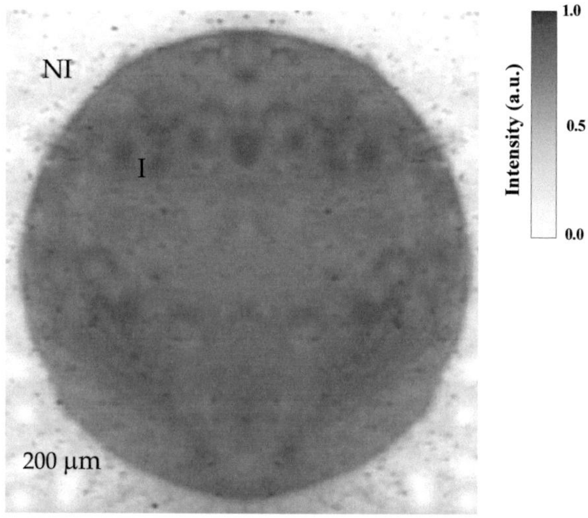

Fig. 14. Optic image of non-irradiate (NI) and irradiate (I) region.

Fig. 15 shows the PL spectra of the PPV film before (0 min) and after (35 min) the photo-irradiated exposure, using an Ar+ laser (458 nm) at 200 mW/cm^2 in atmospheric conditions. The PL-intensity increases 250% after being exposed to light, without significant changes in the spectral line shape. The inset in Fig. 15 shows the normalized PL-intensity enhancement, which reaches saturation for an exposition time of 30 min. This figure shows that the zero-phonon peak at about 516 nm (2.4 eV) and the vibronic progression in the low energy spectral range are unchanged in both spectra. During the irradiation, the ratio between the intensities of zero-phonon and the first-phonon replica peak (553 nm) is approximately constant and equal to 0.29. This value indicates a high electron-phonon coupling (Huang Rhys parameter), characteristic of the structural disorder, impaired by the molecular random packing of the cast deposition technique.

The presence of chemical structure degradation is evident in the absorption of the UV-Vis range and infrared (IR) measurements reported by Gobato et al., 2002. After light exposure, the absorption spectrum is blue shifted and the IR spectrum presents a major increase in 1690 cm^{-1} peak, identified as the carbonyl group (C=O) incorporation (Barford & Bursill, 1997; Chandross et al., 1994; Friend, et al., 1997; Onoda et al., 1990). This defects incorporation indicates chemical changes in PPV main chains, raising the PPV HOMO-LUMO band gap, with a reduction in the effective conjugation degree of the PPV.

Fig. 16 shows AFM images obtained by using the tapping mode for the non-irradiated (Fig. 16a) and irradiated (Fig. 16b) regions, indicated as NI and I in the Fig. 14, respectively. The simple visual analysis of these images shows a significant difference between both regions. The non-irradiated region shows a minor number of peaks with height greater than 25 nm, distributed randomly on the surface. The irradiated region presents a considerable increase in the number of peaks, with height greater than 25 nm, distributed more homogeneously on the surface. In the 10 μm AFM image scale, the non-irradiated region presents peaks with a diameter larger than irradiated region, as observed in the macroscopic image in Fig. 14.

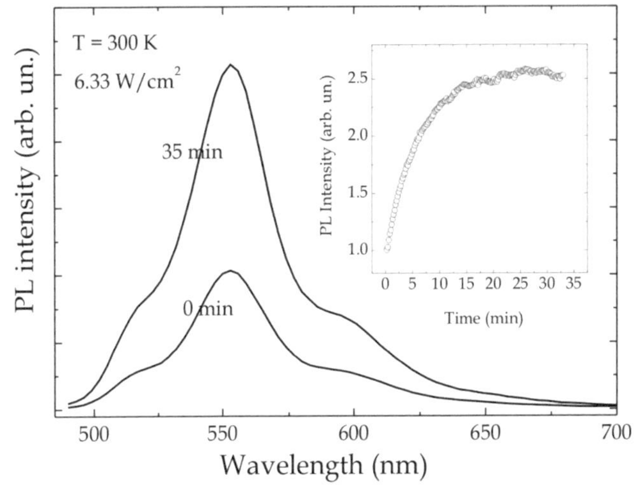

Fig. 15. PL spectra before (0 min) and after (35 min) photo-irradiation at 200 mW/cm² for a PPV cast film under environmental conditions. The insert shows the time evolution of the PL-intensity.

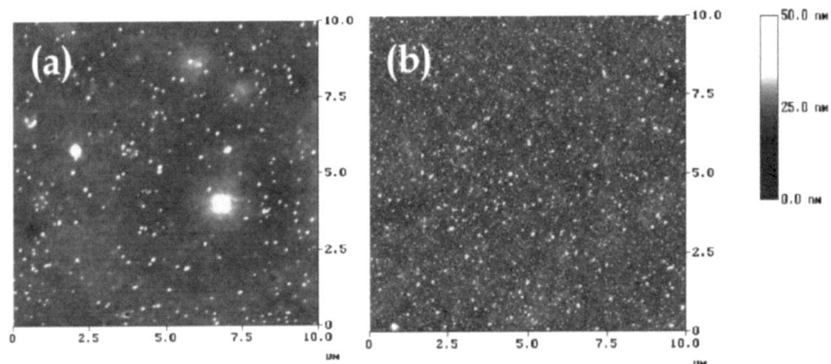

Fig. 16. AFM image in tapping mode of non-irradiate (a) and irradiate (b) regions.

The roughness is 4.0 nm in the non-irradiated and 2.8 nm in the irradiated region. These findings include surface changes induced during the laser exposure of the PPV film, but it is not enough to conclude that the irradiated surface is more homogeneous than the non-irradiated region. The surface analysis of polymeric films using a subjective AFM image analysis and the roughness parameter is difficult, once the phase image does not show additional information like in films composed by polymeric blends.

The new methodology proposed in this work is based on the quantitative analysis of AFM images using a statistical study of the peak height distribution and distance between peaks. A substantial increase in the total number of peaks, with the irradiated region presenting approximately 5 times more peaks (1.6×10^4) than the non-irradiate (3.3×10^3) is observed. Fig. 17 shows the distribution of peaks height in the irradiated (Fig. 17a) and non-irradiated (Fig. 18b) region. In both regions, the major contribution to the histogram is in the range between 6 and 14 nm. For these initial data, we can observe in the irradiated region that the peaks height fraction, superior to 12 nm and 18 nm in the non-irradiated region, are less than 1% of the total

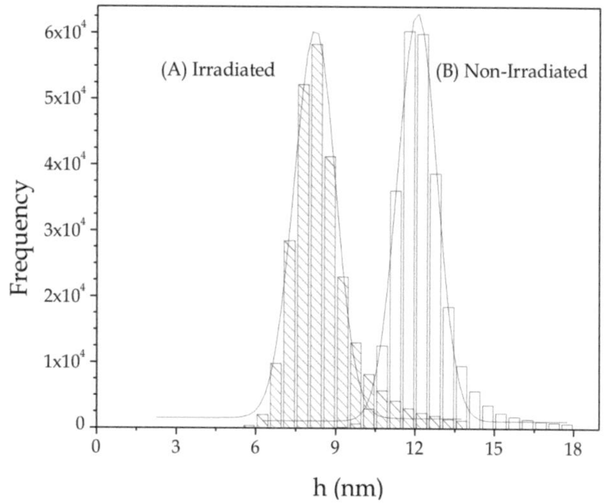

Fig. 17. Histogram of peak heights in irradiated (a) and non-irradiated (b) regions. Gaussian function was used to fit the distribution.

peaks on both surface. The histograms in Fig. 17 were adjusted using a Gaussian curve. The distribution width is narrow, with 1.5 nm in both regions, centralized at 12.1 nm and 8.2 nm, and the maximum is 1.2×10^5 and 1.0×10^5 for non-irradiated and irradiated regions, respectively. These findings show that the surfaces are not changed dramatically, with a small tendency to increase the surface disorder with the decrease of the peak height. In comparison with the AFM image in Fig. 16, we observe that the major contribution to the surface is centralized in the region with low peak heights (<15 nm), which is not perceived visually.

The distribution of the distance between peaks is plotted in Figure 18. Histograms were adjusted using a Gaussian curve and presented similar parameters. Distances between peaks, in both regions, are centralized at 5.1 μm and the width at 6.3 μm. The main difference is the amplitude of the distribution, which is one order greater for the irradiated region. These findings corroborate with the data obtained in Fig. 18, and the principal contribution to the surface morphology is in a nanometer scale that it is not perceived by the visual analysis of the AFM image in Figure 16.

When the PPV casting films were photo-irradiated for a short time, we observed an increase in the intensity of photoluminescence is observed. On the other hand, a long exposure to irradiation decreased the intensity. Both effects can be connected to the degradation process in the structure of the polymeric chain. By using a photo-oxidation process, we produced a diffraction grating in the PPV film. Fig. 19a shows the AFM diffraction grating image where alternating irradiated and non-irradiated regions can be observed. Similar surface relief grating formation has been observed in azopolymeric films, in which a driving force that depends on the gradient of the electric field and the local plasticization assisted by the trans-cis-trans photoisomerization is the mechanisms responsible for the molecular movimentation (Bian et al., 1999; Kumar et al., 1998). However, in the present work, the surface relief grating observed is due to a polymeric chain photo-oxidation, which causes changes in the PPV photo-luminescence, as mentioned before. This result can be used in light emitting display to store the information on the surface.

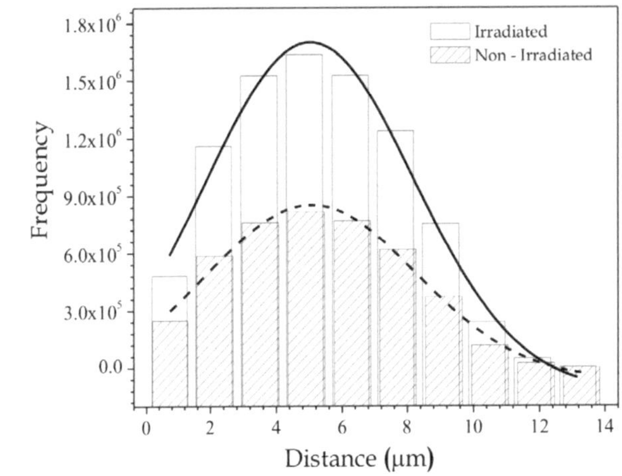

Fig. 18. Histogram of the distance between peaks in non-irradiated and irradiated regions. A Gaussian function was used to fit the distribution.

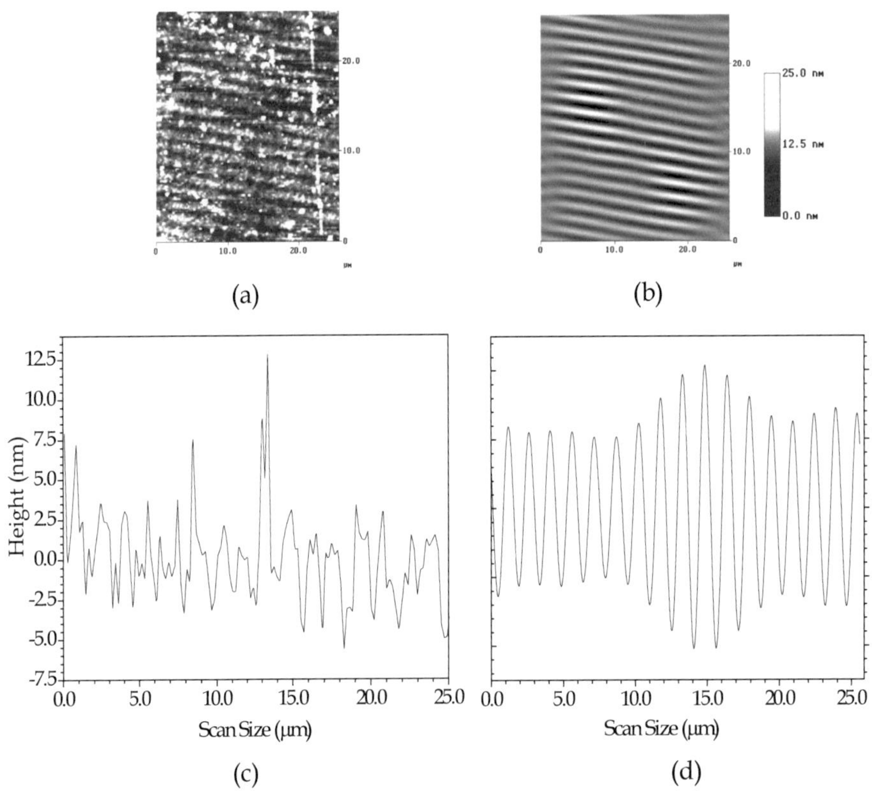

Fig. 19. (a) shows the AFM diffraction grating image alternating between irradiated and non-irradiated regions and (b) the AFM diffraction image obtained after the Fourier transform application explained in the experimental section. Figure (c) and (d) show the transversal section of the original AFM image (a) and the image obtained after the Fourier filtering (b), respectively.

Fig. 19a was obtained by carrying out the Fourier transforming explained in the experimental section. After the Fourier transforming shown in Fig. 19a, two bright points in the second and fourth quadrant were obtained. These points correspond to a frequency period of 1.6 μm. These values are related to the well-defined grating periodicity observed in the real image. These frequencies correspond to a regular standard of undulation in the real image. The grating periodicity found through the imaging processing method (1.6 μm) is similar to the one calculated in the experimental section (1.5 μm), revealing the reliability of the adopted image analyses. Figures 19c and 19d show the transversal section of the original AFM image and the image obtained after the Fourier filtering. The peak-peak height obtained is ~ 1.3nm for the original image and is 1.55nm for the Fourier filtered image.

The peak-valley height of the grating inscribed in this film through the photo-oxidation process as determined by AFM, is approximately 7.9 nm. This value is smaller than those usually obtained for surface relief grating in azo-doped films, which in general present amplitude of modulation of around 100 nm. Furthermore, surface relief grating formation is, in general, polarization dependent, being more efficient for p-polarized than for s–polarized light. No indication of such polarization dependence was observed in our experiments. This result leads to the conclusion that the grating formation process here presented here is probably a bulk process, in agreement with previous evidences.

In summary, the quantitative analyses show that the surface morphology is not affected significantly during the laser exposure process at low power. We observe that the visual analysis of the FM image is not sufficient since important features of the microscopic morphology for homogeneous polymeric PPV films cannot be distinguished visually. The use of an image statistical analysis and specific software showed the importance of quantifying the morphological parameters, the heights of pixels and the distances between peaks. By using this methodology, the image interpretation can replace the usual subjective visual analysis. The decrease in the σ_{RMS} parameter verified by using specially developed software indicates that the surface is not affected significantly and can be used to deposit metallic electrodes in the fabrication of polymeric light-emitting devices.

4. Conclusion

Surface statistical analysis of homoplymer films using height AFM images proved to be important to quantify the topological structure and to identify surface type. Thickness, mechanical modification, and photo-blanch effects on the surface topology of casting PPV film were explored in the present chapter, using first and second order statistical analysis. The simple observation of AFM images or height histograms is not able to infer the changes on the surface topology due to physical or chemical modifications. Through casting PPV films, we explore all potentiality of the mathematical tools introduced in section 2. In summary, the height distribution $p(h)$ moments describe the type of the surface, normal surfaces and surface covered with bumps and pits, or a combination of them. The second statistical order describes the typical distance until the heights and the effects on, for example, scatter lights are correlated. Additional analysis, using the distance between adjacent peaks, can inform about surface homogeneity. It is important to emphasize that the development of new procedures or techniques to analyze AFM images is essential to improve the knowledge on surface topology and mechanical, electrical, and optical properties of luminescent polymers. The study is not limited to a matter of scientific investigation. It became a matter of technological interest with applications in organic electronics.

5. Acknowledgment

This work had the financial support of FAPEMIG, CNPq, CAPES and INEO/MCT (Brazil). The authors are grateful to Professor Cléber Mendonça from *Instituto de Física de São Carlos* (USP, Brazil) for the use of the Institute's experimental facilities and to Carlos Alberto Rodrigues from *Departamento de Ciências Exatas* (UEFS, Brazil) for the use of the SPIA computing program facilities. The authors are grateful to Hugo Santos Silva a student from the *Grupo de Espectroscopia de Materias* (UFU, Brazil), for the text layout.

6. References

Alliprandini, P.; da Silva, G. B.; Barbosa Neto, N. M.; Silva, R. A. & Marletta, A. (2009). Induced Secondary Structure in Nanostructured Films of Poly(p-phenylene vinylene). *Journal of Nanoscience and Nanotechnology*, Vol.9, No.10, (Octuber, 2009) pp. 5981-5989, ISSN 1533-4880

Assender, H.; Bliznyuk, K. & Porfyrakis, V. How Surface Topography Relates to Materials' Properties. *Science*, Vol.297, No.5583, (August, 2002), pp. 973-976, ISSN 0036-8075

Bar, G. & Thomann, Y. (1998). Characterization of the Morphologies and Nanostructures of Blends of Poly(styrene)-block-poly(ethene-co-but-1-ene)-block-poly(styrene) with Isotactic and Atactic Polypropylenes by Tapping-Mode Atomic Force Microscopy. *Langmuir*, Vol.14, No.5, (February, 1998), pp. 1219–1226, ISSN 0743-7463

Barford, W. & Bursill, R. J. (1997) Theory of molecular excitons in the phenyl-based organic semiconductors . *Chemical Physics Letters*, Vol.268 , No.5-6 (April,1997), pp. 535-540, ISSN 0009-2614

Bennett, J. M. & Mattsson, L. (1999). *Introduction to surface Roughness and Scattering*, Optical Society of America, ISBN 1-55752-609-5, Washington DC, USA

Bian, S.; Williams, J. M.; Kim, D. Y.; Li, L.; Balasubramanian, S.; Kumar, J. & Tripathy, S. (1999). Photoinduced surface deformations on azobenzene polymer films. *Journal of Applied Physics*, Vol.86, No.8,(July, 1999), pp. 4498-4508, ISSN 0021-8979

Binnig, G.; Quate, C. F. & Gerber, Ch. (1986). Atomic Force Microscope. *Physical Review Letters*, Vol.56, No.9, (March, 1986), pp. 930–933, ISSN 0031-9007

Chandross, M.; Mazumdar, S.; Jeglinski, S.; Wei, X.; Vardeny, Z. V.; Kwock, E. W. & Miller, T. M. (1994). Excitons in poly(para-phenylenevinylene). *Physical Review B: condensed matter and materials physics*, Vol. 50, No. 19, (November, 1994), pp. 14702–14705, ISSN 1098-0121

Costa, L. D. F.; Rodrigues, C. A.; Souza, N. C. D. & Oliveira, O. N. Statistical Characterization of Morphological Features of Layer-by-Layer Polymer Films by Image Analysis. *Journal of Nanoscience and Nanotechnology*, Vol.3, No.3, (June, 2003), pp. 257-261, ISSN 1533-4880

Cimrová, V.; Remmers, M.; Neher, D. & Wegner, G. (1996). Polarized light emission from LEDs prepared by the Langmuir-Blodgett technique. *Advanced Materials*, Vol.8, No.2, (February, 1996), pp. 146–149, ISSN 1521-4095

Friend, R.H.; Denton, G.J.; Halls, J.J.M.; Harrison, N.T.; Holmes, A.B.; Köhler, A.; Lux, A.; Moratti, S.C.; Pichler, K.; Tessler, N.; Towns, K. & Wittmann, H.F. (1997). Electronic excitations in luminescent conjugated polymers. *Solid State Communications*, Vol.102, No.2-3, (April, 1997), pp. 249-258, ISSN 0038-1098

Gobato, Y. G.; Marletta, A.; Faria, R. M.; Guimarães, F. E. G.; de Souza, J. M. & Pereira, E. C. (2002). Photoinduced photoluminescence intensity enhancement in poly(p-phenylene vinylene) films. *Applied Physics Letters*, Vol.81, No.5, (June, 2002), pp. 942-944, ISSN 0003-6951

Gua, X.; Raghavana, D.; Nguyenb, T.; VanLandinghamb, M.R & Yebassaa, D. (2001). Characterization of polyester degradation using tapping mode atomic force microscopy: exposure to alkaline solution at room temperature. *Polymer Degradation and Stability*, Vol.74, No.1, (September, 2001), pp. 139-149, ISSN 0141-3910

Holzera, W.; Penzkofera, A.; Pichlmaiera, M.; Bradleyb, D.D.C W. & Blauc J. (1999). Photodegradation of some luminescent polymers. *Chemical Physics*, Vol.248, No. 2-3, (December, 1999), pp. 273-284, ISSN 0301-0104

Kumar, J.; Li, L.; Jiang, X. L.; Kim, D.-Y.; Lee, T. S. & Tripathy, S. (1998). Gradient force: The mechanism for surface relief grating formation in azobenzene functionalized polymers. *Applied Physics Letters*, Vol.72, No.17, (February, 1998), pp. 2096-2098, ISSN 0003-6951

Magonov, S. N. & Whangbo M.-H. (1996). *Surface Analysis with STM and AFM: Experimental and Theoretical Aspects of Image Analysis*, Wiley-VCH, ISBN 978-3527293131, Weinheim, Federal Republic of Germany

Marletta, A.; Gonçalves, D.; Oliveira, O. N.; Faria, R. M. & Guimarães, F. E. G. (2000). Rapid Conversion of Poly(p-phenylenevinylene) Films at Low Temperatures. *Advanced Materials*, Vol.12, No.1, (January, 2000), pp. 69–74, ISSN 1521-4095

Marletta, A.; Vega, M. L.; Rodrigues, C. A.; Gobato, Y. G.; Costa, L. F. & Faria, R. M. (2009). Photo-irradiation effects on the surface morphology of poly(p-phenylene vinylene) films. *Applied Surface Science*, Vol.256, No.10, (March, 2010), pp. 3018–3023, ISSN 0169-4332

Marti, O.; Stifter, T.; Waschipky, H.; Quintus, M. & Hild, S. (1999). Scanning probe microscopy of heterogeneous polymers. *Colloids and Surfaces A: Physicochemical and Engineering Aspects*, Vol.154, No.1-2, (August, 1999), pp. 65-73, ISSN 0927-7757

Onoda, M; Manda, Y.; Iwasa, T.; Nakayama, H; Amakawa, K & Yoshino, K. (1990). Electrical, optical, and magnetic properties of poly(2,5-diethoxy-p-phenylene vinylene). *Physical Review B: condensed matter and materials physics*, Vol.42, No.18, (December, 1990), pp. 11826–11832, ISSN 1098-0121

Palasantzas, G. (1993). Roughness spectrum and surface width of self-afFine fractal surfaces via the K-correlation model. *Physical Review B: condensed matter an materials physics*, Vol.48, No.19, (May, 1993), pp. 14472-14478, ISSN 1098-0121

Simpson, G. J.; Sedin, D. L. & Rowlen, K. L. (1999). Surface Roughness by Contact versus Tapping Mode Atomic Force Microscopy. *Langmuir*, Vol.15, No.4, (January, 1999), pp. 1429–1434, ISSN 0743-7463

Therézio, E. M.; Piovesan, E.; Anni, M.; Silva, R. A.; Oliveira, O. N. & Marletta, A. (2011) Substrate/semiconductor interface effects on the emission efficiency of luminescent polymers. *Journal of Applied Physics*, Vol.110, No.4, (August, 2011), pp. 044504-044509, ISSN 1089-7550

Therézio, E. M.; Piovesan, E.; Vega, M. L.; Silva, R. A.; Oliveira, O. N. & Marletta, A. (2010). Thickness and annealing temperature effects on the optical properties and surface morphology of layer-by-layer poly(p-phenyline vinylene)+ dodecylbenzenesulfonate films. *Journal of Polymer Science Part B: Polymer Physics*, Vol.49, No.3, (February, 2011), pp. 206-213, ISSN 1099-0488

Thomas, T. R. (1999). *Rough Surfaces,* Imperial College Press, ISBN 978-1860941009, London, England

Weisenhorn, A. L.; Maivald, P.; Butt, H. J. & Hansma P. K. (1992). Measuring adhesion, attraction, and repulsion between surfaces in liquids with an atomic-force microscope. *Physical Review B: condensed matter and materials physics*, Vol.45, No.19, (May, 1992), pp. 11226–11232, ISSN 1098-0121

Zhao, Y.; Wang, G.-Ch. & Lu, T. M. (2001). *Characterization of amorphous and crystalline rough surface: principles and applications,* Academic Press, ISBN 0-12-475984-X, San Diego CA, USA

4

Polyamide-Imide Membranes of Various Morphology – Features of Nano-Scale Elements of Structure

S.V. Kononova[1], G.N. Gubanova[1],
K.A. Romashkova[1], E.N. Korytkova[2] and D. Timpu[3]
[1]*Institute of Macromolecular Compounds, Russian Academy of Sciences, St. Petersburg*
[2]*Institute of Silicate Chemistry, Russian Academy of Sciences, St. Petersburg*
[3]*Petru Poni Institute of Macromolecular Chemistry, Romanian Academy, Iasi*
[1,2]*Russia*
[3]*Romania*

1. Introduction

Polyamide-imides (PAI), having a valuable set of properties due to the presence in the main chain amide and imide groups of atoms, in addition to resistance to chemical environments and to processability at elevated temperatures, are very attractive materials for the formation of coatings, films and membranes of different applications [1]. Interest in polymers of this class periodically arose throughout several decades in various scientific schools of the world. First of all, a possibility of preparing from them isolating materials for microelectronic was considered [2, 3]. Workings out materials produced in industrial scale from polymers of commercial marks Torlon ® (Solvay Advanced Polymer), Tecamax ® (Graftech Industries, Inc.), Durimide ® (Fujifilm Electronic Materials, Inc.), Tecator ® (San Diego Plastics, Inc.) are known, and also along with them unique laboratory researches [4-6]. One of the most successful applications of polyamide-imides may be formation of materials for membranes, particular for membranes of composite type [7-8]. On a basis of polyamide-imides either their polymeric or polymer-inorganic compositions, membranes of a wide spectrum of classification groups, and consequently, different morphology are offered. In particular, microporous membranes with nano-scale pores in a skin-layer are discussed [9-11]. Investigations of the flat or the hollow fiber membranes, porous or dense non-porous films and composite multilayer structures are known. Last two of the listed above morphological types of films are widely used as diffusion membranes, these are membranes containing one or more non-porous separating layers. For example, on a basis of polyamide-imides with trade-mark Torlon ®, asymmetric and composite diffusion membranes for gas separation or pervaporation processes were developed [12-13]. Polyamide-imides have appeared selective for a wide spectrum of the separating problems solved in a pervaporation [14-16]. In particular, the film from Torlon 4000T has been offered for use at pervaporation separation of isomers (xylene isomers [17]).

At the same time, the spectrum of PAI applications is limited, that is connected with specificity of synthesis of these polymers. For preparation of membranes with a specific morphology, it is necessary to choose an appropriate strategy to optimize the synthesis conditions for obtaining PAI so that the polymer could be formed into a material with desired properties. So, in case the need for self-supporting films, optimizing the synthesis conditions should result in the first place, to the preparation of stable polymer of high molecular weight with narrow molecular weight distribution. Complex requirements of chemical and thermal stability and mechanical properties needed to form flexible and strong self-supporting films satisfy PAI samples, obtained by polycondensation in solution at low temperature. Conditions of mentioned above reaction have been optimized at the Institute of Macromolecular Compounds Russian Academy of Sciences (IMC RAS) [18]. Main feature of the reaction consists in introduction in the reactionary environment at the lowered temperatures monomers, one of which contains in the structure an imide fragment. Such method gives the chance not only to obtain polymer with a necessary complex of properties, but also to avoid imidization as an additional stage of the process. Proposed in 1971 by researchers from the Romanian Academy of Sciences and optimized by chemists of IMC RAS method has found followers and has been successfully used to this day, as in Romania [19-20], and in Russia [21-23].

2. Asymmetric porous films based on polyamide-imides in the membranes of various morphological types

In focus of the authors of this paper were always materials with unique transport properties, or other characteristics, that allow the use these materials in the high-performance composite membranes. In previous our publications it has been shown, that with use of polyamide-imides as active (in the course of separation) materials of supporting or intermediate layers, membranes high-effective for separation of mixtures of gases or liquids can be generated. So, in [24-26] were considered gas separation properties of the composite membranes containing microporous supporting layers from polyamide-imides of a similar chemical structure, differing with a diamine fragment. In these membranes after the formation of structural components with the certain morphologies, the conditions of highly selective transport through the support material were achieved. As a result of this, the support from the category of passive substrates, which are responsible only for the mechanical strength of the composite material as a whole, passed into the category of active ones. The choice of the polyamide-imide for formation of such membranes became basic, as well as conditions preparing microporous films on their basis. On Fig.1 the schematic image of the typical multilayer composite membrane and microphotographs of low-temperature fractures of porous PAI membranes (supports in composite membranes) prepared in identical conditions of phase - inversion process from PAI's with similar chemical structure, but differing in diamine component, are presented.

Variations in the morphological features of microporous PAI films demonstrated in Fig. 1 correlate with their transport properties: for example, N_2 permeability in these PAI–films series is $(4 \div 95) \cdot 10^{-3}$ $cm^3/(cm^2\ s\ (cm\ Hg))$. This level of permeability shows the presence of cross-porous structure in all asymmetric films discussed above. The morphologies of all supports are similar: the finger-shaped conical macropores traverse the support, and the cone voids in cross section decrease towards the upper surface of the support (skin layer). The walls of the macropores have a network - morphology with a pore size of 50 – 100 nm. All supports, concerning SEM photographs in Fig.1, have almost perfectly smooth flat surfaces with pores in skin layer of less 20 nm in diameter.

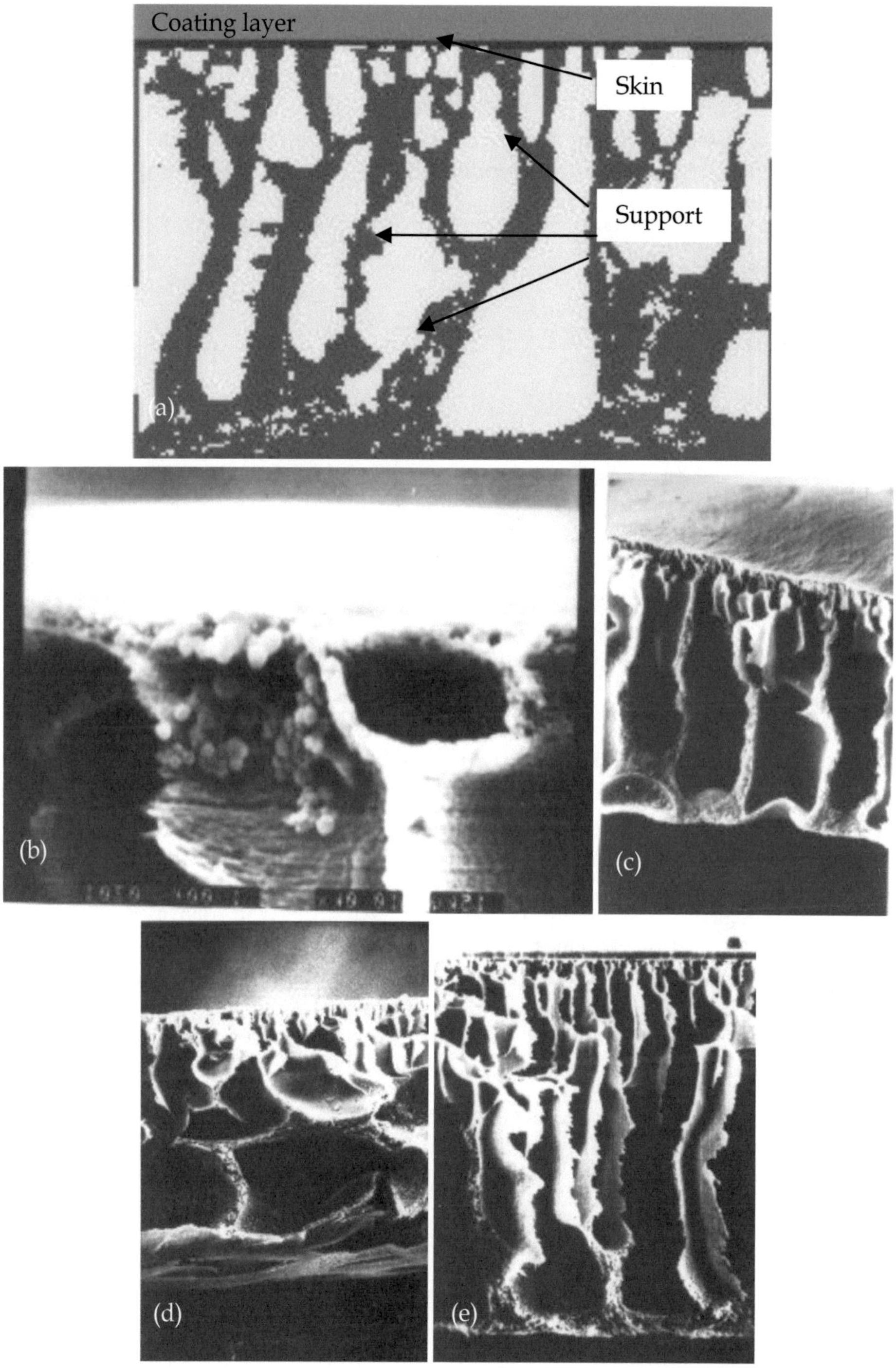

Fig. 1. (a) schematic representation of the cross section of a composite membrane; (b,c,d,e) microphotographs of low-temperature fractures of porous PAI's membranes prepared in the present work using water as non-solvent [26].

These features of morphology are illustrated well also by Fig. 2. In this Figure microphotographs of low-temperature brittle fracture (cross-section) and the top surface of the asymmetric porous film (support) are placed. Film was obtained by a method of wet formation from polyamide-imide synthesized at the Institute of Macromolecular Compounds of RAS from phthalimidobenzenedicarbonyl dichloride and diaminediphenyl oxyde (PAI-1) by low-temperature polycondensation in solution [18].

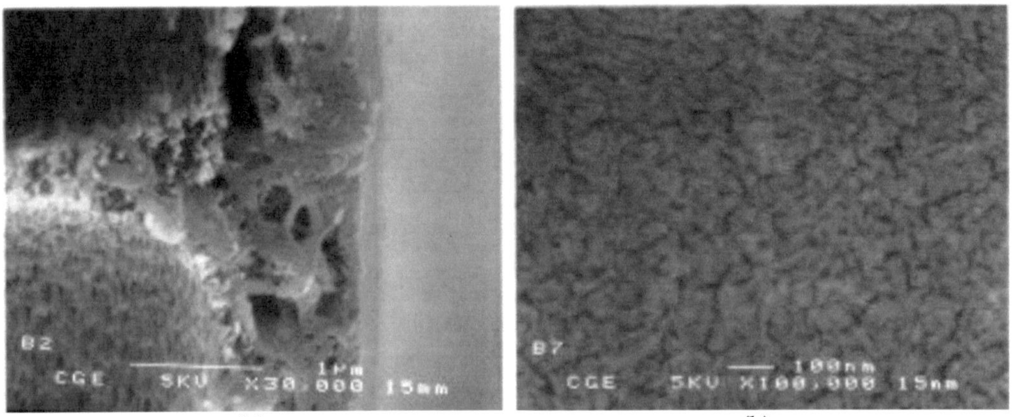

(a) (b)

Fig. 2. SEM photographs of PAI-1 microporous support: (a) low-temperature fracture, (b) surface of skin-layer.

Formation of composite membrane by a method of coating the polymeric solution on a surface of skin-layer of the asymmetric support resulted, as is noted above, to formation of intermediate layers of complicated morphology, and in some cases - outwardly homogeneous layers of difficult structure. In [27] it is shown, that composite membranes presented in a Fig. 3 have diffusion separating layer, unlike porous support in their structure. On a surface of a skin-layer of the same PAI support (a microporous asymmetric film) from different coating polymers (polyvinyltrimethylsilane, polypropylmethacrylate) thin separating layers of various morphology can be generated in identical conditions [27]. Thus, obviously expressed "pores" and holes in coating layers are not through, and the internal intermediate layers generated on a skin-layer of support answer for the transport properties of composite membranes discussed in work [27].

It was demonstrated [26] with an example of poly(2,2,3,3,4,4,5,5-octafluoro-n-amylacrylate)/ polyamide-imide composite membranes that highly selective diffusion layers, particularly of low-selective elastomers, can be realized on the support by optimization of the chemical structure of PAI, i.e., by varying the nature and (or) concentration of adsorption sites on the support surface and, therefore, its energy characteristics. This allows make diffusion membranes the separation characteristics of which differ from those predicted theoretically on the basis of the resistance model, so that they surpass the selectivity of all individual polymers composing the composite membrane. In similar conclusions about an essential role of internal intermediate layers resulted the analysis of characteristics of multilayer composite membranes, effective at separating pervaporation of liquids [28]. In this case the material of a porous support layer of PAI-1 played even more significant role and influenced essentially transport properties of a membrane.

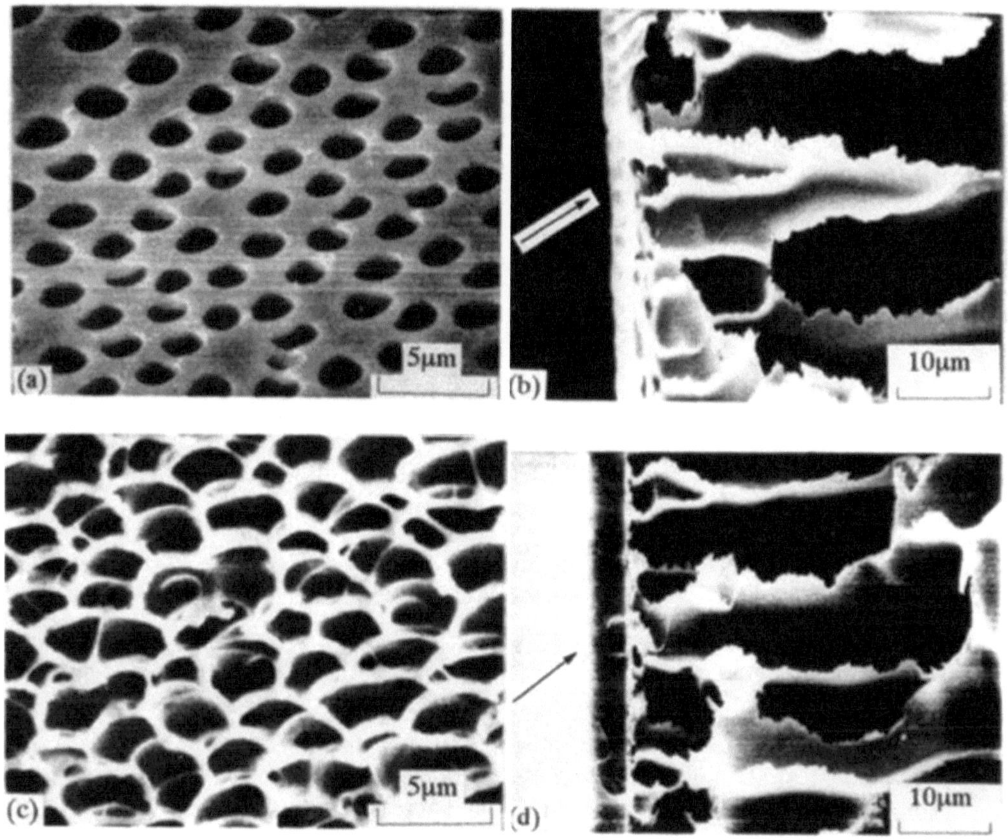

Fig. 3. SEM photographs of the surfaces (a, c) and of the cross-sections prepared by low-temperature cleavage (b, d) of the PVTMS-PAI-3 (a, b) and PPMA-PAI-3 (c, d) composites membranes. The arrow shows the upper surface.

We have studied the asymmetric porous membranes with ultra-thin diffusion layers of PAI-1 and also two - and multi-layered composite structures containing microporous polymer layers with a similar but looser morphology. These layers are characterized by penetrating porous structure with nano-sized pores in the skin layer and the nearby - layers. Unlike gas separation when glass transition temperature of coating polymer, the size of pores of a skin-layer, and also a parity of thickness of various layers of a composite membrane are the matter of principle, in a case of pervaporation last factor can have defining role for realization of transport properties of PAI, forming a microporous support of a membrane. As well as in a case of gas separation, the essential contribution to transport properties of multilayer membranes brings the boundary layer formed on a surface of a skin-layer of a microporous PAI film.

It is possible to consider the multilayer composite membranes presented on Fig. 4 as simple, but demonstrative examples. In this Figure SEM photograph (Fig. 4,a) of a highly effective multilayer composite membrane is resulted, in which on a surface of skin-layer of a microporous film intermediate layer of polydimethylamine ethylmethacrylate (PDMA) and a top-layer of polysiloxane are formed. Presented in [28], the transport properties of the

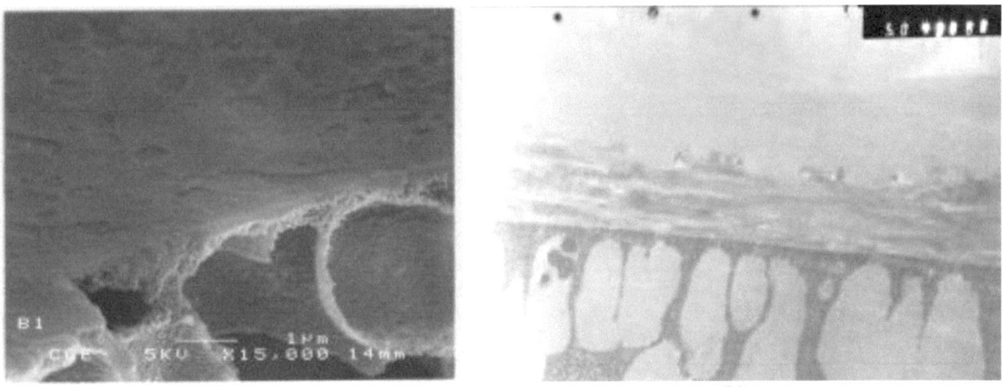

(a) (b)

Fig. 4. Microphotographs of (a) the cross-sections prepared by low-temperature cleavage of multilayer composite membrane polysiloxane/PDMA/PAI-1, (b) ultra-thin cut of composite membrane with the coating layer of poly-γ-benzyl-L-glutamate.

membrane can be considered it one of the most effective membranes for the separation of mixtures of methanol - methyl tertiary butyl ether in different concentrations. However the Fig.4a shows a presence in a membrane structure of a faltering rough polysiloxane layer on border polysiloxane - air. It is possible to assume, that internal membrane layers boundary with a support top-surface provide diffusion separation.

The similar situation is created at formation of a composite membrane poly-γ-benzyl-L-glutamate (PBG)/PAI, the microphotograph of ultra-thin cut of which is presented in a Fig. 4,b. Despite formation of the thin intermediate layer having roughness, and external emptiness on border with air, the composite membrane is steady in pervaporation conditions and shows high selectivity at separation of mixtures of toluene and n-heptane [29].

Possibly, in a boundary region between polymers of various layers of a multilayer composite membrane, where interactions of adsorbed molecules with effective adsorption nodes on a surface of adsorbate (skin-layer polymer) are possible, in the process of the membrane formation a intermediate layer of the difficult structure is formed. This layer contains areas of the raised density, sites of domain character, and also the difficult architecture, consisting of coating polymer or both polymers. The morphology of the intermediate layer depends on conditions of formation of a composite. The laws found in works under discussion have led to necessity of detailed research of morphological, structural, physical and chemical characteristics of all layers in membranes, and first of all – of surfaces of skin-layers in PAI asymmetric films.

3. The surface morphology of the skin layer of asymmetric PAI film depending on the conditions of its preparation

In [30] the special attention has been given to PAI-1, containing diphenyloxyde fragments in structure of a molecular unit. PAI-1 concerns to moderate hydrophilic polymers. Phase separation at contact of its viscous solution in N-MP with strong for that polymer non-solvent – water – occurs with the high speed. Therefore the asymmetric microporous film with a skin-layer which thickness of 50 - 60 nm is formed. The SEM photographs of low-temperature fractures and surface of skin-layers of asymmetric microporous PAI-1 films presented in [30] show obvious distinctions in density of packing of polymer in surface layers (top and bottom), such that the average size of a pores formed in them differ ~ in 100 times.

Comparison of the SEM photograph of the skin layer surface of PAI-1 asymmetric film, shown in Fig. 2b, and the data of atomic force microscopy (Figs. 5a, 5b) leads us to conclude that fiber-like polymer chain agglomerates are oriented mainly in one direction of the film plane (presumably along the direction of doctor blade motion during pre-formation of membrane) to form a complex texture whose fragments go under the surface plane or rise above it. The observed pattern resembles a knitted fabric where caves (voids) with different sizes are formed. It is difficult to reveal among the latter penetrating pores and estimate their average diameter; however, it is obvious that the average pore diameter does not exceed 15 nm [30].

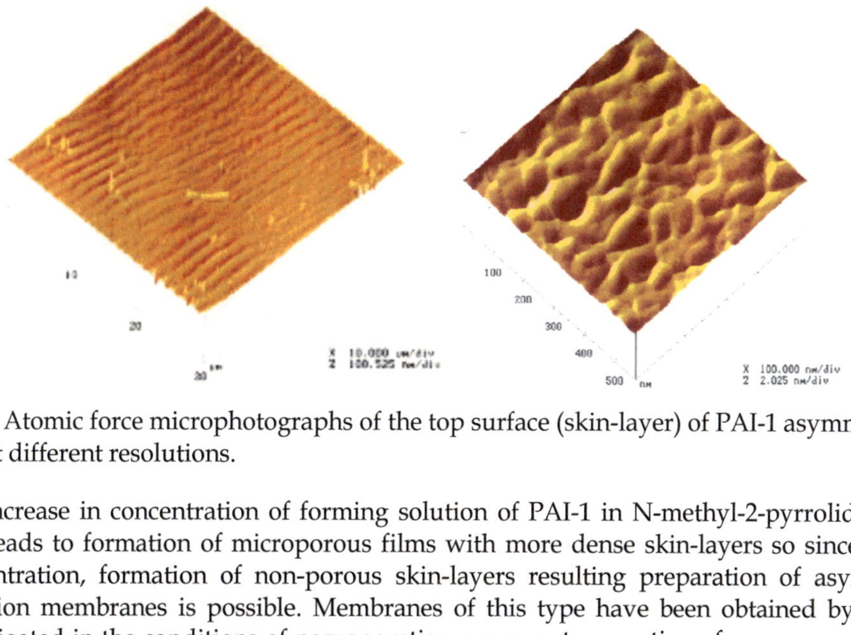

Fig. 5. Atomic force microphotographs of the top surface (skin-layer) of PAI-1 asymmetric film at different resolutions.

The increase in concentration of forming solution of PAI-1 in N-methyl-2-pyrrolidone (N-MP) leads to formation of microporous films with more dense skin-layers so since certain concentration, formation of non-porous skin-layers resulting preparation of asymmetric diffusion membranes is possible. Membranes of this type have been obtained by us and investigated in the conditions of pervaporation process at separation of aqueous-organic or organic liquids mixtures [31]. In our publications distinctions of physical, chemical and structurally - morphological characteristics of surface layers of asymmetric PAI-1 membranes [22], received in various conditions of phase-inversion process are shown.

It is of interest to compare data of atom-force microscopy of surfaces of skin-layers of asymmetric films on the basis of PAI-1, prepared from solutions with various PAI-1 concentrations, and also of a surface of non - porous dense film from this polymer. Mentioned the latter was obtained by a pouring technique with following slow removal of solvent from a solution during heating. In a Fig. 6 are presented SEM - photographs of low-temperature fractures of asymmetric diffusion PAI-1 membrane with non-porous dense skin-layer (Fig. 6a, 6b) and also the atom-force images (Fig. 6c, 6d) of a surface of this membrane in comparison with data of an atom-force microscopy for non-porous PAI-1 film (Fig. 6e, 6f). Formation conditions of asymmetric membranes (Fig. 3 and Fig. 6) differed only with concentrations of PAI-1 formation solutions.

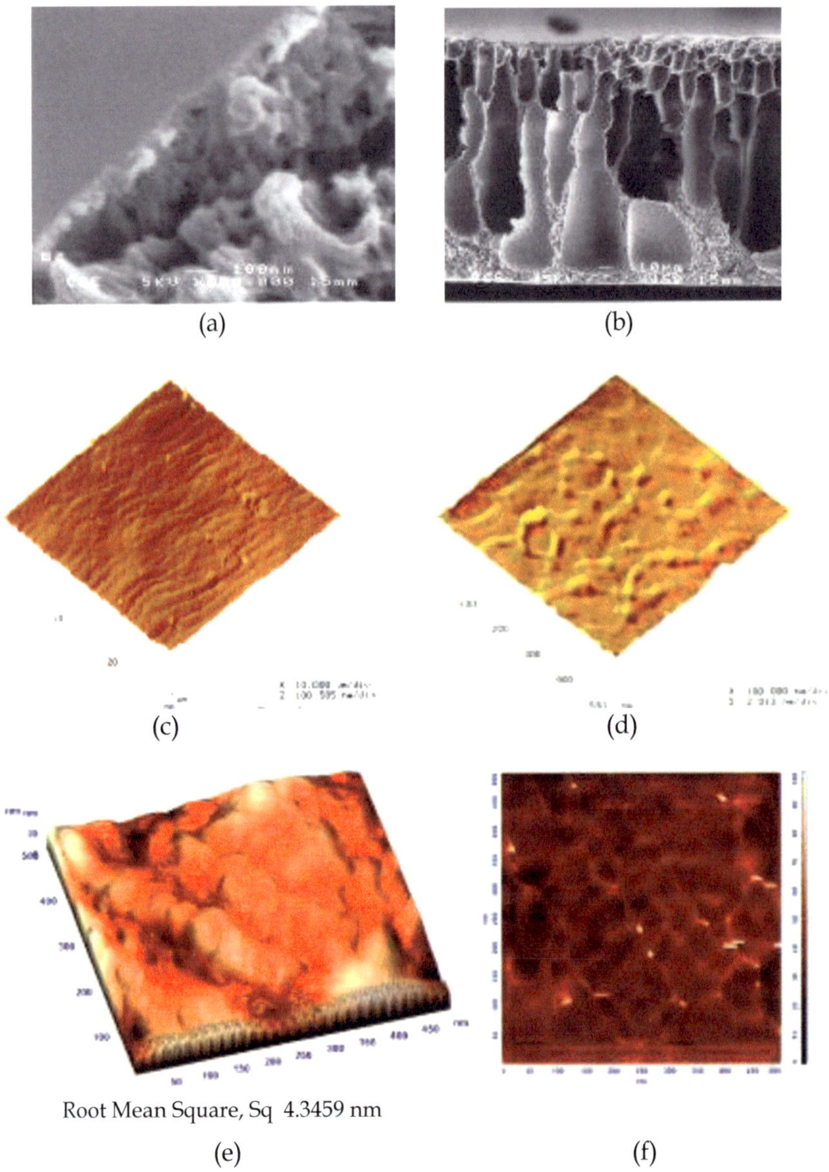

Fig. 6. (a, b) SEM - photographs of low-temperature fractures and AFM images of (c,d) asymmetric diffusion PAI-1 membrane and (e, f) non-porous dense film based on PAI-1 (e - 3D image, f - phase contrast image).

According to data of the atom-force microscopy, presented on Figs. 5 and 6, surfaces of all considered films are characterized by presence of the oriented polymeric domains. Their average size increases at transition from an ultrafiltration membrane (porous skin-layer) to diffusion one, that depends on increase in concentration and viscosity of a forming solution. Logical would be to expect the greatest smoothing of a surface at transition to non-porous polymeric film. On the contrary, in the case of PAI-1 film, most rough surface layer with the large oriented domains coming from a surface plane is observed.

Obviously, such unexpected effect is connected with ways of preparation of discussed samples. In case of asymmetric films of complex morphology of cross-section (Figs. 2a and 6a, 6b) the skin-layer is formed in process of "meeting head-on" of both the polymer solution and the precipitant (water), in the conditions of wet formation. At the same time at film formation on a surface of a smooth inorganic substrate (glass) there is a slow removal of solvent mainly in a direction, perpendicular the top surface. Polymer of top layer in these conditions "takes a great interest behind solvent" and takes a great interest behind it in that measure in which viscous polymeric gel formed already allows.

The similar picture can be observed at studying of a surface morphology of films of another polyamide-imide of PAI-2 synthesized by use of 3,5-diaminobenzoic acid as a diamine reagent in polycondensation. In a Fig.7 are presented AFM 3D image and phase contrast image to the top surface (it is generated on border polymer/air) of non-porous film of PAI-2. The film surface layer also is characterized by presence of the directed domains of polymer, but the smaller size, than at PAI-1. Carboxylic groups containing in a polymeric chain of PAI-2, more hydrophilic than PAI-1, takes a great interest in evaporating solvent so effectively that the craters which presence illustrates phase contrast image are formed.

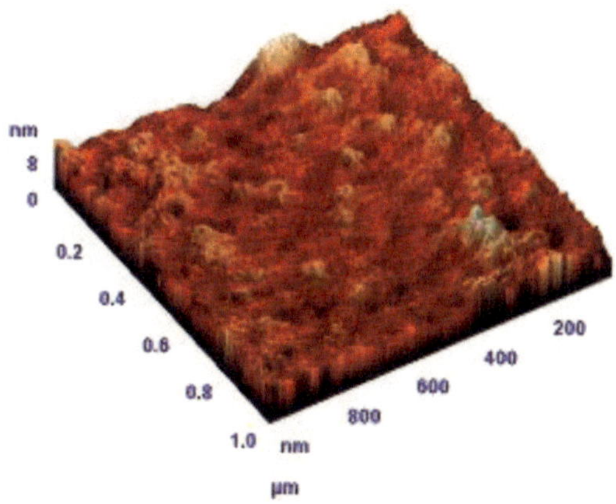

(a)

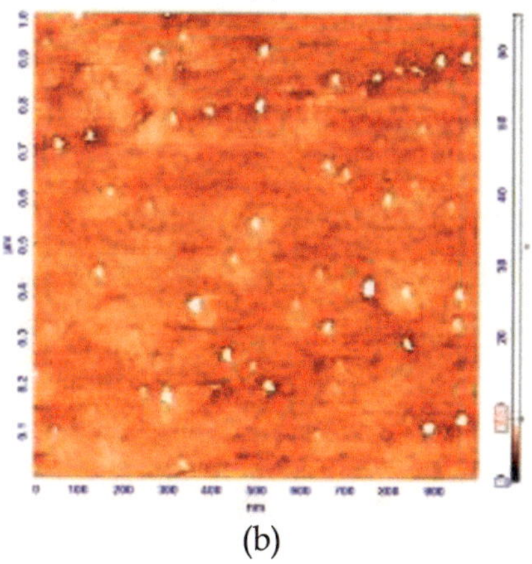

(b)

Fig. 7. AFM images of non-porous dense film based on PAI-2 (a - 3D image, b - phase contrast image).

There is a question what reason of the formation of the oriented PAI domains, «as weaved» from polymeric chains. The results of X-ray researches showing are known, the structures of dense PAI films and asymmetric PAI-supports, as a whole, do not contain sites with notable degree of heterogeneity; fragments with high degree of crystallinity are absent [32-34]. In PAI-1 films obtained by a pouring on a solid basis of a solution of polymer with its subsequent drying, always there is a structure which is characterized by orientation of chains in a film plane (Fig. 8).

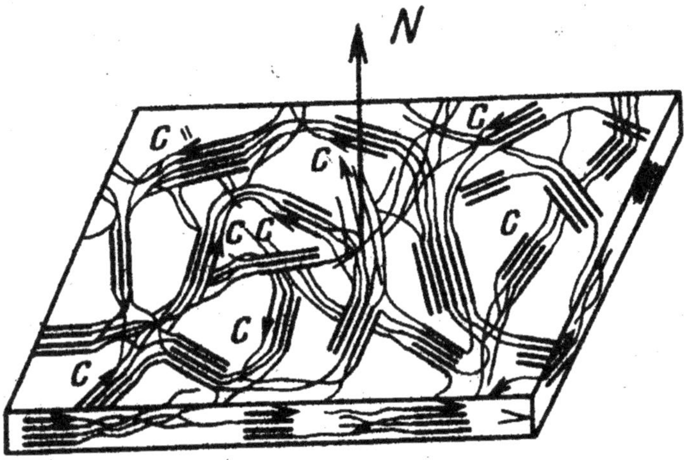

Fig. 8. Scheme of an arrangement of the ordered structures of macromolecules relative to the normal to a film surface [34].

In case of a film formation by pouring of a viscous solution of polymer on a surface of a glass plate with use of a doctor blade, it is necessary to expect a primary arrangement of the ordered fragments in a direction of movement of a knife, as illustrate Figs. 6 and 7. Thus, received by a poring method on a surface of a smooth inorganic basis PAI films are characterized by the ordered structure, such that polymeric chains most typical allocate in a film plane, and in case of additional focusing influence are extended in the primary direction set by external forces. In case of wet formation of a microporous film the front of precipitant interferes with movement of chains of polymer in a direction of removal of steams of solvent therefore more smooth surface is formed, than in case of non-porous film.

4. Thermo-physical characteristics of PAI films

The assumptions discussed above are in agreement with the data of thermo-gravimetric analysis (TGA) of PAI-2 samples (curve 2, Fig.9). The Fig. 9 shows, that speed of weight loss increases for PAI-2 since 200 °C, and at 600 °C the weight loss attains 47 % of initial weight.

TGA/DSC investigations were performed simultaneously with analysis of volatile products, which rise during thermal treatment of PAI-2. In Fig.10 3D FTIR diagram in coordinate – temperature, wave length and intensity is presented. It was supposed that three main processes are observed here: release of water in temperature region 25-550 °C, removal of N-MP solvent in temperature region 250-420 °C and the beginning of thermal degradation (release of CO_2, starting with 400 °C).

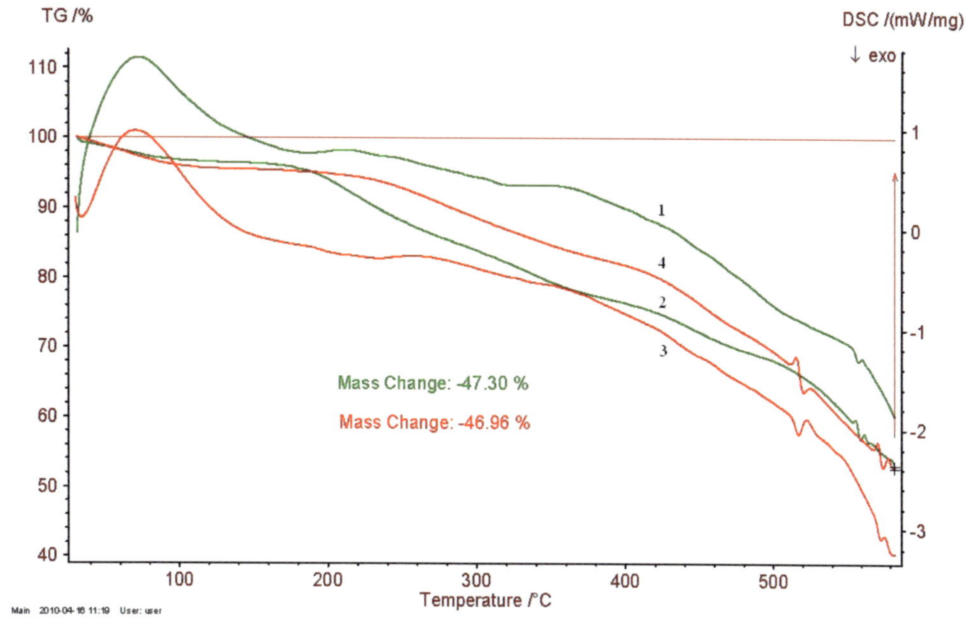

Fig. 9. TGA (2, 4) and DSC (1, 3) curves for PAI-2 (1, 2) and PAI-2 composite (3, 4) (2 $_{wt}$% of nanotubes) non-porous films.

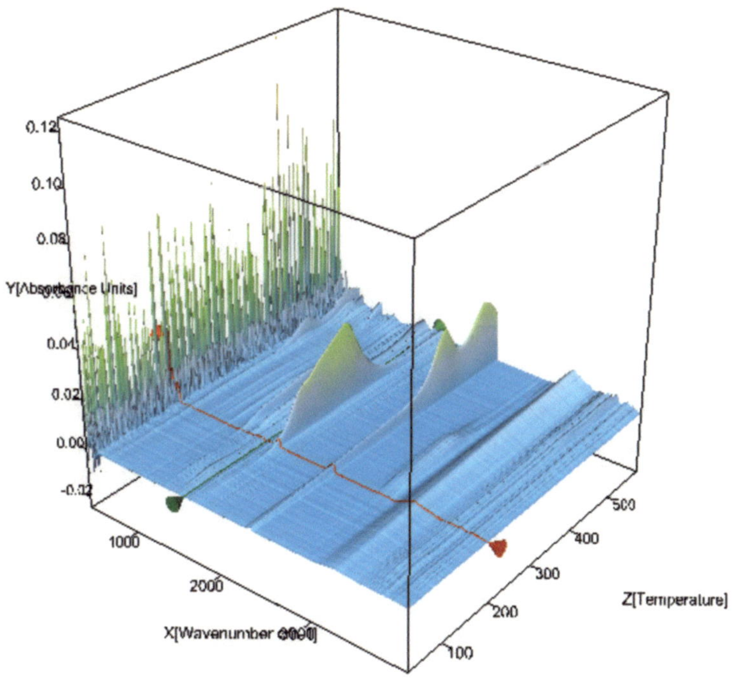

Fig. 10. 3D FTIR spectroscopy data for PAI-2 (simultaneously registered with TG-DSC).

The TGA curves related to the neat polymer (PAI-2) and the PAI-2-NT composite with the NT concentration of 2 $_{wt}$% are presented in Fig. 9 (curves 2, 4). For PAI-2 (curve 2), two steps of the mass loss can be recognized: within (20-100) °C (the mass loss is about 5 $_{wt}$%) and at 180 °C. The first step relates to the water evaporation, that confirmed by DSC data for polymer (curve 1, Fig 9). Endothermic peak is obviously observed on the DSC curve in this region. The next step of weigh loss (curve 2) relates to the N-MP evaporation out of the inner layers of the sample (the boiling temperature of N-MP is 202 °C) and followed by thermal degradation of the sample.

The same behavior we observed in the PAI-2-NT composite (curve 4), but the beginning of the second step of mass loss is shifted to more high temperature (220 °C) compared with neat polymer matrix. It is interesting to note that in the region 200-400 °C the TGA curve for PAI-2-NT disposes above the TGA curve for PAI-2 matrix and mass loss does not exceed 20 $_{wt}$% at 400 °C. This fact might be connected with NT's influence. Thus, introduction of NT's in polymer matrix promotes increasing thermal stability of PAI-2 and at the same time prevent of N-MP solvent and water removal from polymer-NT composite film.

5. Nanotubes with a chrysotile structure

The information on a chemical composition and structure of nanotubes, used in the given work is necessary for finding-out of the nature of the phenomenon discussed. Chrysotile $Mg_3Si_2O_5(OH)_4$ is the predominant fibrous form of serpentine. The crystal morphologies of the serpentines include cylindrical or conical rolls (chrysotile), planar structure (lizardite) and corrugated structure (antigorite). The structure of the chrysotile has been reviewed by Wicks and O'Hanley (1988) [35]. The first note concerning the tubular structure of chrysotile was by Pauling (1930) who suggested that the Mg-analogue of kaolinite should have a curved structure because of the misfit between the octahedral and tetrahedral sheets [36]. Later investigators confirmed this hypothesis and the tubular structure of chrysotile was demonstrated by electron diffraction and transmission electron microscopy (TEM) [37].

Chrysotile consists of sheets of tetrahedral silica in a pseudo-hexagonal network joined to a brucite layer in which Mg is in octahedral coordination with the apical oxygen of the SiO_4 layer and additional hydroxyl groups. The mismatch of the smaller lateral dimension of the SiO_4 sheet with respect to the $Mg(OH)_2$ layer is accommodated by the concentrically or spirally curled cylindrical chrysotile structure. This misfit results in a strain that can be relieved by curling of the double layer with tetrahedral part on the inner surface and the octahedral part on the outer surface [39]. As calculated by Whittaker (1955), the misfit is completely compensated at the ideal radius-of-curvature of 8.8 nm. There are several varieties of chrysotile, they differ mainly in the stacking of the double sheets and in the direction around which the sheets are rolled. Clinochrysotile and ortho-chrysotile are rolled around [100]. Para-chrysotile is rolled around the b axis. About 10 layers, each one 0.73 nm thick, constitute the wall thickness of the cylindrical rolls. The rolls possess hollow cores with a diameter of 4-5 nm because the layers energetically cannot withstand too tight a curvature. Figs. 11d and 12 present the schemes showing the incommensurate conformation between the smaller silica tetrahedra (light) and the sheet of darker $Mg(OH)_2$ octahedral.

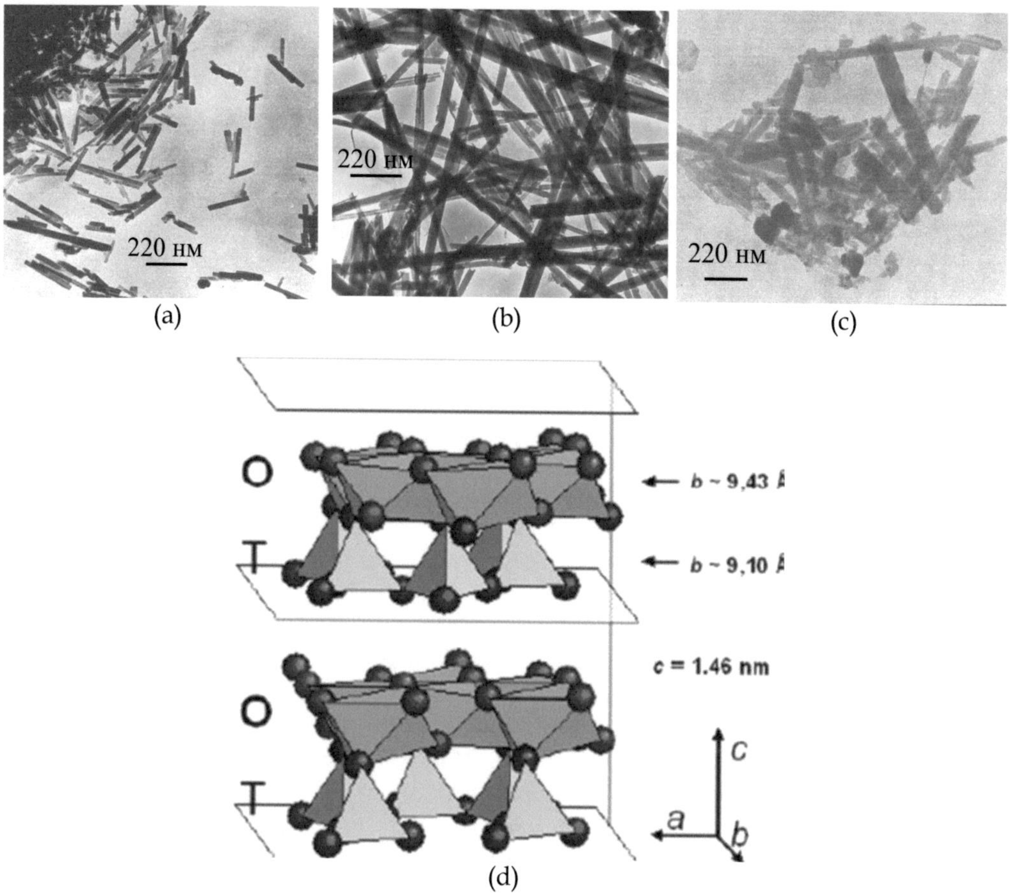

Fig. 11. (a, b, c) TEM photographs of $Mg_3Si_2O_5(OH)_4$ nanotubes of different sizes and morphologies [38]; (d) lattice of chrysotile (scheme 1) [37].

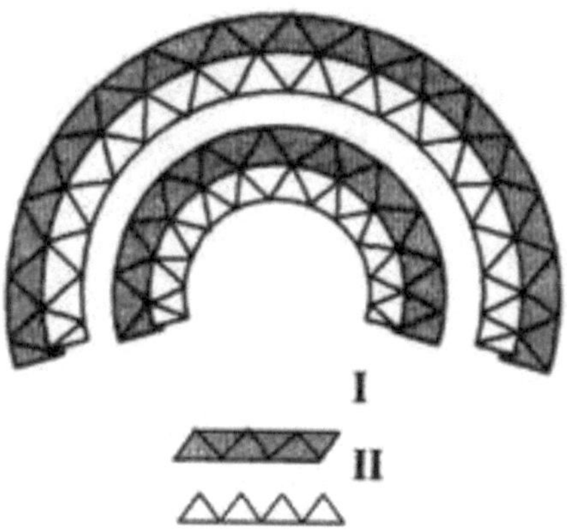

I - octahedral sheet
II - tetrahedral sheet

Fig. 12. Lattice of chrysotile (scheme 2) [40].

In the work, the following inorganic fillers were used in the composites: nanotubes (NT) of $Mg_3Si_2O_5(OH)_4$ composition with a chrysotile structure. These NT were produced by hydrothermal treatment of various precursors: magnesium oxide, silicon dioxide, and $MgSiO_3$ with NaOH solutions at NaOH concentration of up to 3 $_{wt}$% at temperatures of (250 - 450) °C and pressures of 30-100 MPa by the procedure described in [41, 42]. Nanotubes of composition $Mg_3Si_2O_5(OH)_4$ were used, with a cylindrical morphology and following dimensions: outer diameter d_o = 20-25 nm, inner diameter d_i = 4 nm, and length L = 500 -1000 nm (NT).

Thus, introduction of described above nanotubes in a polymeric matrix of PAI means introduction of particles with a crystalline structure possessing the big area of a surface and having on their external surfaces reactionary-capable hydroxyl-groups.

6. Morphological peculiarities of PAI-NT composite films

In the paper [38] the morphology and mechanical and transport properties of the composites formed were analyzed, which enabled optimization of their synthesis conditions. It was shown an increase in permeability to liquids (pervaporation) upon introduction of inorganic nanotubular additives in PAI-1. These properties of PAI-1-NT composite material obtained are very different from those of typical composites consisting of polymer and inorganic filler. As a rule, introduction of inorganic particles into a polymeric matrix results in that an inorganic phase impermeable to gases and liquids is formed in the matrix and the mass-transfer zone becomes narrower. As a result, the permeability of non-porous films falls [43]. A study of the transport properties of the PAI-1-NT composite films (homogeneous membranes) during pervaporation demonstrated the following. Introduction into the polymeric matrix of porous inorganic particles with tubular structure without their additional treatment (chemical or orienting) leads to an increase in the flux of water across the film membrane. This effect is the most pronounced in the case of a good compatibility of components of the composite material. The results of tests of PAI-1 and PAI-1-NT films in dehydration of an aqueous ethanol solution (48 $_{wt}$% ethanol) in the course of pervaporation at 40°C are shown that as the content of nanotubes in the composite increases from 2 to 10 $_{wt}$%, the permeability of the material to polar liquids, such as water and ethanol, grows. Introduction of 2 $_{wt}$% NT into PAI-1 film yields a composite for which the separation factor of the water/ethanol mixture even somewhat increases as compared with that for the homogeneous PAI-1 film.

In [44] it is also shown by us, how pervaporation properties of PAI-2-NT differ from corresponding properties of PAI-1-NT and also of a homogeneous PAI-1 and PAI-2 films. As is known, distinction in pervaporation characteristics of samples can be connected with influence of several factors among which are considered to be the most important two factors: diffusion and sorption ability [45]. According to diffusion-sorption model, primary transport of water and polar liquids through discussed samples testifies about hydrophylic properties of studied polyamide-imides which increase at transition from PAI-1 to PAI-2, containing carboxylic groups in a polymeric chain. Introduction of nanotubes with hydroxylic groups on their surface only enhances effect of wetting ability of material. At the same time, introduction into a polymeric matrix of objects with crystal structure reduces an area of carrying over penetrating liquids through a composite film. Thereof followed expect permeability decrease under the relation to all penetrants, in comparison with films of base polymer. Nevertheless, permeability on water only increases, and in case of PAI-2-NT exceeds permeability of film PAI-1-NT more than 5 fold, and PAI-2 film is twice more permeable by water than PAI-1. It is necessary to notice, that permeability of samples with NT in relation to a little polar liquids remains on a low level. Hence, speech does not go about substantial growth of free volume in a material or about occurrence of micro-defects in it as a result of introduction inorganic nanoparticles though it is impossible to deny their function of spacers. Apparently, NT's in a matrix of the semi-rigid polymer having ability to strong intermolecular interactions, fill free spaces which are arise at formation of non-porous film, not moving apart polymer chains in the visible extent. It is shown [38, 46] in our works, that the increase in quantity of nanotubes entered into a matrix PAI, according to the law of negation of negation, results, since certain quantity, to allocation superfluous NT on a surface of a polymeric film. Possibly, in this system intermolecular interactions of semi-fixed chains have appeared stronger, than interactions on border polymer - an inorganic phase.

It was of interest to investigate, how structural-morphological characteristics of a polymeric matrix PAI can change as a result of introduction in it NT's presented above.

On Fig. 13 data of atom-force microscopy are presented, obtained at surface research of nanocomposite film PAI-1-NT containing 2 $_{wt}$% of NT. A surface more smooth, than in case of film PAI-1. On the image a presence of nanotubes in near-by-surface layers of a polymeric matrix is visible. This result will well be coordinated with the information received at the analysis of ultra thin cross-sections of samples of type discussed, which photo is presented on Fig. 14 (a, b).

The Fig. 14 shows that nanotubes are distributed in polymeric matrix PAI-1 non-uniformly, in most cases by groups in quantity to 10 pieces on group. In some places NT approach to a surface, but are kept by polymer. With increased concentration of filler to 5 $_{wt}$%, NT's are going out on a surface of a composite film that illustrates Fig. 15.

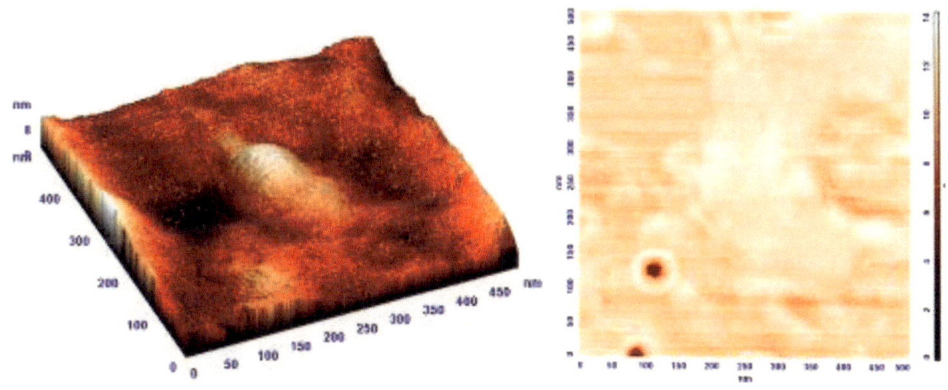

Root Mean Square, Sq 2.11508 nm
Peak-to-peak, Sy 15.6719 nm

(a) (b)

Fig. 13. 3D image (a) and (b) phase contrast image (b) for composite based on PAI-1 (2 $_{wt}$% of NT).

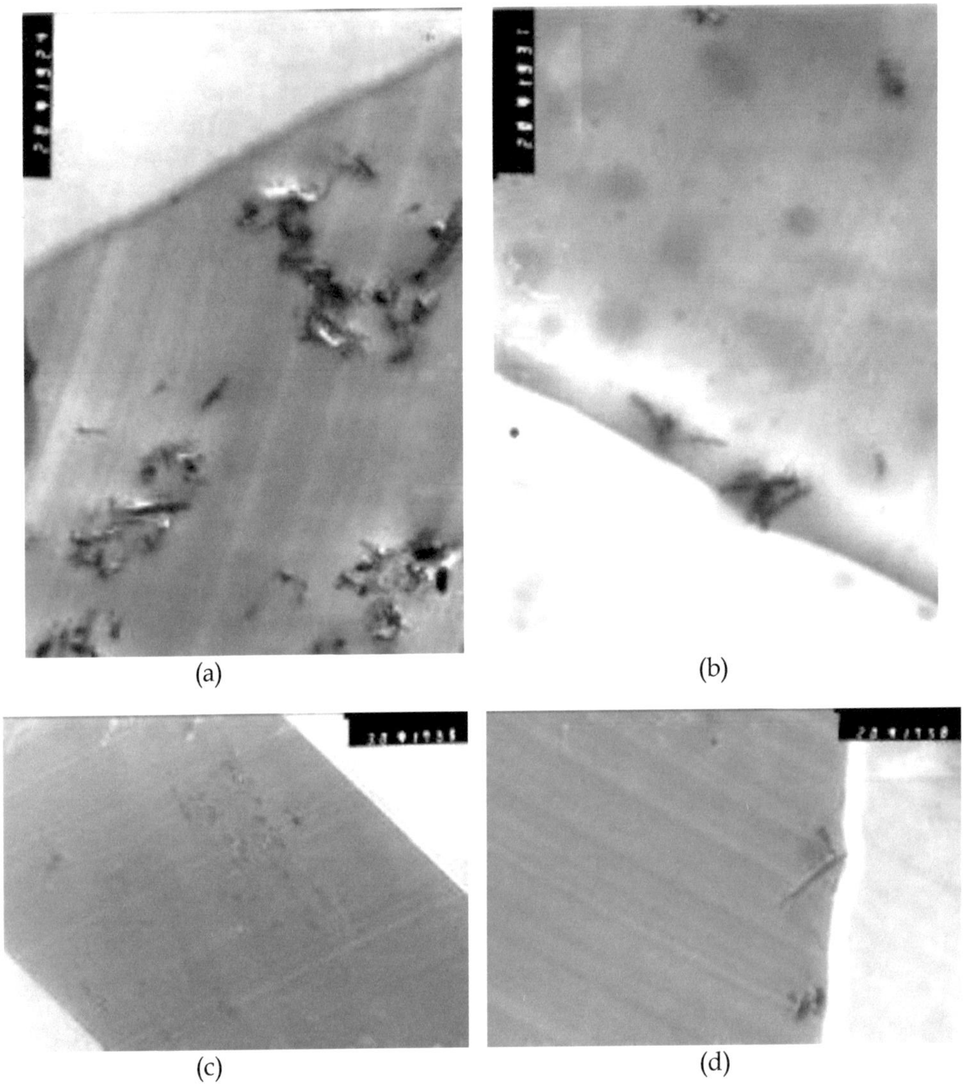

Fig. 14. Microphotographs of ultra-thin cross-sections for (a, b) PAI-1-NT (2 $_{wt}$%) and (c, d) PAI-2-NT (2 $_{wt}$%). Magnifications: (a, b, d) 20000, (c) 30000.

The analysis of images Fig. 15 leads to a conclusion, that at going of NT out on a surface of film adhesion between NT and a polymer matrix (that is well visible on the three-dimensional image and on profile) is broken, that negatively affects on transport properties and on values of durability of a composite [38]. In case of PAI-2-NT (2 $_{wt}$%), nanotubes are distributed in a film in more regular intervals by small groups. Near to a surface they are strongly kept by polymer that illustrates a Fig. 14 (c, d). The topology of a surface of composite film PAI-2-NT (2 $_{wt}$%) essentially differs from that at a surface of film PAI-2. If at AFM - image of surface of PAI-2 film craters from a solvent exit are shown (Fig. 7), the surface of composite PAI-2-NT (2 $_{wt}$%) has absolutely other appearance - the same relief from more or less similar hills that illustrates a Fig. 16.

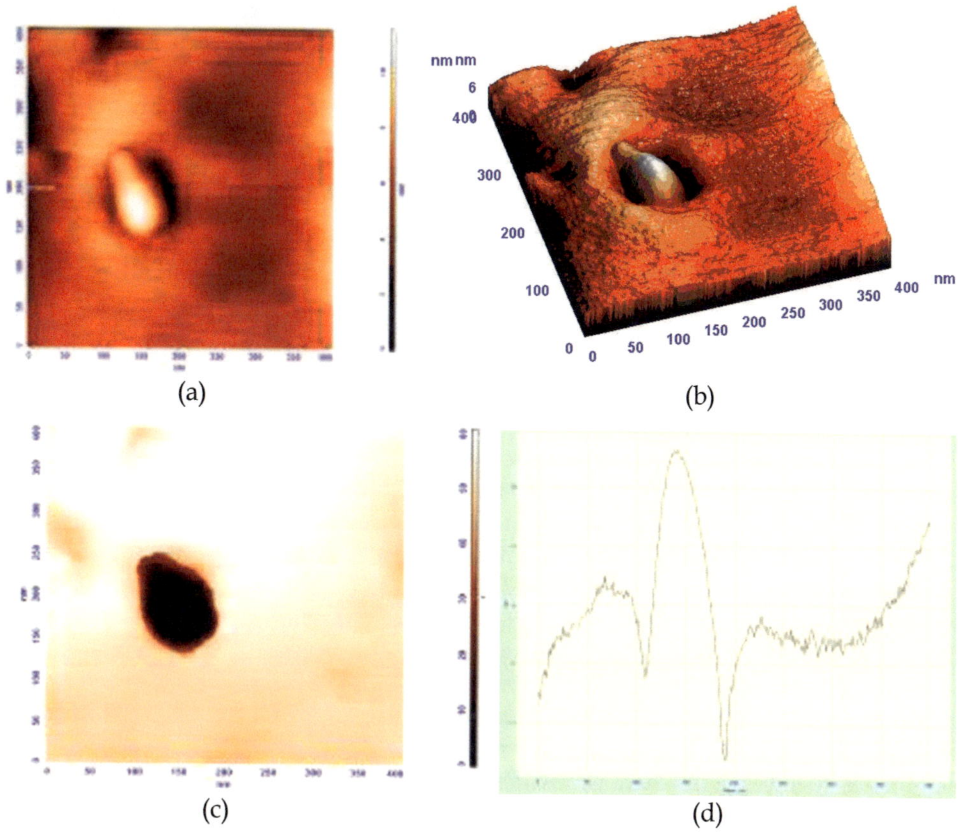

Fig. 15. AFM – images of magnesium hydrosilicate NT on the surface of polymer matrix at 5 $_{wt}$% of nanofiller content: topography (a), 3D image (b), phase contrast (c) and profile (d).

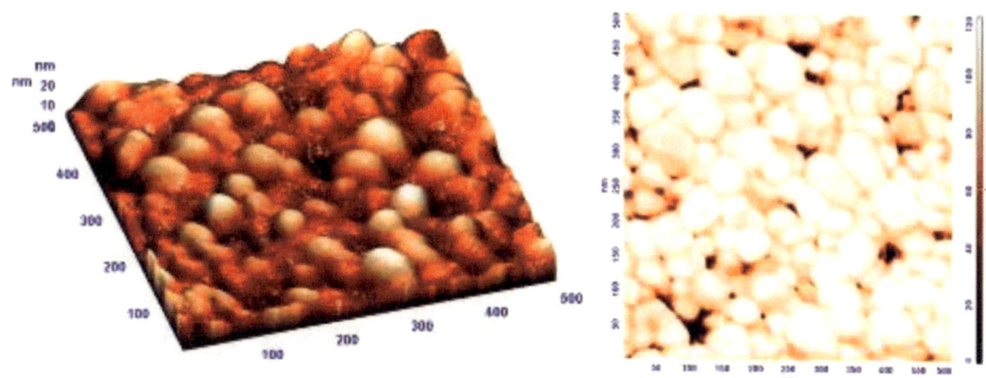

Root Mean Square, Sq	3.02961 nm
Peak-to-peak, Sy	20.1248 nm

(a) (b)

Fig. 16. 3D image (a) and phase contrast image (b) for composite film based on PAI-2 (2 $_{wt}$% of NT).

On microphotographs of surface of PAI-2-NT top layer nanotubes are not visible, that confirms phase contrast image. In this case strong adhesion between nanoparticles and polymer thanks to interaction of hydroxyl groups (on the surface layers of nanotubes) and carboxylic polymer groups leads to that all nanotubes settle down in a polymeric film, and the surface relief is formed at allocation of solvent at a stage of heating of a film (Figs 9,10). Distinctions in pervaporation characteristics of PAI and PAI-NT films lead to thought that the reason of the primary accelerated transport of water through PAI-NT not differences in hydrophilic properties of samples can be only. Apparently, located in small groups in a polymeric film, nanotubes form areas of the raised permeability on water [44].

7. Conclusions

The present publication urged not only to show primary features of polyamide-imides as polymers for formation of a wide spectrum of actual materials of different morphology, but also to note once again a role of interphase border in formation of materials for composite membranes. Being formed on border polymer/polymer, polymer/air or polymer / inorganic filler, these boundary regions are capable to influence essentially on distribution of transport streams of penetrants and unexpectedly to change characteristics of diffusion membranes. Structural and morphological properties of the composite boundary layers, as well as the number of active sites of adsorption in them are characteristics that are extremely difficult to study because of problems obtaining the necessary data. Apparently, it is necessary to come nearer, methodically and carefully investigating the general morphology of membranes in a complex with their properties. Apparently, in order to get closer to the result, we need to methodically and thoroughly investigate the general morphology of the membranes together with their properties. In this context, atomic force microscopy was indispensable method, which in combination with electron microscopy, thermo-physical methods and techniques for evaluating the transport properties allow for a fresh look at the processes in the composite membranes.

8. Acknowledgments

The authors are grateful to I.L. Potokin (State Research Institute of Ultrapure Biopreparations, Russia), M. Schlossig-Tiedemann and H. Kamusewitz (GKSS Research Center, Germany) for their help in performing experiments on membrane characterization

9. References

[1] Guchhait PK, Ray A, and Maiti S. Processable Heat-Resistant Polymers. XV. Synthesis and Properties of Polyamideimides from N-(p-carboxy Phenyl) Trimellitimide and p,p '- Di(amino Phenyl) Methane. *J. Macromol. Sci.-Chem.* 1983; 20: 93- 108.

[2] Agnihotri RK. Polyimides in lithography. *Polymer Eng. And Sci* 1977; 17: 366-371.

[3] Frank W. Harris et al. Polyimides used as microelectronic coatings. Patent number: 7074493; Jan 28, 2000.

[4] Choi KY, Yi MH. Synthesis and characterization of N-alkylated polyamidoimides. *Angewandte macromolekulare chemie* 1994; 222: 103-109.

[5] Wang Y, Chung TS, Wang H, Goh SH. Butanol isomer separation using polyamide-imide/CD mixed matrix membranes via pervaporation. *Chemical engineering science* 2009; 64: 5198-5209.

[6] Kononova SV, Kuznetsov YuP, Apostel R, Paul D, Schwarz HH. New polymer multilayer pervaporation membrane. *Angewandte Macromoleculare Chemie* 1996; 237: 45 - 53.

[7] Li FY , Li Y, Chung T-S, Chen H, Jean YC, Kawi S. Development and positron annihilation spectroscopy (PAS) characterization of polyamide imide (PAI)–polyethersulfone (PES) based defect-free dual-layer hollow fiber membranes with an ultrathin dense-selective layer for gas separation. *Journal of Membrane Science* 2011; 378: 541- 550.

[8] Yoshikawa M, Higuchi A, Ishikawa M, Guiver MD, Robertson GP. Vapor permeation of aqueous 2-propanol solutions through gelatin/Torlon® poly(amide imide) blended membranes. *Journal of Membrane Science* 2004; 243: 89–95.

[9] Setiawan L, Wang R, Li K, Fane AG. Fabrication of novel poly(amide - imide) forward osmosis hollow fiber membranes with a positively charged nanofiltration-like selective layer. *Journal of Membrane Science* 2011; 369: 196–205.

[10] Sun SP, Wang KY, Peng N, Hatton TA, ChungT-S. Novel polyamide-imide/cellulose acetate dual-layer hollow fiber membranes for nanofiltration. *Journal of Membrane Science* 2010: 363: 232–242.

[11] Sun SP, Wang KY, Rajarathnam D, Hatton TA, Chung T-S. Polyamide-imide Nanofiltration Hollow Fiber Membranes with Elongation-Induced Nano-Pore. *AICHE Journal* 2010; 56: 1481-1494.

[12] Zhang Y, Wang R, Yi S, Setiawan L, Hu X, Fane A G. Novel chemical surface modification to enhance hydrophobicity of polyamide-imide (PAI) hollow fiber membranes. *Journal of Membrane Science* 2011; 380: 241-250.

[13] Wang Y, Jiang L, Matsuura T, Chung TS, Goh SH. Investigation of the fundamental differences between polyamide-imide (PAI) and polyetherimide (PEI) membranes for isopropanol dehydration via pervaporation. *Journal of Membrane Science* 2008; 318: 217-226.

[14] Higuchi A, Yoshikawa M, Guiver MD, Robertson GP. Vapor Permeation and Pervaporation of Aqueous 2-Propanol Solutions through the Torlon ® Poly(amide imide) Membrane. *Separation Science and Technology* 2006; 40: 3419.

[15] Kreiter R, Wolfs DP, Engelen CW, van Veen HM, Vente JF. High-temperature pervaporation performance of ceramic-supported polyimide membranes in the dehydration of alcohols. *Journal of Membrane Science* 2008; 319: 126–132.

[16] Wang Y, Jiang L, Matsuura T, Chung TS, Goh SH. Investigation of the fundamental differences between polyamide-imide (PAI) and polyetherimide (PEI) membranes for isopropanol dehydration via pervaporation. *Journal of Membrane Science* 2008; 318: 217-226.

[17] Chafin R, Lee JS, Koros WJ. Effects of casting and post casting annealing on xylene isomer transport properties of Torlon - 4000T films. *Polymer* 2010; 51: 3462-3471.

[18] Gusinskaya VA, Koton MM, Batrakova TV, and Romashkova KA. Polyamidoimides based on symmetric and asymmetric dicloride imido acids. *Vysokomol. Soedin.* 1976; 18(A): 2681-2686.

[19] Hamciuc C, Carja I-D, Hamciuc E, Vlad-Bubulac T, and Musteata V-E. Silica-containing fluorinated poly(amide-imide) hybrid films. *High Performance Polymers 2011*; 23: 362–373.

[20] Bruma M, Sava I, Daly WH, Mercer FW, Belomoina NM, Negulescu II. Synthesis and Properties of New Polyphenylquinoxaline-imide-amides. *Journal of Macromolecular Science* 1993; 30: 789-799.

[21] Gusinskaya VA, Churganova SS, Koton MM., et al. Aromatic Polysulfone Imides and Membranes based on Them. *Zh. Prikl. Khim.* 1984; 57: 1819.

[22] Kononova SV, Kuznetsov YP, Romashkova KA, and Kudryavtsev VV. Interrelation between Preparation Conditions and Structure of Asymmetric Membranes Based on Poly((diphenyl oxide amido)-N-phenylphthalimide). *Polymer Science* 2006; 48(A): 967–973.

[23] Kononova SV, Romashkova KA, Gofman IV, Kremnev RV, Kruchinina EV, and Svetlichnyi VM. Aromatic Polysulfone Imides and Membranes Based on Them. *Russian Journal of Applied Chemistry* 2009; 82: 1033–1040.

[24] Kononova SV, Kuznetsov YP, Sukhanova TE, Belonovskaya GP, Romashkova KA, and Budovskaya LD. Gas separation Properties of Composite Membrane: Effect of the Support on the Structure of a Selective Skin Layer. *Polymer Science* 1993; 35: 229-234.

[25] Shchukarev AV, Kononova SV, Kuznetsov YP. XPS study of polymer molecules orientation in polyamide - imide films, Prepr. of the 1996 4th European technical symposium on polyimides and high performance polymers. *STEPI 4*, U. Montpellier 2, France, 13 - 15 May, 1996;

[26] Kononova SV, Kuznetsov YP, Shchukarev AV, Ivanova VN, Romashkova KA, and Kudryavtsev VV. Structure and Gas Separation Properties of Composite Membranes with Poly(2,2,3,3,4,4,5,5-octafluoro-n-amyl Acrylate) Cover Layer. *Russian Journal of Applied Chemistry* 2006; 76: 791-799.

[27] Kononova SV, Kuznetsov YP, Sukhanova TE, Belonovskaya GP, Romashkova KA, and Budovskaya LD. Gas separation properties of composite membtranes: effect of the support on the structure of a selective skin layer. *Polymer Science* 1993; 35: 229-234.

[28] Kuznetsov YP, Kononova SV, Kruchinina EN, Romashkova KA, Svetlichnyi VM, Molotkov AV. Pervaporation membranes for separation of methanol/methyl tert. butyl ether mixtures. *Russian Journal of Applied Chemistry* 2001; 74: 1302.

[29] Kononova SV, Kremnev RV, Baklagina YG, Volchek BZ, Vlasova EN, Shabsels BM, Romashkova KA, Romanov DP, Arkhipov SN, Bogomazov AV, and Uchytil P. Interrelation between the Structural and Transport Properties of Pervaporation Membranes with Diffusion Layers Based on Poly-γ-Benzyl-L-Glutamate. *Crystallography Reports* 2011; 56: 530–535.

[30] Kononova SV, Kruchinina EN, Romashkova KA, Potokin IL, Shchukarev AV, and Kudryavtsev VV. Phase-Inversion Gradient-Porous Films on the Basis of

Polyamidoimides Derived from Phthalimidobenzenedicarbonyl Dichloride and Various Diamines. *Russian Journal of General Chemistry* 2010; 80: 1977–1985.

[31] Kuznetsov YP, Kononova SV, Romashkova KA, Kudryavtsev VV, Gusinskaya VA. Asymmetric polymer pervaporation membrane. Patent, Russian Federation, № 2126291, 20.02.1999, priority 26.11.1996.

[32] Sidorovich AV, Svetlichnyi VM, Baklagina YuG, Gusinskaya VA, Batrakova TV, Romashkova KA, Goikhman MYa. Thermomechanical Properties and structure of blends and copolymers of polyamidoimides. *Vysokomol. Soedin.* 1989; 31(A): 2597-2602.

[33] Gusinskaya VA, Baklagina YuG, Romashkova KA, Batrakova TV, Kuznetsov YuP, Koton MM, Sidorovich AV, Mikhailova NV,Nasledov DM, Lyibimova GN. Features of Formation of Supermolecular Structure in Polyamidoimide Systems. *Vysokomol. Soedin.* 1988; 30(A): 1316-1320.

[34] Baklagina YuG, Sidorovich AV, Urban I, Pelzbauer Z, Gusinskaya VA, Romashkova KA, Batrakova TV. Supermolecular structures of polyamidoimide's films. *Vysokomol. Soedin.* 1989; 31(B): 38-42.

[35] Wicks FJ., O'Hanley. Serpentine minerals: structure and petrology. *Rev.Mineral.* 1988; 19: 91-159.

[36] Pauling L. The structure of chlorites. *P Natl Acad Sci.* USA 1930; 16: 578-582.

[37] Piperno S, Kaplan-Ashiri I, Cohen SR, Popovitz-Biro R, Wagner HD, Tenne R, Foresti E, Lesci IG, Roveri N. Characterization of Geoinspired and Synthetic Chrysotile Nanotubes by Atomic Force Microscopy and Transmission Electron Microscopy. *Advanced Functional Materials* 2007; 17: 3332-3338.

[38] Kononova SV, Korytkova EN, Romashkova KA, Kuznetsov YuP, Gofman IV, Svetlichnyi VM, and Gusarov VV. Nanocomposite Based on polyamidoimide with Hydrosilicate Nanoparticles of Varied Morphology. *Russian Journal of Applied Chemistry* 2007; 80: 2142-2148.

[39] Whittakker EJW. A classification of cylindrical lattices. *Acta Crystallogr.* 1955; 8: 571-574.

[40] Levi KJT., Veblen DR. "Eastonite" from Easton, Pennsylvania: A mixture of phlogopite and a new form of serpentine. *Amer. Miner.* 1987; 72: 113.

[41] Korytkova, E.N., Maslov, A.V., Pivovarova, L.N., Drozdova, I.A., and Gusarov, V.V. Formation of $Mg_3Si_2O_5(OH)_4$ Nanotubes under Hydrothermal Conditions, *Glass Phys. Chem.(Engl. Transl.)* 2004; 30: 51-55.

[42] Korytkova, EN, Maslov, AV, Pivovarova, LN, Polegotchenkova, YuV, Povinich VF, and Gusarov VV. Synthesis of nanotubular $Mg_3Si_2O_5(OH)_4$ - $Ni_3Si_2O_5(OH)_4$ hydrosilicates at elevated temperatures and pressures. *Inorg.Mater.(Engl.transl.* 2005; 41: 743-749.

[43] Yamaguchi T, Miyazaki Y, Nakao S, et al. Membrane design for pervaporation or vapor permeation separation using a filling-type membrane concept. *Ind. Eng.Chem. Res.* 1998; 37: 177-184.

[44] Kononova SV, Korytkova EN, Maslennikova TP, Romashkova KA, Kruchinina EV, Potokin IL, and Gusarov VV.Polymer-inorganic nanocomposites Based on Aromatic Polyamidoimides Effective in the Processes of Liquids Separation. *Russian Journal of General Chemistry* 2010; 80:1136-1142.

[45] Mulder M. Basic Principles of Membrane Technology. Dordrecht: Kluwer, 1991.

[46] Gubanova GN, Kononova SV, Vylegzhanina ME, Sukhanova TE, et al. Structure, Morphology, and Thermal Properties of Nanocomposites Based on Polyamide Imide and Hydrosilicate Nanotubes. *Russian Journal of Applied Chemistry* 2010; 83: 2175-2181.

5

Characterization of Complex Spintronic and Superconducting Structures by Atomic Force Microscopy Techniques

L. Ciontea[1], M.S. Gabor[1], T. Petrisor Jr.[1], T. Ristoiu[1], C. Tiusan[1,2] and T. Petrisor[1]
[1]*Technical University Cluj-Napoca, Material Science Laboratory, Napoca*
[2]*Institut Jean-Lamour, UMR7198 CNRS-Nancy Université, Vandoeuvre les Nancy*
[1]*Romania*
[2]*France*

1. Introduction

Within this chapter we would like to address two main applications of Atomic Force Microscopy techniques. The first one illustrates the use of the Atomic Force Microscopy for the optimization of the morphological properties in multilayer stacks dedicated to spintronic devices, in which the electric current flows perpendicular to the layers (current-perpendicular-to-plane CPP geometry). The second part of our chapter presents the use of the Magnetic Force Microscopy as a tool for the micro-magnetic characterization of magnetic thin film dedicated to interface systems with high temperature superconductors. The performances of both spintronic and superconductor/ferromagnet interface devices are directly related to the optimal structural and magnetic properties of the constituent thin films. During the process of the complex sample growth, the structural, morphological and magnetic characterization represents one of the most important steps. Various ex-situ techniques are involved for characterization at microscopic and macroscopic scale. For the multilayer stacks, which require flatness and continuity of the constituent layers, the near field microscopy techniques represent one of the most commonly used tools. The standard atomic force microscopy (AFM) allows to extract precise information about the thin film surface topology in terms of roughness and morphology. This information is furthermore correlated with the crystallographic properties of the layers determined by diffraction techniques (X-Ray, electrons…). The magnetic force microscopy (MFM) operating mode provides a complete characterization of the micromagnetic properties of a magnetic film. This analysis at a microscopic scale is often correlated to the macroscopic magnetic properties measured using the magnetometry techniques.

2. The use of AFM technique for the characterization of multilayer stacks for spintronic devices

After the discovery in 1989 of the giant magnetorezistive effect in magnetic multilayers by A. Fert and P. Grunberg (Baibich et al 1988, Binash et al 1989), the spintronics became one of the most attractive research fields from both fundamental and applicative point of view. The spintronic devices, composed by alternating magnetic and nonmagnetic multilayer structures, represent nowadays one of the major issues of the sensors and data storage industries (Wolf et al 2001). The main functional property of a spintronic device is based on the skilful manipulation of the spin of the electrons carrying the charge current.

In the current-perpendicular-to-plane (CPP) spintronic devices, the interfacial roughness has to be minimized in order to get flat and abrupt interfaces between the different layers. This avoids the intermixing and provides a perfect continuity with minimum thickness fluctuations of spacers sandwiched between adjacent layers. We chose to show here the characterization steps by AFM in a complex magnetic tunnel junction multilayer stack using as magnetic electrodes the full Heusler Co_2FeAl film and the single crystal MgO as insulating barrier. Therefore, the magnetic tunnel junction is constituted of the MgO insulating barrier sandwiched between two magnetic metallic films (Tiusan et al 2007) (transition metal and Heusler alloy). The electric current crosses the insulator by quantum tunneling whose thickness is typically less than 12 monolayers. Therefore, one can easily understand that the extreme control of the interfacial roughness is a key parameter for proper operating the device. Analysis of our samples by AFM provides the surface topography of the layers. This has been correlated with their crystalline structure analyzed by X ray diffraction techniques. Our analysis allowed a deep understanding of the growth mechanisms by the sputtering of the layers constituent of the complex multilayer stack.

The Co_2FeAl (CFA) Heusler alloy has a cubic structure with a lattice parameter close to 0.573 nm. For this reason, in order to facilitate the epitaxial growth of the Heusler alloy, one can use single crystalline MgO(001) substrates. The MgO has a crystalline structure belonging to the Fm-3m space group, essentially formed by the interpenetration of two face centered cubic sub-lattices containing Mg and O atoms. Taking into account the CFA and MgO lattice parameters, the CFA should normally grow on MgO with the epitaxial relation: CFA(001)[110]||MgO(001)[100], which implies a 45° in-plane rotation of the CFA lattice with respect to the MgO one. In this configuration the lattice mismatch between CFA and MgO is $(\sqrt{2}/2a_{CFA} - a_{MgO})/a_{MgO} = -3.8\%$, which is adequate to promote epitaxy. However, in order to better adapt the lattice mismatch between the epitaxial film and the single crystalline substrate one can employ a buffer layer. We explored this possibility by using an epitaxial Cr film as buffer. Cr has a body centered cubic crystalline structure belonging to the Im-3m space group. The lattice parameter is 0.2884 nm and if one assumes a Cr(001)[110]||MgO(001)[100] epitaxial relation of Cr on MgO, the lattice mismatch is -3.15%. Moreover, considering a cube-on-cube epitaxy of CFA on Cr, i.e. CFA(001)[100]||Cr(001)[100] (see figure 1), the lattice mismatch between CFA and Cr is only -0.7%, and as a consequence, Cr is well suited to be used as a buffer layer for the epitaxial growth of CFA films on MgO(001) substrates.

2.1 Chromium buffer layer

Although CFA Heusler alloy thin films were also grown on un-buffered MgO (001) single crystal substrates, we will first discuss the epitaxial growth of Cr (001) buffer layer. The process of epitaxial growth of CFA on MgO is essentially similar to that of Cr on MgO, even

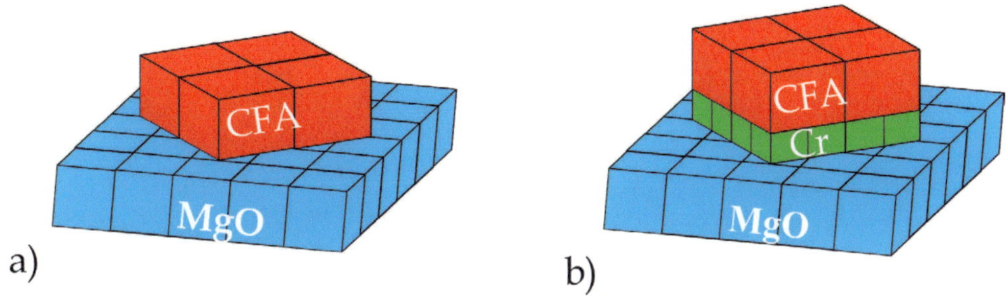

Fig. 1. Schematic representation of the (a) CFA and (b) Cr/CFA bilayer epitaxy on MgO(001) single crystalline substrate.

though it is more complicated due to the complex chemical ordering within the Heusler alloy. Therefore, we will first treat, in extensive details, the growth of Cr on MgO and use the drawn conclusions when presenting the epitaxial growth of CFA on MgO.

The Cr films were deposited using standard DC magnetron sputtering on polished MgO (001) substrates in a system having a base pressure better than $4x10^{-9}$ torr. The Ar working gas pressure was maintained at 1.0 mtorr during sputtering and the deposition rate was around 0.1nm/s. The substrate temperature during growth was varied between room temperature (RT) and 600 °C.

High angle specular X-Ray diffraction scans were performed to test the orientations present in the different Cr films. figure 2 shows a typical XRD symmetric scan for the Cr film grown at a substrate temperature of 400°C showing only the (002) film reflection. Regardless of the growth temperature, no evidence of other orientation, except for (002) was found. The inset in figure 2 shows the evolution of the Cr (002) peak with the growth temperature. A large decrease of the Cr (002) peak intensity and position shift can be observed for the sample grown at RT relative to the high temperature deposited ones. This suggests the occurrence of an important change in the structure at atomic level for this sample. Figure 3 shows the evolution of the in-plane and out-of-plane lattice parameters as a function of the growth temperatures extracted from symmetric and asymmetric X-Ray diffraction experiments. A 3.2% expansion of the out-of-plane lattice parameter from the bulk value is observed for the sample grown at RT. This growth with the out of-plane lattice expansion is not accompanied by clear in-plane lattice contraction, the small variation of the in-plane lattice parameter from the bulk value being within the measurement error limits. Nevertheless, a small increase of the in-plane lattice parameter can be argued. This behavior is consistent with the in-plane constant expansion due to the tensile strain in the growth plane produced by the lattice mismatch between MgO and Cr (-3.15%).

The out-of-plane lattice distortion is rather difficult to explain. A similar growth temperature dependence of the lattice parameter was observed in triode sputtered V (001) epitaxial films on MgO (001) substrates (Huttel et al., 2005). A complex experimental analysis accompanied with *ab-initio* simulations (Huttel et al., 2007) showed that this low temperature out-of-plane lattice expansion is a metastable state and that it is most likely due to the inhomogeneous disorder created by the bombardment of the growing film by neutral atoms reflected off the target. Another possible mechanism for the lattice expansion is the residual stress of the thin films produced by sputtering at low pressures (Thornton &

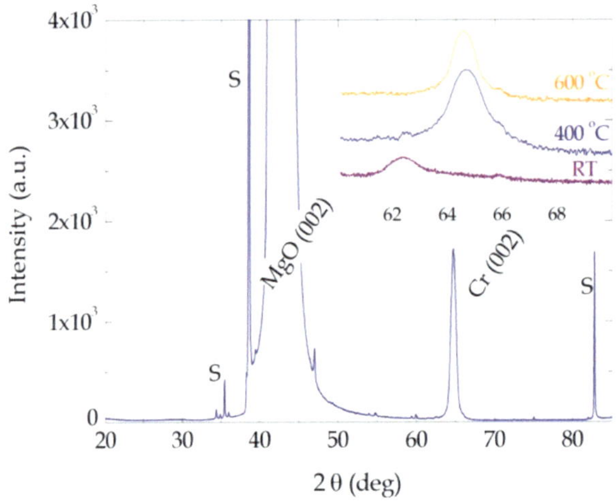

Fig. 2. High angle symmetric scan for the MgO (001)//Cr (16.5nm) film grown at 400°C, showing the presence of Cr (002) reflection, in addition to the MgO (002) one. The peaks marked with S are substrate refections corresponding to the Cu Kα and WLα wavelengths. The inset shows the Cr (002) peak for films deposited at different temperatures. The vertical dashed line marks the position of the Cr bulk 002 reflections. The scans are shifted vertically for better visibility.

Hoffman, 1989). At low sputtering pressures, because of the poor thermalization, there is an increase in high energy atoms arriving at the substrate and, due to the momentum transfer, Cr atoms are forced to into spaces too small to accommodate them under existing thermal equilibrium conditions and, as a result, an out-of-plane lattice expansion takes place (Hsieh et al., 2003). Following these assumptions, one should expect that a substrate temperature increase drives the system to its most stable state relaxing the strains and disorder (Huttel et al., 2007). This is indeed the case and for the sample grown at 400 °C a much smaller strain is observed, the out-of-plane lattice parameter being contracted by 0.12% while the in-plane lattice parameter expanded by 0.18% and, consistent with the in-plane tensile stress to the Cr to MgO lattice mismatch. Moreover, for the sample grown at 600 °C the strain is totally relaxed, most likely by formation of misfit dislocations, and the in-plane and out-of-plane lattice parameters are being consistent with the bulk values.

As expected, the epitaxial quality of the Cr(001) films improves with increasing the growth temperature. The full width at half maximum (FWHM) of the rocking curves around Cr (002) and (011) reflections as a function of temperature are depicted in figure 3b. The figure indicates a FWHM of 0.74° for the (002) reflection and of 0.91° for the (011) reflection, for the sample deposited at 600 °C. These values are in agreement with the previously reported ones (Harp. & Parkin, 1994; Harp. & Parkin, 1996; Fullerton, 1993) and demonstrate a high degree of epitaxy.

The main conclusion from the structural point of view is that it is possible to obtain epitaxial Cr(001) thin films on MgO(001) substrates by DC sputtering starting from room temperature and with improved crystalline quality at higher growth temperatures. Conventionally, it is considered that the epitaxial growth of Cr on insulating substrates by sputtering can be achieved only when the deposition takes place at elevated temperatures (Harp. & Parkin,

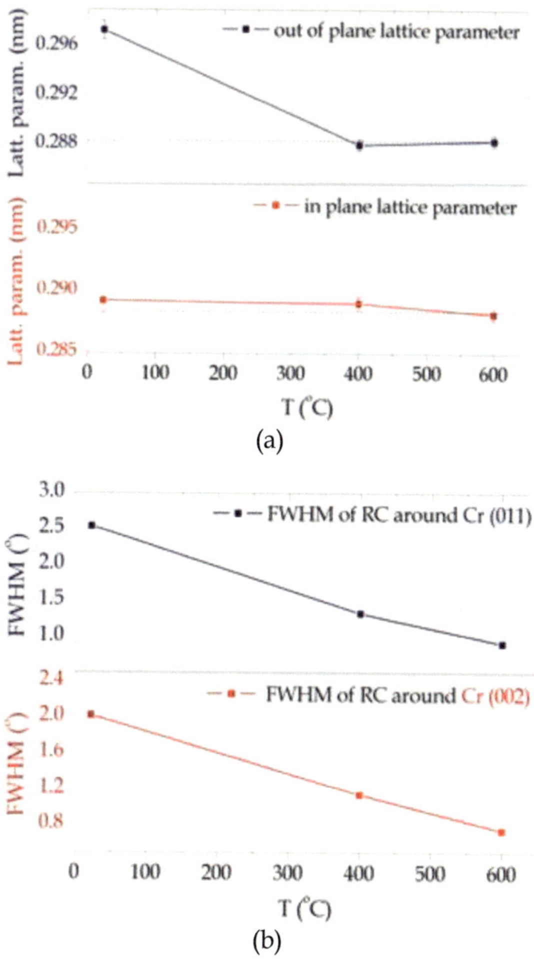

Fig. 3. (a) Evolution of the in-plane and out-of-plane lattice parameters of the Cr films with the growth temperature and of the (b) full width at half maximum of the rocking curves around Cr symmetric (002) and asymmetric (011) reflections.

1994; Harp. & Parkin, 1996; Fullerton et al., 1993a, 1993b), while our findings show a lower growth temperature required for epitaxy. This behavior is most likely related to our specific growth conditions which imply a relative low Ar pressure during deposition. In contrast to evaporation methods, sputtering deposition methods are characterized by an impulse transfer from accelerated energetic particles to the surface atoms of the target. The energy distribution of the sputtered species has a maximum around 5-10eV, presenting a high energy tail with a low percentage of high energy sputtered particles as well, that can increase by decreasing the Ar pressure (Depla & Mahieu, 2008). It has been shown that these energetic particles can modify the growth kinetics trough the formation of a high density of nucleation centers resulted from their energetic impact on the sample surface (Kalff et al., 1997). The maximum of the sputtered particles energy distribution can be shifted to lower energies by the thermalization process (Meyer et al., 1981), i.e. collisions with the working gas atoms between the target and the substrate. The thermalization process is controlled by the total gas pressure in the sputtering chamber. Thus, by decreasing the Ar pressure, the thermalization is reduced, so the energy of the atoms arriving at the substrate increases, and eventually, they will have enough energy to undergo diffusion on the surface to a high-binding energy site at a lower surface temperature. The higher mobility of the ad-atoms, together with the increase in the density of nucleation centers, would explain why lower growth temperatures are required for epitaxy to occur under our specific growth conditions.

Atomic force microscopy (AFM) was used to investigate the effect of the deposition temperatures on the films surface morphology. Figure 4 shows the AFM images for Cr films grown on MgO (001) at different temperatures. The deposition of Cr at room temperature yields a very flat featureless surface with a root mean square roughness (RMS) of 0.14 nm. As the temperature is increased to 400 °C and 600 °C, temperature dependent morphological features start to form. The surface of the 400 °C grown sample exhibits 3D rounded islands, while the surface of the 600 °C grown sample display larger, well defined and regulated ones, with the sides parallel to the substrate [110] and to the [1-10] directions. At the same time, the RMS increases from 0.45 nm for the film deposited at 400 °C to 1.25 nm for the film grown at 600 °C.

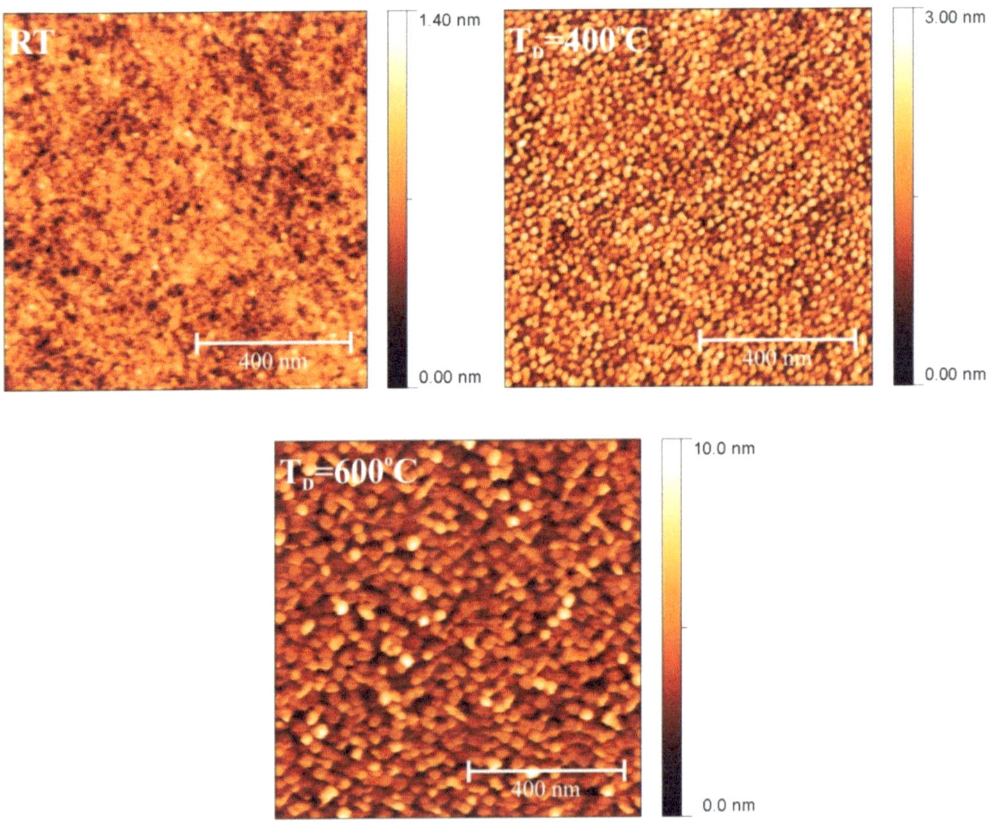

Fig. 4. AFM images (1x1 μm^2 scans) showing the morphology of the MgO(001)//Cr(001) films grown at different temperatures.

It is a common feature for the mid to late transition metals that in conditions of thermodynamic equilibrium not to wet an oxide surface (Campbell, 1997). However, our growth conditions are far from thermodynamic equilibrium and in this case the film growth process is governed by kinetic effects. In a simplified picture layer by layer growth occurs when additional islands start to form on the surface after the previous layer has coalesced. Therefore, in order to prevent nucleation before coalescence, the *ad-atoms* have to be mobile enough to reach the island edges and to have enough energy to overcome the Ehrlich-Schwoebel energetic barrier (Ehrlich & Hudda, 1966; Schwoebel, 1969) encountered by the *ad-atom* upon descending a step. In the special case of non-wetting metals on oxides, the energetics has two essential features: the strong attraction to the edge of an island due to the lateral metal-metal bonding (Ei) and the energy difference (ΔE) between the absorption energy of a metal atom a oxide site and on a metal site (Campbell, 1997; Ernst et al., 1993). If the energy difference (ΔE) exceeds the lateral metal-metal bonding energy (Ei), then there will be no stable sites at the edges of the 2D islands and 3D clusters will form starting from the lowest coverage. On the other hand, if the energy difference (ΔE) is smaller than the lateral metal-metal bonding energy (Ei), the ad-atoms will stick to the edges of the 2D islands, provided that thermal energy is high enough for the ad-atom migration across the surface and for the down-stepping to occur.

Moreover, the up-stepping activation barriers, which is approximately equal to the lateral metal-metal bonding energy (Ei) have to be large as compared to the thermal energy, so that the thermal thickening of the islands to be kinetically disallowed. This pseudo layer-by-layer growth mode usually leads to the formation of 2D islands below a critical coverage, after which additional layers grow in a layer-by-layer mode on top of these islands.

The critical coverage is directly related to the density of 2D islands which is, in turn, linked to the density of nucleation centers. In sputtering deposition, the low percentage of high energy sputtered particles that can reach up to hundreds of eV of energy leads to the formation of a high density of nucleation centers (Kalff, 1997). This will eventually increase the critical coverage giving rise to the formation of at continuous layers at relative low growth temperatures. The arguments above, of kinetically governed growth, can explain the observed morphology of the Cr film deposited at room temperature (see figure 4). As the growth temperature increases, the ad-atoms have enough energy to surface diffuse overcoming the up-stepping energetic barrier, allowing the system to move in the actual thermodynamic equilibrium configuration of 3D islands on the MgO surface (see figure 4). The sample grown at 600 °C exhibits square shaped 3D crystalline clusters with the sides parallel to the substrate [110] and [1-10] directions proving the 45° in-plane rotation epitaxy (see figure1) of Cr on MgO. Another interesting feature of the 3D islands is the narrow size distribution. This aspect is connected with strain and strain relief mechanism. In lattice mismatched epitaxy, after an island is formed, the misfit strain relaxation in the island causes a strain concentration at the edges which increases monotonically with increasing island size. Since the ad-atoms tend to diffuse from the high strain sites to lower strain sites, the strain concentration at the edges will translate into an additional kinetic barrier for the ad-atoms to diffuse to the islands, thus the island growth rate is slowed down as the island size increases, leading to the formation of homogenously sized islands (Chen & Washburn, 1996).

The main result from the X-Ray Diffraction and Atomic Force Microscopy studies is that crystallinity improvements of Cr films deposited on MgO comes with the cost of surface morphology quality (see figures 5, 6 and 7). Since epitaxy is favored at high temperatures, while at surfaces al low temperatures it is difficult to select deposition temperatures that allow for both high degree of epitaxy and flat surfaces. However, the sample deposited at room temperature exhibits very good morphologic properties, while preserving the epitaxy. Therefore, to overcome the challenge of growing Cr films with both flat surface and high degree, one can choose to deposit the films at room temperature and to subsequently perform high temperature annealing stages. Figure 5 shows AFM images for two samples one deposited at room temperature, and the other one deposited at room temperature and post-annealed in vacuum at 600 °C for 20 minutes. As indicated by the images, the post-annealing process preserves and even improves the surface morphology, the root mean square roughness being reduced from 0.17 nm down to 0.12 nm after annealing. As indicated by the X-Ray diffraction measurements (see figure 3), the as deposited film shows a rather large tetragonal lattice distortion that is associated with the growth process. After annealing the film at 600 °C the distortion is relaxed and the in-plane and out-of-plane lattice parameters regain the bulk value. The epitaxial quality of the film increases with annealing, as reflected by the decrease of the FWHM of the rocking-curve around the (002) reflection from 2.54° to 1.51°. The crystallinity also increases after annealing, the mean crystallite size reaching a value comparable with the thickness of the film (16.5 nm). Figure 6 shows a High Resolution Transmission Electron Microscopy (HRTEM) image of a Cr film deposited on MgO at room temperature and subsequently annealed at 600 °C. The image confirms the high quality single crystalline nature of the Cr film.

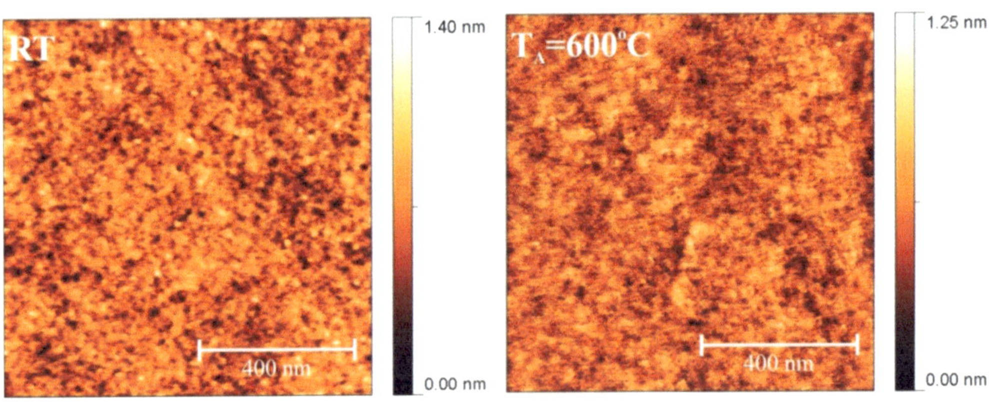

Fig. 5. AFM images (1x1μm^2 scans) showing the morphology of the MgO(001)//Cr(001) films grown at room temperature and post annealed at 600 °C.

In conclusion, we showed that in order to obtain smooth epitaxial Cr films on MgO(001), by sputtering deposition, a two step process must be employed. First, the deposition must take place at low temperatures, to obtain textured films with a flat surface morphology. Secondly, in order to improve the structural properties of the films, a high temperature annealing stage must be performed. Following this procedure, the flat surface morphology is preserved, while the structural properties of the films are improved.

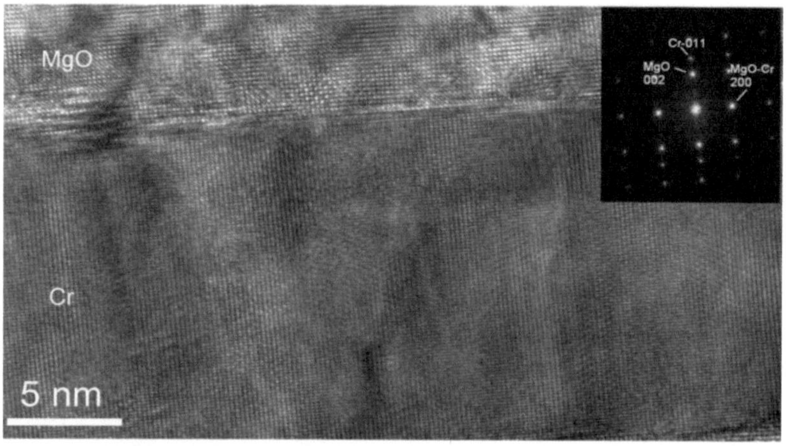

Fig. 6. High Resolution Transmission Electron Microscopy image of an MgO//Cr film deposited at room temperature and annealed at 600°C, confirming the single crystalline growth of Cr on MgO (by courtesy of E. Snoeck CEMES, TOULOUSE).

2.2 Co_2FeAl (CFA) epitaxial films

In a first stage we deposited CFA Heusler alloy thin films on un-buffered MgO(001) single crystal substrates by RF and DC sputtering. As in the case of Cr films, before deposition, the substrates were degassed in-situ at 600°C for 20 minutes. After cooling down to room temperature, a 5 nm thick MgO coating layer was deposited on the substrate. The CFA films with a thickness of 50 nm were sputtered at room temperature from a stoichiometric target ($Co_{50\%}Fe_{25\%}Al_{25\%}$) at 30 W under an Ar pressure of 1 mtorr.

As seen in the case of the Cr buffer layer, depositing the films at high temperatures will increase the crystalline properties but also will degrade the surface morphology through the formation of large 3D clusters. In order to test if this is also valid for CFA we have grown two types of films: one consisting of films deposited at high temperatures and the other of films grown at room temperature and subsequently high temperature vacuum annealed. The surface morphology of the layers was studied by Atomic Force Microscopy and the results are depicted in figure 7.

As seen from the figure 7 the films deposited at high temperatures have a granular structure with increasing grain size for higher deposition temperatures. In the case of the films grown at room temperature and ex-situ post annealed the surface is very flat and featureless, this characteristic being maintained even after annealing at 600 °C. Figure 8 shows the evolution of the Root Mean Square (RMS) surface roughness parameter and of the Maximum Peak-Valley (M_{PV}) distance for CFA films deposited at various substrate temperatures and for CFA films grown at room temperature and ex-situ vacuum annealed. In the case of the annealed samples the RMS and the MPV remain at low values regardless of the annealing temperature. Still, a minimum of RMS of 0.13 nm and M_{PV} of 1.48 nm is obtained for the sample annealed at 400 °C, with a small increase of RMS to 0.18 nm and of the M_{PV} to 2.8 nm for the layer annealed at 600°C. In case of the samples deposited at various substrate temperatures, the RMS and the M_{PV} distance show a monotonous increase from RMS=0.13 nm and M_{PV} =1.48 nm, for the RT deposited sample, to RMS=8.3 nm and M_{PV} =64 nm, for the films grown at 600°C. For the latter, the M_{PV} distance is even larger than the expected film thickness (50 nm). As seen form figure 7 the film consists of very large 3D clusters and most likely that the layer is not even continuous. This is expected since in conditions of thermodynamical equilibrium the metallic films have the tendency not to wet an oxide surface (see discussion in the previous paragraph).

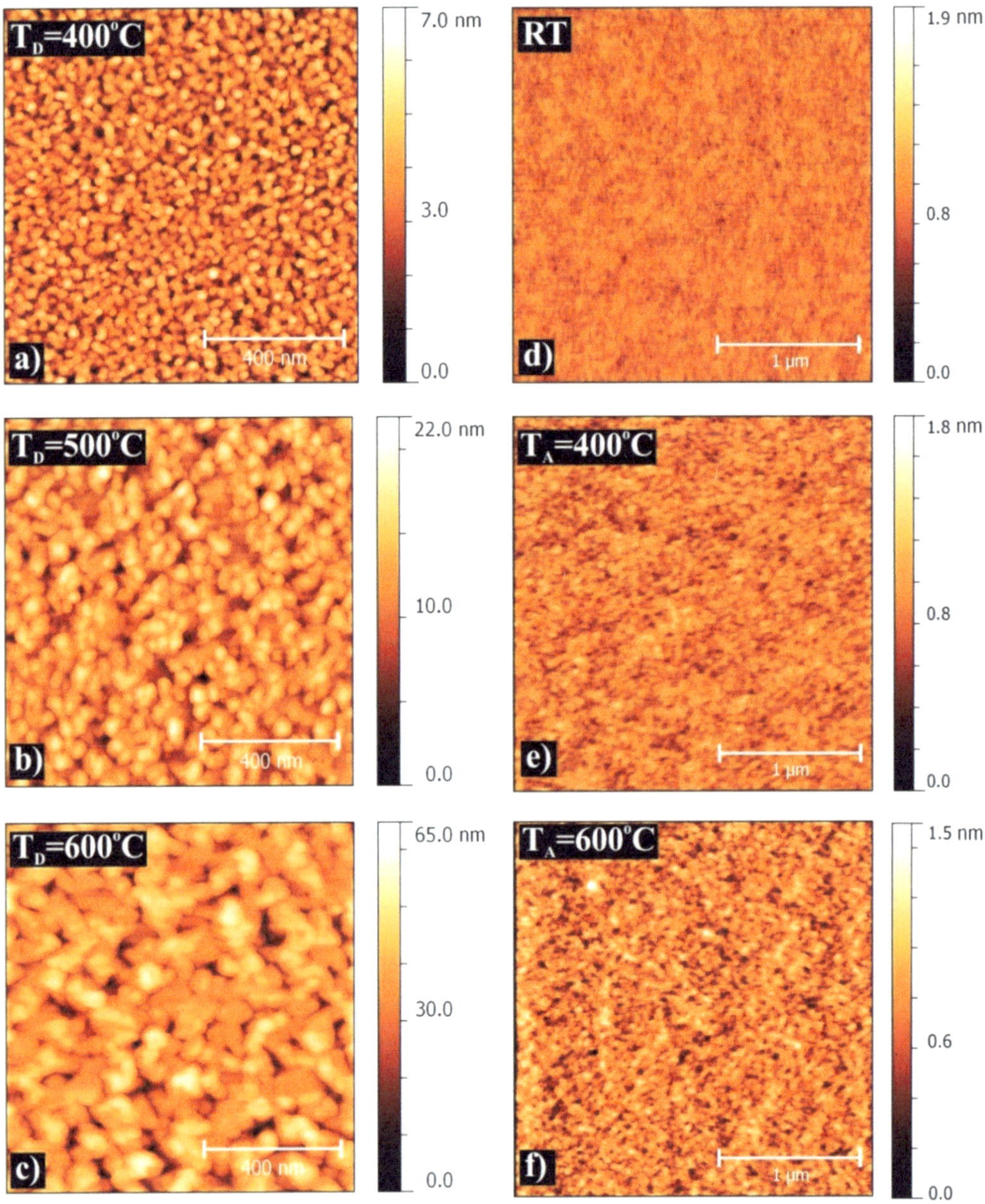

Fig. 7. AFM images showing the surface morphology of CFA films deposited at high temperatures (a), (b) and (c) and deposited at room temperature (d) and ex-situ vacuum annealed (e) and (f).

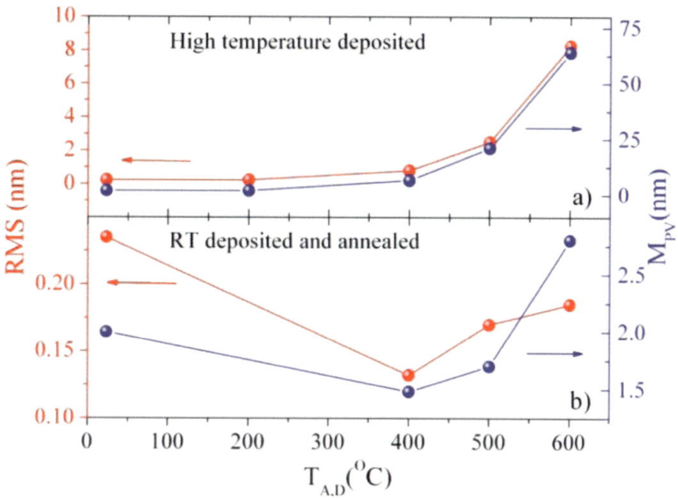

Fig. 8. Evolution of the Root Mean Square (RMS) surface roughness parameter and of the Maximum Peak-Valley (M_{PV}) distance as a function of temperature for (a) CFA films deposited at various substrate temperatures and (b) for CFA films grown at room temperature and ex-situ vacuum annealed.

Our analysis indicated that, due to the granular morphology, the CFA films deposited at high temperatures are not compatible with the manufacturing of magnetic tunnel junctions. Therefore, our efforts have been focused on the room temperature deposited and subsequently annealed samples. The $2\theta/\omega$ specular X-Ray diffraction measurements (see figure 9 (a) indicated that the CFA films are epitaxial even when deposited at room temperature and that the crystalline quality of the films increases with annealing (see figure 9 b), the best crystalline properties being obtained for the film annealed at 600 °C. However, the morphological properties of the film annealed at 600 °C are somewhat degraded relative to the one annealed at 400 °C, making the choice of the optimal annealing temperature relative difficult, since both the crystalline and morphological properties of the CFA films are of extreme importance for the further development of the magnetic tunnel junction multilayer stack.

In addition to the films deposited directly on MgO we have grown epitaxial CFA films using a Cr buffer layer. The buffer layer, with a thickness of 20 nm, was deposited as described in the previous section. To ensure a high crystalline quality of the buffer layer, after growth the Cr film was annealed *in-situ* at 600 °C. After cooling down to room temperature the CFA films, with a thickness of 50 nm, were deposited by RF sputtering under 1.0 mtorr of Ar working pressure. As in the case of the films grown directly on MgO, the $2\theta/\omega$ specular X-Ray diffraction measurements (see figure 10 a) indicated that the CFA films are epitaxial even when deposited at room temperature and that the crystalline quality of the films increases with annealing (see figure 10 b), the best crystalline properties being obtained for the film annealed at the highest temperature.

An interesting feature is that the values of the FWHM for the samples deposited on the Cr buffer layer are smaller as compared to the ones recorded for the films deposited directly on MgO (figure 9). This is a direct consequence of the smaller lattice misfit between CFA and

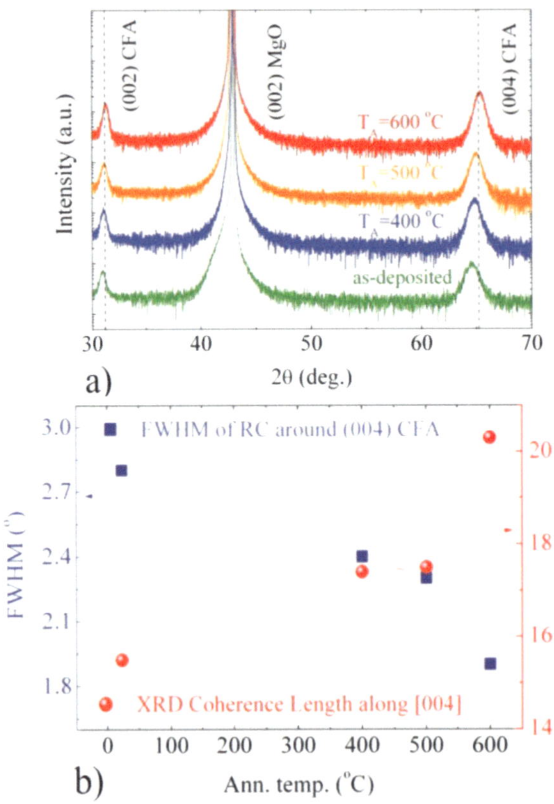

Fig. 9. (a) Specular 2θ/ω X-Ray diffraction patterns for the CFA films as a function of the annealing temperature; (b) FWHM of the rocking curve around the (004) CFA reflection and of the X-Ray diffraction coherence length along [004] direction versus the annealing temperature.

Cr (-0.7%), the one than between CFA and MgO (-3.8%). If the lattice misfit is higher, the interfacial strain energy also increases. This energy is normally relaxed by the formation of dislocations in the film, which, of course, degrade the crystallinity of the epilayer. This is also sustained by the evolution of the mean size of the X-Ray coherently diffracting domains as a function of annealing temperature. The mean size has the same behavior, as in the case of MgO deposited CFA films (figure 9), correlating perfectly with the decrease of the FWHM, but the values are larger for the Cr buffered films. The increased size of the coherently diffracting domains indicates that the epitaxial perfection of the CFA lattice is spread over larger distances, the density of lattice dislocations being lower than in the case of the films grown on MgO.

The surface morphology of the CFA films deposited by RF sputtering on Cr buffered MgO substrates was investigated by Atomic Force Microscopy. Figure 11 shows AFM images for the CFA films as-deposited at room temperature and post-annealed at 600 °C for 20 minutes. One can see that the surface of both films are very flat and featureless. For the as deposited film, the RMS roughness parameter is around 0.1 nm and slightly smaller for the annealed one. The M_{PV} distance decreases from 1.7 nm, for the as-deposited film, to just less than 1 nm for the

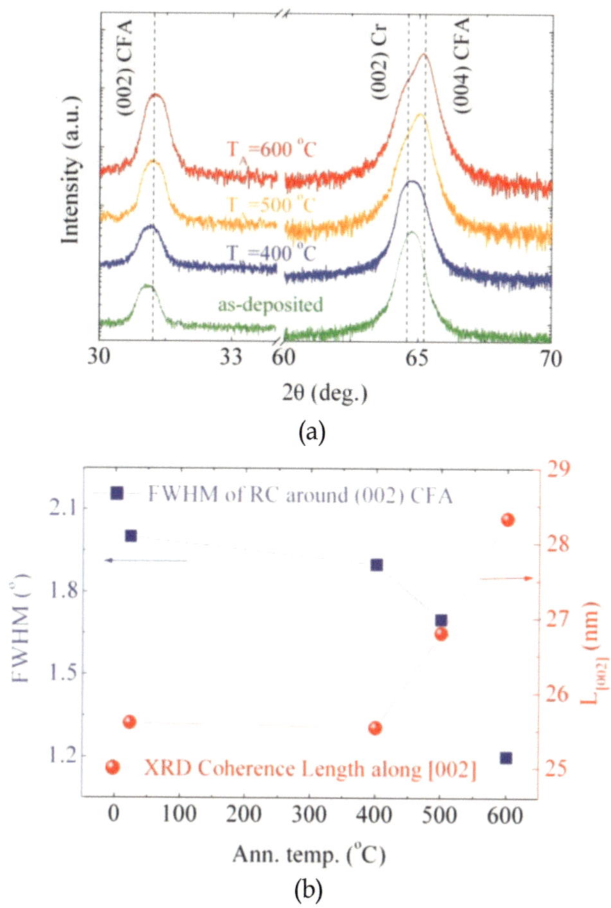

Fig. 10. (a) Specular 2θ/ω X-Ray diffraction patterns for the Cr/CFA films as a function of the annealing temperature; (b) FWHM of the rocking curve around the (002) CFA reflection and of the X-Ray diffraction coherence length along [002] direction versus the annealing temperature, indicating the improvement of the crystallinity of the films with annealing.

annealed one. These results indicate that, contrary to the case of the CFA films deposited directly on the MgO substrate, both the crystalline properties and the surface morphology of the Cr buffered CFA film improve with the increase of the annealing temperature.

The growth of a high quality tunnel barrier in a magnetic tunnel junction is in general a very difficult task since the tunnel barrier must be very thin, typically 1-2 nm, and pinhole free in order to avoid metallic shorts in the junction. This implies that the lower electrode of the junction to be extremely flat, which is indeed the case of the Cr buffered CFA film annealed at 600°C. Using this optimized lower electrode we have grown the subsequent MgO tunnel barrier and the upper CoFe electrode. The AFM measurements have indicated a very flat morphology for the tunnel barrier (see figure 12 a), with a RMS roughness around 0.2 nm, which ensure the continuity of the barrier, as also suggested by the HRTEM images in figure 12 b. Moreover, the HRTEM images indicate that the whole magnetic tunnel junction stack is epitaxial.

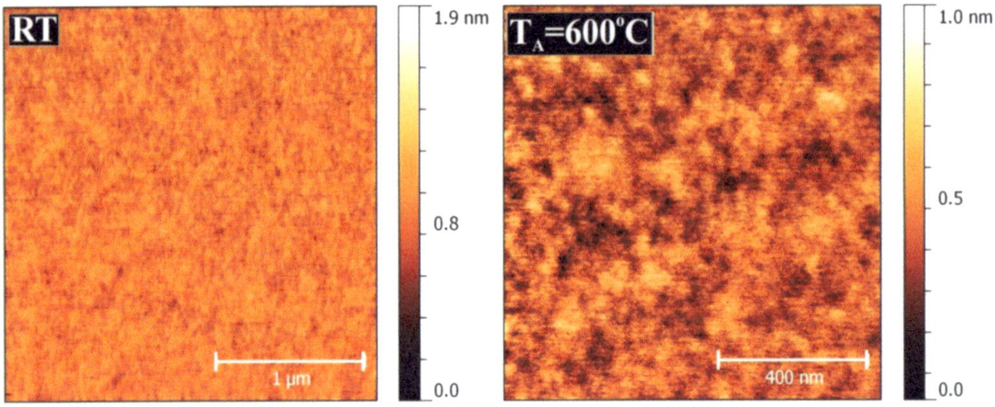

Fig. 11. AFM images of Cr buffered CFA films as-deposited at room temperature and post-annealed at 600°C for 20 minutes.

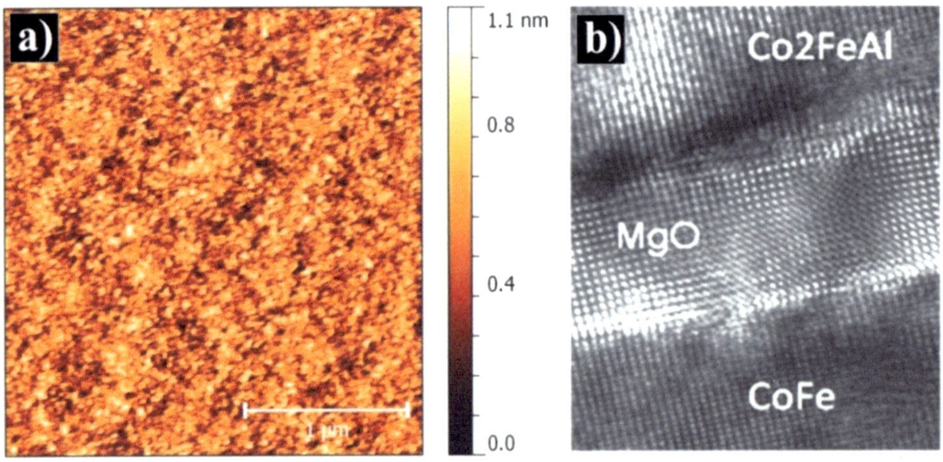

Fig. 12. (a) AFM image on the MgO//Cr(15 nm)/CFA(20 nm)/MgO(2.2 nm) stack indicating the morphology in the tunnel barrier; b) High Resolution Transmission Electron Microscopy image of the magnetic tunnel junction indicating the crystallinity and continuity of the barrier.

Based on this analysis, we succeeded to grow and pattern by optical lithography magnetic tunnel junctions which give significant tunnel magnetorezistive effects at room temperature.

3. The use of AFM/MFM techniques for the study of hybrid interface systems between high temperature superconductors and magnetic structures

Concerning the second part of the chapter, we present the magnetic force microscopy as a powerful tool to get deep insight in the micro-magnetic properties of a magnetic thin film interfaced with a high temperature superconductor layer. More recently, a new hybrid research domain has emerged: superconducting spintronics, dealing with systems constituted by alternating superconducting and magnetic layers. The aim of these novel systems is to drive new functionality of the constituent layers by taking profit on the mutual interfacial interaction between the superconductor and the magnetic film. For instance, vortices and the anisotropy of the electronic transport in the superconductor can be controlled by the stray-fields emerging from a modulated magnetic structure (Carneiro 2007, Hoffman et al 2008, Karapetrof et al 2009). The engineering of the micromagnetic properties in continuous or patterned magnetic thin films represent therefore a powerful tool to control the electronic transport in the superconductor. We illustrate the results concerning the modulation of the magnetic structure of Permalloy thin films via the thickness of the film. We follow the evolution of the magnetization from in-plane to out-of-plane by increasing the thickness and we present a direct correlation between the micromagnetic properties, measured by MFM, and the magnetic properties at macroscopic scale (magnetization hysteresis curves) measured by standard magnetometry. Furthermore, we exemplify the possibility to engineer the magnetic/micromagnetic properties in mezoscopic size magnetic objects patterned by optical lithography. Single objects and arrays of circular shape of mezoscopic dots with lateral size in the micron range, are considered. Alternatively, nanoscopic magnetic objects are elaborated using polystyrene balls shadowing/lift-off technique.

In the past decade numerous research groups have addressed the problem of magnetic pinning in superconductors (Aladyshkin et al., 2009). The relevance of this subject lies in the possibility of attaining large pinning forces of vortices in type II superconductors, that are temperature independent, through the interaction between the vortex flux and the magnetization of a magnetic structure, *i.e.* ferromagnetic films, ferromagnetic micro-, or nanostructures.

Large scale applications of superconductivity, such as energy transport, superconducting magnets, engines, generators, etc., require superconducting materials with a critical current density, J_c, greater than 1 MA/cm^2 and high irreversibility lines. In order to fulfill these requirements, the only way is to avoid the vortex motion, which is responsible for the energy dissipation in a type II superconductor. Up to date, the method used to block the vortex movement is their pinning on normal impurities present in a superconductor, the so-called *normal core pinning*. This type of pinning relies on the tendency of the vortex normal core to attach itself to regions in the superconductor where superconductivity is suppressed, in an attempt to minimize the overall system energy. For this reason, normal zones within the superconductor are artificially created in different ways. These zones coexist with the natural occurring normal regions arising from inherent growth defects, such as grain boundaries, dislocations and the presence of secondary phases. Several ways of externally inducing normal regions in superconductors have been demonstrated. They include: irradiation by swift heavy ions, artificially introduced regular arrays of holes, artificially introduced nanoparticles that create columnar defects in superconducting films, or surface decoration of the substrate upon which the superconducting film is grown (Civale et al., 1991; Augieri et al., 2010; Mele et al.,2006; Sparing et al., 2007). However, classic non-magnetic pinning becomes ineffective at high temperatures making it difficult to imagine practical applications of HTS materials.

In this context, magnetic pinning, originating from the Zeeman interaction between a magnetization and the magnetic flux of the SC vortices, may be a valid alternative for effective vortex pinning applications, as the pinning potential created by a magnetic structure, in this case displaying a stripe domain structure varying along the *x* axis, is expressed as

$$U_{mp} = \Phi_0 M(x) d_s \tag{3}$$

d_s being the superconducting layer thickness and $M(x)$ the magnetization value, which, neglecting the domain wall contributions, takes the values of $\pm M_0$ (Bulaevskii et al., 2000). Even though the above expression of the pinning potential is calculated for a particular case of magnetic configuration, the point it stands for may be generalized to any magnetic distribution, and it is that magnetic pinning potential is independent of temperature. Of course, the Curie temperature of the ferromagnet has to be high enough so that there is no significant variation of the magnetization with temperature, in the superconducting regime of the SC film. This fact makes this type of pinning as the ideal case of superconducting vortex pinning, suitable for high temperature superconductors. Equating the magnetic pinning force, calculated from the pinning potential (3), as U_{mp}/l, with l as the domain width, to the Lorentz force acting on the vortex line, stemming from the transport current in the SC layer, $J\Phi_0 d_s/c$, gives a rough estimate of the critical current

$$J_c \sim cM_0 / l. \tag{4}$$

It can be seen that within the model proposed, the critical current density, J_c, increases as the perpendicular component of the magnetization, M_0 increases, justifying the use in magnetic pinning experiments FM structures that exhibit a perpendicular anisotropy.

In the following paragraphs we give an account of the results obtained on some magnetic systems exhibiting out-of-plane anisotropy investigated for possible magnetic pinning applications of superconducting vortices. They include permalloy continuous thin films, permalloy micronic disks and cobalt nanostructures. A particular emphasis is placed on the magnetic characterization of the above structures by means of magnetic force microscopy (MFM) and the information this analysis may give regarding the magnetic pinning characteristics of the respective investigated systems.

3.1 Permalloy thin films

In the case of Permalloy, Py, ($Ni_xFe_{(1-x)}$, x=19-21 at.%) thin films, perpendicular anisotropy originates from a negative magnetostriction constant correlated with an in- plane tensile strain of the films. In this case, the anisotropy constant is given by the expression (Saito et al., 1964):

$$K_U = \frac{3}{2}\lambda_L \sigma \tag{5}$$

where λ is the magnetostriction constant and σ represents the strain of the film.

The orientation of the magnetization within Py thin films is determined by the competition between the perpendicular anisotropy energy term and the magnetostatic energy. Since the uniaxial anisotropy is not strong enough to determine the perpendicular orientation of the magnetization with respect to the film surface, the resulting configuration corresponds to an out-of-plane deviation of the magnetization. The domain pattern stabilized by this orientation is comprised of stripes in which the magnetization points alternatively, outwards and inwards with respect to the film surface, Figure 13. Because the deviation angle, θ, of the magnetization is less than 90°, this domain configuration is referred to as a *weak stripe* domain pattern. A consequence of the presence of weak stripe domains is the so called *rotatable anisotropy*. This phenomenon consists in the orientation of the domain stripes along the direction of a relatively weak, in-plane applied magnetic field, and the persistence of this orientation after the field is removed. The field magnitude is typically of few hundreds of Oe. The two properties of thin Py films recommend them for fundamental studies of magnetic pinning in superconductors, as they allow for the preparation of two different states with respect to the effects of vortex pinning, as illustrated in Figure 14. These two states consist in the parallel and perpendicular orientation of a current density, J, which is injected into a superconducting $YBa_2Cu_3O_7$ (YBCO) strip, with respect to the orientation of the magnetic stripe domains of a Py film deposited on top. As it can be seen, the Lorentz-type force acting on the superconducting vortices is directed perpendicularly to the current density. For the perpendicular configuration, Figure 14a, the vortices will not be acted upon by any pinning force as the vortex motion takes place along the stripe domains where the magnetization is constant. This configuration is referred to as a *vortex guide* configuration. On the other hand, in the parallel configuration, Figure 14b, the force due to the current density acts on the vortices so as to move them across the magnetic domains. In this case the vortex motion will be impeded by the presence of the periodic modulation of the magnetization. Thus, the vortices will by pinned by a periodic magnetic pinning potential. In order to be able to fabricate and study such systems, a careful analysis of the magnetism of the Py thin films is required. For this purpose we have fabricated several polycrystalline thin Py films by means of dc magnetron sputtering, deposited on Si (111) single crystal substrates. The thickness of the films was varied between 270 and 720 nm. The as deposited films were analyzed by means of vibrating sample magnetometry (VSM) and magnetic force microscopy (MFM).

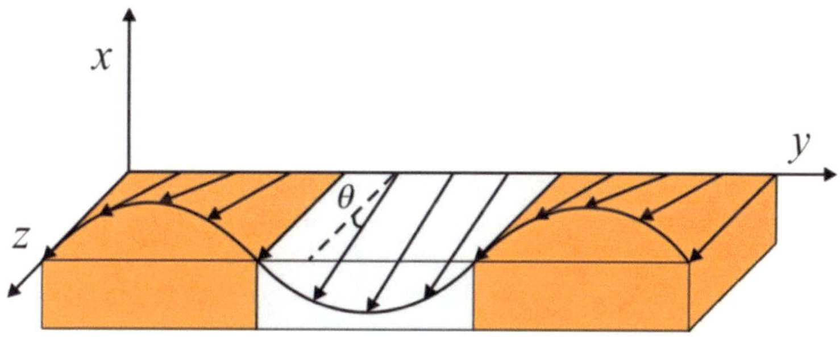

Fig. 13. Schematic representation of the spin distribution within the weak stripe domain configuration, as proposed by Chikazumi (Chikazumi, 1997). It is to be noted that the out-of-plane component of the magnetization has a sine variation along the y direction. The out-of-plane deflection angle of the spins is denoted by θ.

The magnetization dependence on the externally applied field, Figure 15, has the characteristic shape of the films that exhibit weak stripe domains. The magnetic field was applied parallel to the film surface plane. The linear decrease of the magnetization from saturation is observed with a small hysteresis. The decrease is due to the out-of-plane rotation of the film magnetization, corresponding to the weak stripe formation. The

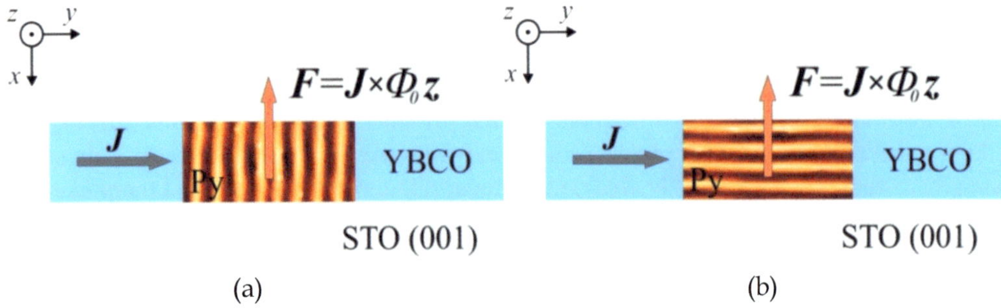

Fig. 14. Schematic representation of the vortex guide (a) and the periodic pinning potential (b) configurations of a Py/YBCO heterostructure.

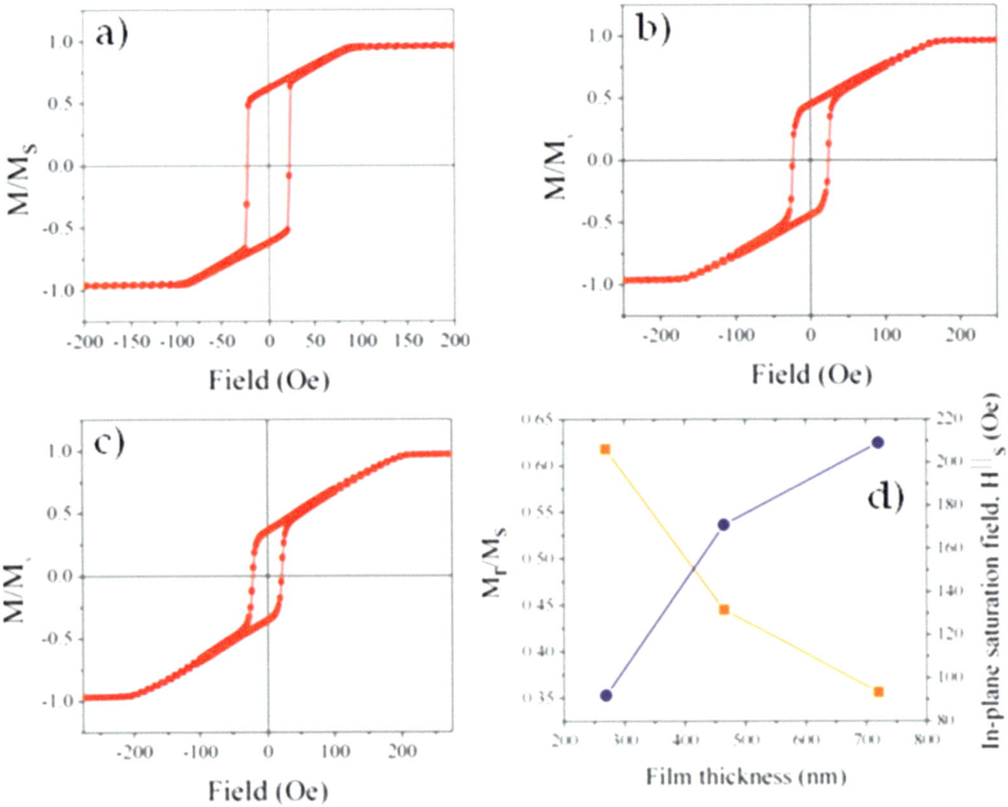

Fig. 15. Hysteresis loops of the (a) 270 nm, (b) 646 nm and (c) 730 nm thick Py thin films. The shape of the loop is typical for films having weak stripe domain configuration; (d) Variation of the normalized remanent magnetization, M_r/M_s, and of the in-plane saturation field, $H||_{sat}$, as a function of Py film thickness.

hysteresis on the other hand, may be explained by the existence of a slight difference in the stripe domain pattern in the two branches.

In order to test the expected rotatable magnetic anisotropy, the cycles were performed applying the magnetic field in two different perpendicular directions. No modification of the loops were observed, proving thus the existence of rotatable anisotropy. Figure 15d presents the thickness dependence of two characteristic magnetic quantities, the magnetization at remanence, M_r, and the in-plane saturation field, $H_{||sat}$. As the film thickness increases, the magnetostatic energy of surface charges decreases, allowing for the perpendicular anisotropy to manifest by an increase of the deviation angle of the spins, θ, with respect to the film surface plane. This increased deviation leads to a decrease of the in-plane magnetization remanence. Consequently, the larger magnetization perpendicular to the film plane leads to a higher in-plane saturation field. With respect to our goal to fabricate and study hybrid interface SC/FM systems for vortex pinning, higher θ values are expected to have a more pronounced influence on the superconducting properties of the bottom SC layer. Concomitantly, as analytical calculations have shown (Saito et al., 1964, Chikazumi, 1997, Murayama, 1966), an increase of film thickness also leads to an increase of stripe periodicity, *i.e.* domain width. In the following we present the results of MFM measurements on the film domain structure and how it is affected by the film thickness.

The domain structure of the films having thicknesses of 270 nm, 460 nm and 720 nm are shown in Figure 16. As it can be seen, the expected stripe domain configuration is present. Extracting a profile from the above images allows for an evaluation of the stripe wavelength, λ, and accordingly, of the domain width, $d=\lambda/2$, Figure 17. The domain width variation as a function of thickness has also been calculated according to the three models (Saito et al., 1964, Chikazumi, 1997, Murayama, 1966). For the calculation, the physical parameters entering the domain width expression were taken from (Ben Youssef et al., 2004), so that M_s=826 emu/cm^3, the out-of-plane anisotropy constant, $K_U=5\times10^4$ erg/cm^2, while the exchange stiffness constant was taken to be $A=1\times10^{-6}$ erg/cm. Comparing the measured results with the three models, it is noted that the initial model of stripe domains, proposed by Saito et al. (Saito et al., 1964), in which the spin direction varies perpendicularly to the stripes in a *zig-zag* as a function of position, is the worst approximation to the domain width value. A sine variation (Chikazumi, 1997) of the spin direction yields a better approximation of the domain width. Yet, the best results are found using the model proposed (Murayama, 1966). Because of the fact that their model allows for the highest freedom in spin ordering, allowing for a variation of the spins with respect to all the coordinate axes, they are able to take into account the presence of closure domains at film surface. As confirmed by numerical simulations run on low out-of-plane anisotropy Py thin films (Vlasko-Vlasov et al., 2008, Ben Youssef et al., 2004), the model proposed by Murayama is the closest in describing the actual physical picture in these systems, and thus yields the most satisfactory result in predicting the domain width dependence on the film thickness.

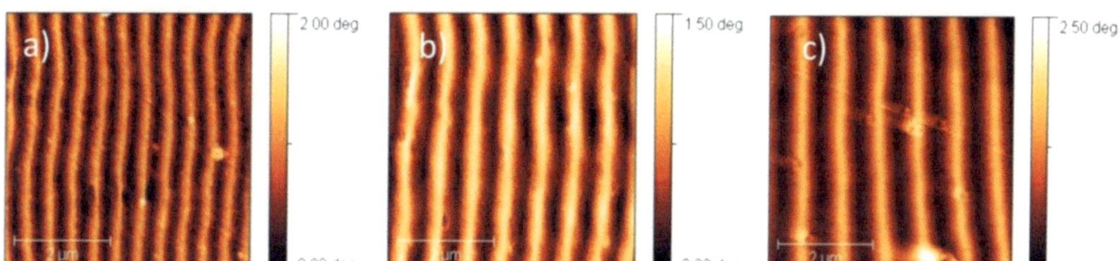

Fig. 16. 5 μm×5 μm MFM images of the (a) 270 nm, (b) 460 nm and (c) 720 nm thick Py films.

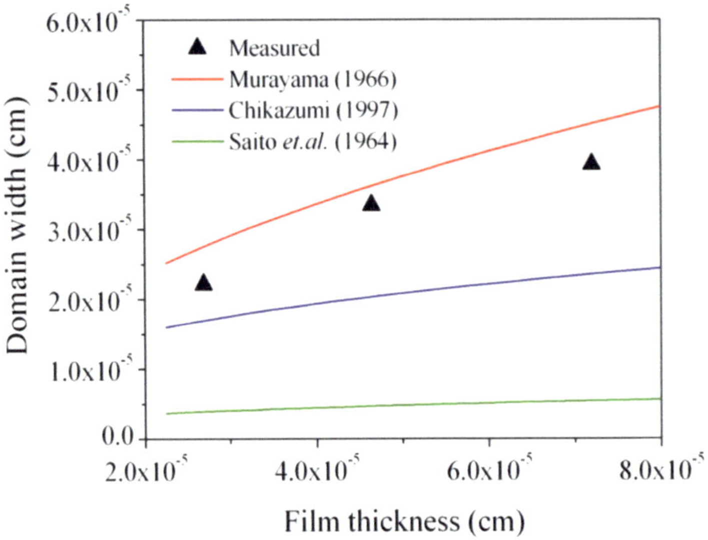

Fig. 17. Domain width of the 270 nm, 460 nm and 720 nm thick Permalloy films. Domain width variation as a function of film thickess, calculated according to the three different models presented in the text.

As far as the rotatable magnetic anisotropy is concerned, Figure 18 shows the orientation of the domain stripes in a remanent state after the application of an in-plane magnetic field along different directions. The value of the applied field was set at 650 Oe in order to ensure saturation of all the studied films. As it can be observed, the stripe domains align themselves along the direction of the applied magnetic field and do not change their orientation after the field is removed. The rotatable anisotropy feature of low out-of-plane anisotropy magnetic thin films is essential for the study of the influence of a periodic magnetic field on the superconducting properties of SC thin films, as it allows for a systematic modulation of the stray magnetic field produced by the FM layer (Belkin et al., 2008).

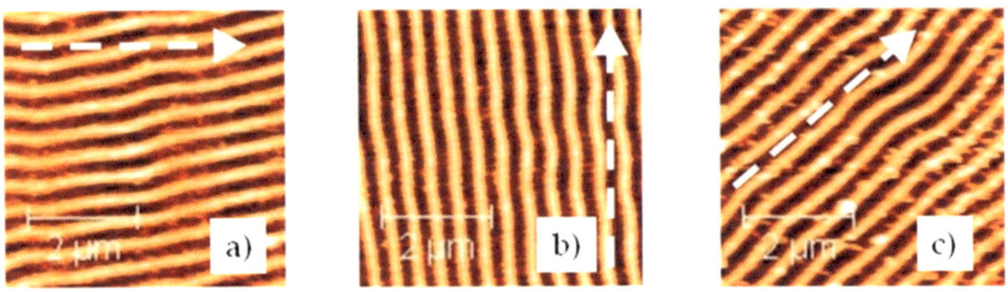

Fig. 18. Rotatable anisotropy. Different magnetic stripe orientations correspond to the direction of the applied magnetic field for a 460 nm thick Py film.

3.2 Magnetic micro,- and nanostructures

Periodic stripes of alternating "up" and "down" orientation of magnetization with respect to the film surface plane are expected to provide a strong pinning of the SC vortices when these alternations are oriented perpendicularly to the direction of the vortex motion. Magnetic micro,- and nanostructures, such as regular dot arrays, provide an additional degree of freedom to the modulation, as they are also subject to a topographic periodicity, given by the specific pattern of the arrays defined by lithographic techniques. In these sense, even the sole presence of such structures in the vicinity of SC films may give rise to a pinning potential of the vortices, as demonstrated by Hoffmann et al. (Hoffmann et al., 2000) in their study of a Nb film deposited on top of a regular array of non-magnetic Ag dots. The flexibility in defining different configurations of dot lattices coupled with their respective magnetic configurations produce *commensurability effects*.

These effects consist in resonant changes of the magnetoresistance of the SC layer, $\rho(H_{ext})$, with the appearance of equidistant minima having a period in H determined by lattice constant of the magnetic dot arrays. Also, these minima are visible in the field dependence critical current density, $J_c(H_{ext})$, or in the magnetization curves, $M(H_{ext})$. These effects are explained considering a strong magnetic pinning potential exerted by magnetic dots on the SC vortices, coupled with a structural match between the dot and vortex lattices. Thus, when the condition that an integer number of vortices per magnetic unit cell is fulfilled, a maximum number of vortices are pinned, producing resistivity minima or J_c maxima of the SC film. The field dependence of these effects is explained by the fact that the vortex lattice parameter is field-dependent. Because of the fact that the vortex lattice is triangular strong commensurability effects are produced by triangular dot lattice. Lateral structuring of the magnetic films into micro or nanostrucutres also influences their magnetic configuration, so that the magnetic state of the film will be altered when it is patterned into dot sutructures. Jubert and Allenspach (Jubert and Allenspach, 2004) constructed a phase diagram of the magnetic states of circular nanometric dots as a function of their thickness and diameter, based on micromagnetic simulations. By varying the lateral size and thickness of the dots one is able to stabilize different magnetic states: in-plane single domain (IPSD), vortex (V) and out-of-plane single domain (OPSD). Carneiro (Carneiro, 2004) calculated the interaction energy between a SC vortex and an in-plane and out-of-plane magnetized dipole showing that different attraction interaction profiles exist between the dipole and the vortices in the two cases. While in the case of out-of-plane magnetic dipoles the largest pinning potential arises at the center of the dipole, in the case of an in-plane dipole configuration pinning is present at the edges of the object where the variation of the magnetization is highest. Larger, micronic, dots that may be able to accommodate a multi-domain magnetic configuration are also interesting in the study of FM/SC structures as their net magnetic moment may be tuned by magnetizing the sample in a field lower than the saturation field (Aladyshkin et al., 2009). Also larger dots may be able to stabilize "giant" vortices that are able to carry more than one flux quanta (Aladyshkin et al., 2009).

In view of the above arguments, it can be seen that within the context of FM/SC heterostructures, magnetic micro,- and nanostructures present a special importance as they can produce effective magnetic pinning that can be modulated both by the dot lattice, as well as their magnetic state. In this sense we have fabricated micronic Py disks having diameters of 10, 5 and 1μm. Also, Co nanostructures were grown using an alternative lithographic process that involves the use of self-assembled polystyrene nano-spheres as a shadow mask.

3.2.1 Permalloy microstructures

Figure 19 presents the magnetic images of circular, 350 nm thick Py dots having a diameter of (a)10, (b)5 and (c)1 μm. For the 10 μm dots, the diameter is well above the stripe domain width, estimated to be around 300 nm, according to the model proposed by Murayama (Murayama, 1966). As a consequence, the disks exhibit a well defined striped domain pattern. The stripes that are situated at the center of the disk have the same orientation and are very much similar to the case of a continuous film. As the stripes are situated closer to the edge of the microstrucutres, they tend to curve along the side of the dots, so as to prevent any magnetic charge build-up at the edges, and so to reduce the magnetostatic energy of the system. These *closure domain* - type structures have a limited spatial extension with respect to the overall area of the dots. As the diameter of the dots is reduced to 5 μm, even though this dimension is still considerably higher than the domain width, the stripe pattern is almost entirely pertaining to the closure domains. As a consequence, the MFM images of the 5 μm dots is composed of alternating bright and dark concentric circles. Again, locally the parallel stripe order is maintained. However, this is not the only domain configuration that can be observed. Another domain pattern that is observed, resembles the one present in the 10 μm structures. It consists of straight magnetic stripes in the center and elongated circles at the edges as closure domains. It can be noted that the surface "covered" by the closure domains is much larger than in the 10 μm case, reaching 100% for the circular stripe pattern. In view of the evolution that was observed so far, the domain pattern for 1 μm dots is not surprising. As the diameter of the microstructures is reduced towards the limiting value of the stripe width, the competition between the anisotropy and magnetostatic energies results in the formation of stripe domains in the form of concentric rings. For the 1μm disks, the single bright/dark alternation in the MFM images indicates the existence of solely two magnetic domains.

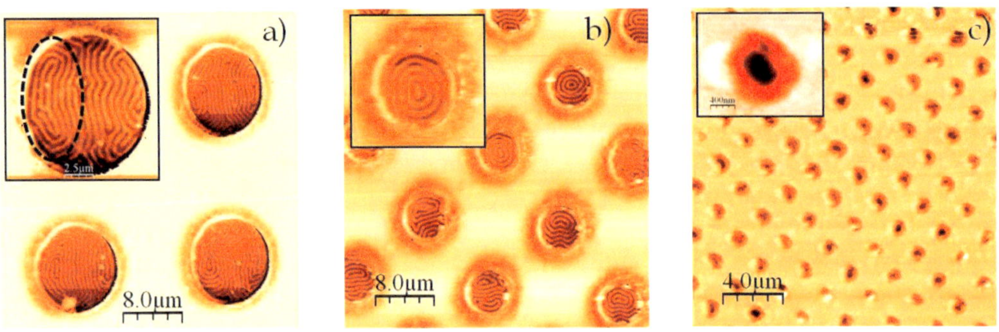

Fig. 19. MFM images of (a)10 μm (40 μm×40 μm), (b) 5 μm (40 μm×40 μm), (c) 1 μm (20 μm×20 μm), Py dots.

3.2.2 Cobalt nanostructures

Co nanostructures were obtained using the polystyrene nanosphere lithography, a method described in detail in (Canpean et al., 2009). 170 nm and 40 nm thick films were deposited on a nanoshpere patterned Si substrate. After lift-off, triangular Co structures remain on the substrate. Even though a large difference in the deposited film thickness was prepared, the height of the nano-structures in the case of the two films did not differ significantly. In the case of the 40 nm thick film, the mean height of the magnetic structures was of 12 nm, while for the thicker 170 nm film, the resulting dots were only 25 nm high. The large difference in the expected value of the height of the dots and the actual one is probably due to the shadowing effect produced by a divergent beam of sputtered atoms and also by the large diameter of the spheres (450 nm), which accentuate these effects. The MFM analyses performed on the two films are shown in Figure 20. For the 25 nm high structures, the dots are in a mono-domain state. The MFM contrsast consists of a dark/bright formation corresponding to an in-plane magnetic dipole structure. If the initial film thickness is decreased, the magnetic configuration of the nanostructures changes, as can be seen in Figure 20b, into a vortex-like spin configuration. The MFM contrast is typical for nanostructures in this configuration, having a bright center and the rest of the dot exhibiting a darker shade. According to the diameter-thickness phase diagram (Jubert and Allenspach, 2004) both dots should have been in the magnetic vortex state, and moreover the behavior predicted by the phase diagram is that for thicker dots the vortex state becomes more stable for a larger range of dot diameters. Our findings point out a different behavior, in the sense that thicker structures are found to be in the single domain state. Although no concrete explanation was found for this trend, we suspect that the particular shape of the dots, in our case triangular, may play an important role in modifying the magnetic phase diagram.

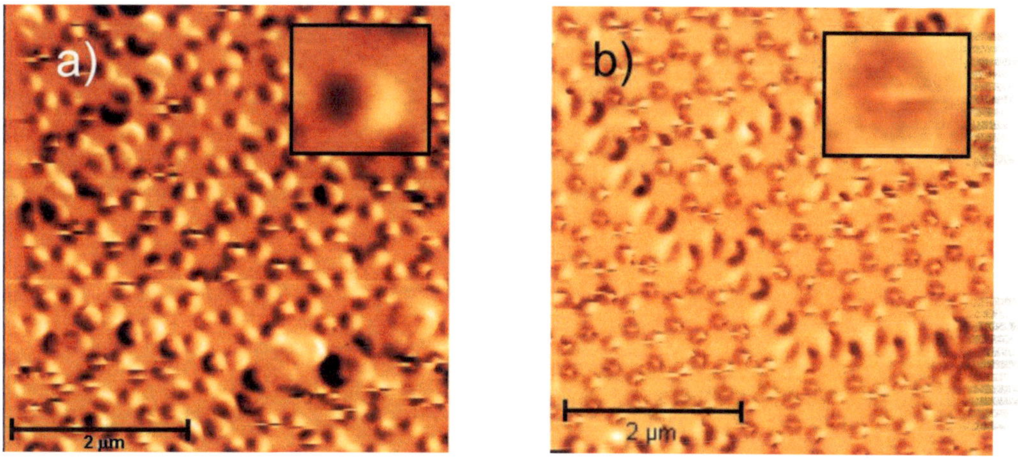

Fig. 20. MFM images of Co nanostructures having different thicknesses: (a) 20 nm (magnetic dipoles) and (b) 12 nm (magnetic vortices).

4. Conclusions

Within this chapter we illustrated the use of the Atomic Force Microscopy techniques as a powerful tool in characterization of complex thin film systems. Usually, when elaborating multilayer thin film stacks, the constituent layers have to be continuous and to present small roughness. Therefore, the AFM characterization will provide important information about the film surface topology in studies related to roughness reduction. Furthermore, high resolution analysis of the film topology can be used in studies of thin film growth mechanisms when the growth parameters (substrate temperature, deposition rate, etc...) are varied. Within the field of the Spintronics, for the elaboration of devices such as the magnetic tunnel junction, the extreme control of roughness of the constituent layers and the continuity of the insulating tunnel barrier is required. The interfacial roughness induces here fluctuations in electronic transport properties (perpendicular to the stack) with detrimental effects on the functional properties of the MTJ device.

Moreover, the Magnetic Force Microscopy operating mode represents a versatile tool for the characterization of the micromagnetic properties of continuous magnetic thin films or mezoscopic size patterned magnetic objects. We have chosen here to illustrate the use of the MFM for characterizing the micromagnetic features of continuous magnetic films and patterned magnetic objects which started recently to be used for magnetic pinning of superconducting vortices in high temperature superconductors.

5. Acknowledgments

This work has been partially supported by CNCSIS-UEFISCSU, project number PNII IDEI No. 4/2010, code ID-106 and by POS CCE ID. 574, code SMIS-CSNR 12467.

6. References

Aladyshkin, A. Yu; Silhanek, A. V.; Gillijns, W. & Moshchalkov, V. V. (2009). Nucleation of superconductivity and vortex matter in superconductor–ferromagnet hybrids, *Superconductor Sciemce and Technology*, Vol.22, No.5, (May 2009), pp. 053001-053049, ISSN 1361-6668 (online)

Augieri, A.; Celentano, G.; Galluzzi, V.; Mancini, A.; Rufoloni, A.; Vannozzi, A.; Angrisani Armenio, A.; Petrisor, T.; Ciontea, L.; Rubanov, S.; Silva E. & Pompeo, N. (2010). Pinning analyses on epitaxial $YBa_2Cu_3O_{7-\delta}$ films with $BaZrO_3$ inclusions, *Journal of Applied Physics*, Vol.108, Issue 6, (September 2010), pp. 063906-063911, ISSN 1089-7550 (online)

Bulaevskii, L. N.; Chudnovsky, E. M. & Maley, P. M. (2000). Magnetic pinning in superconductor-ferromagnet multilayers, *Applied Physics Letters*, Vol.76, Issue 18, (March 2000), pp. 2594-2597, ISSN 1077-3118 (online)

Ben Yousseff, J.; Vukadinovic, N.; Billet, D. & Labrune, M. (2004). Thickness-dependent magnetic excitations in Permalloy films with nonuniform magnetization, *Physical Review B*, Vol.69, Issue 17, (May 2004), pp. 174402-174411, ISSN 1550-235x (online)

Belkin, A.; Novosad, V.; Iavarone, M.; Pearson, J. & Karapetrov, G. (2008). Superconductor/ferromagnet bilayers: Influence of magnetic domain structure on vortex dynamics, *Physical Review B*, Vol.77, Issue 8, (May 2008), pp. 180506(R)-180510(R), ISSN 1550-235x (online)

Canpean, V.; Astilean, S.; Petrisor Jr., T.; Gabor, M. & Ciascai, I. (2009). Convective assembly of two-dimensional nanosphere lithographic masks, *Materials Letters*, Vol.63, Issue 21, (August 2009), pp. 1834-1836, ISSN 0167-577X

Carneiro, G. (2007). Tunable pinning of a superconducting vortex by a magnetic vortex, *Physical Review B*, Vol.75, Issue 9, (March 2007), pp. 094504-094514, ISSN 1550-235x (online)

Campbell, C. (1997). Ultrathin metal films and particles on oxide surfaces: structural, electronic and chemisorptive properties, *Surface Science Reports*, Vol.27, Issues 1-3, (May 1998), pp. 1-111, ISSN 0167-5729

Chen, Y. & Washburn, J. (1996). Structural Transition in Large-Lattice-Mismatch Heteroepitaxy, *Physics Review Letters*, Vol.77, Issue 19, (November 1996), pp. 4046-4049, ISSN 1079-7114 (online)

Chikazumi, S. (1997). *Physics of Ferromagnetism*, Oxford Science Publications, ISBN 0-19-851776-9, New York, United States

Civale, L.; Marwick, A. D.; Worthington, T. K.; Kirk, M. A.; Thompson, J. R.; Krusin-Elbaum, L.; Sun, Y.; Clem, J. R.; Holtzberg, F. (1991). Vortex confinement by columnar defects in $YBa_2Cu_3O_7$ crystals: Enhanced pinning at high fields and temperatures, *Phys. Rev. Lett.*, Vol. 67, Issue 5, (July 1991), pp. 648-651, ISSN 1079-7114 (online)

Depla, D. & Mahieu, S., eds., Reactive Sputter Deposition, Springer Series in Materials Science (2008);

Ehrlich, G. & Hudda, F. G. (1966). Atomic View of Surface Self-Diffusion: Tungsten on Tungsten, *Journal of Chemical Physics*, Vol.44, Issue 3, (February 1966), pp. 1039-1049, ISSN 1089-7690 (online)

Elbaum, L.; Sun, Y.; Clem,J. R. & Holtzberg, F. (1991). Vortex confinement by columnar defects in YBa2Cu3O7 crystals: Enhanced pinning at high fields and temperatures, *Physical Review Letters*, Vol.67, Issue 5, (July 1991), pp. 648-651, ISSN 1079-7114 (online)

Ernst, K. H. ; Ludviksson, A.; Zhang, R.; Yoshihara, J. & Campbell, C. T. (1993). Growth model for metal films on oxide surfaces: Cu on ZnO(0001)-O, *Physical Review B*, Vol.47, Issue 20, (May 1993), pp. 13782-13796, ISSN 1550-235x (online)

Fullerton, E.; Conover, M.; Mattson, J.; Sowers, C. & Bader, S. (1993). Oscillatory interlayer coupling and giant magnetoresistance in epitaxial Fe/Cr(211) and (100) superlattices, *Physical Review B*, Vol.48, Issue 21, (December 1993), pp. 15755-15763, ISSN 1550-235x (online)

Fullerton, E.; Conover, M. J.; Mattson, J.; Sowers, C. & Bader, S. (1993). 150% magnetoresistance in sputtered Fe/Cr(100) superlattices, *Applied Physics Letters*, Vol.63, Issue 12, (July 1993), pp. 1699-1702, ISSN 1077-3118 (online)

Huttel, Y.; Navarro, E. & Cebollada, A. (2005). Epitaxy and lattice distortion of V in MgO/V/MgO(001) heterostructures, *Jounal of Crystal Growth*, Vol.273, Issue 3-4, (November 2004), pp. 474-480, ISSN 0022-0248

Huttel, Y.; Cerda, Y.; Martinez, J. & Cebollada, A. (2007). Role of volume versus defects in the electrical resistivity of lattice-distorted V(001) ultrathin films, *Physical Review B*, Vol.76, Issue 19, (November 2007), pp. 195451-195458, ISSN 1550-235x (online)

Hoffmann, A.; Prieto P. & Schuller, I. K. (2000). Periodic vortex pinning with magnetic and nonmagnetic dots: The influence of size, *Physical Review B*, Vol.61, Issue 10, (March 2000), pp. 6958-6965, ISSN 1550-235x (online)

Hsieh, J.; Li, C.; Wu, W. & Hochman, R. (2003). Effects of energetic particle bombardment on residual stress, microstrain and grain size of plasma-assisted PVD Cr thin films, *Thin Solid Films*, Vol.424, Issue 1, (January 2003), pp. 103-106, ISSN 0040-6090

Harp, G. & Parkin, S. (1994). Seeded epitaxy of metals by sputter deposition, *Applied Physics Letters*, Vol.65, Issue 24, (October 1994), pp. 3063-3066, ISSN 1077-3118 (online)

Harp, G. & Parkin, S. (1996). Epitaxial growth of metals by sputter deposition, *Thin Solid Films*, Vol.288, Issue 1-2, (February 1996), pp. 315-324, ISSN 0040-6090

Jubert, P.-O. & Allenspach, R. (2004). Analytical approach to the single-domain-to-vortex transition in small magnetic disks, *Physical Review B*, Vol.70, Issue 14, (October 2004), pp. 144402-144407, ISSN 1550-235x (online)

Karapetrov, G.; Belkin A.; Novosad V.; Iavarone M.; Pearson J. E. & Kwok W. K. (2009). Adjustable superconducting anisotropy in MoGe-Permalloy hybrids, *Journal of Physics: Conference Series*, Vol.150, Part 5, pp. 052095-052099, ISSN 1742-6596 (online)

Kalff, M.; Breeman, M.; Morgenstern, M.; Michely, T. & Comsa G. (1997). Effect of energetic particles on island formation in sputter deposition of Pt on Pt(111), *Applied Physics Letters*, Vol.70, Issue 2, (January 1997), pp. 182-184, ISSN 1077-3118 (online)

Murayama, Y. (1966). Micromagnetics on Stripe Domain Films. I. Critical Cases, *Journal of the Physical Society of Japan*, Vol. 21, pp. 2253-2266, ISSN 1347-4073 (online)

Mele, P.; Matsumoto, K.; Horide, T.; Miura, O.; Ichinos, A.; Mukaida, M.; Yoshida Y. & Horii, S. (2006). Tuning of the critical current in $YBa_2Cu_3O_{7-x}$ thin films by controlling the size and density of Y2O3 nanoislands on annealed SrTiO3 substrates, *Superconductor Science and Technology*, Vol.19, No.1, (November 2006), pp. 44-50, ISSN 1361-6668 (online)

Meyer, K.; Schuller, I. & Falco, C. (1981). Thermalization of sputtered atoms, *Journal of Applied Physics*, Vol.52, Issue 9, (September 1981), pp. 5803-5805, ISSN 1089-7550 (online)

Saito, N. ; Fujiwara, H. & Sugita, Y. (1964). A New Type of Magnetic Domain in Thin Ni-Fe Films, *Journal of the Physical Society of Japan*, Vol.19, pp. 421-422, ISSN 1347-4073 (online)

Sparing, M.; Backen, E.; Freudenberg, T.; Hühne, R.; Rellinghaus, B.; Schultz L. & Holzapfel, B. (2007). Artificial pinning centres in YBCO thin films induced by substrate decoration with gas-phase-prepared Y2O3 nanoparticles, *Superconductor Science and Technology*, Vol. 20, No.9, (September 2007), pp. S239-S246, ISSN 1361-6668 (online)

Schwoebel, R. L. (1969). Step Motion on Crystal Surfaces. II, *Journal of Applied Physics*, Vol.40, Issue 2, (February 1969), pp. 614-618, ISSN 1089-7550 (online)

Tiusan, C.; Greullet F.; Hehn M.; Montaigne F.; Andrieu S. & Schuhl A. (2007). Spin tunnelling phenomena in single-crystal magnetic tunnel junction systems, *Journal of Physics: Condensed Matter*, Vol.19, No.16, (April 2007), pp. 165201-165235, ISSN 1361-648X (online)

Thornton, J. & Hoffman, D. (1989). Stress-related effects in thin films, *Thin Solid Films*, Vol.171, Issue 1, (April 1989), pp. 5-31, ISSN 0040-6090

Vlasko-Vlasov, V.; Welp, U.; Karapetrov, G.; Novosad, V.; Rosenmann, D.; Iavarone, M.; Belkin, A. & Kwok, W.-K. (2008). Guiding superconducting vortices with magnetic domain walls, *Physical Review B*, Vol.77, Issue 13, (April 2008), pp. 134518-134524, ISSN 1550-235x (online)

6

Influence of Thickness on Structural and Optical Properties of Titanium Oxide Thin Layers

Haleh Kangarlou[1] and Saeid Rafizadeh[2]
[1]*Faculty of Science, Urmia Branch, Islamic Azad University, Urmia*
[2]*Faculty of Engineering, Urmia Branch, Islamic Azad University, Urmia*
Iran

1. Introduction

Many researchers have reported on obtaining TiO_2 thin films, using different methods [1-7]. These extensive studies on TiO_2 have been due to the importance of these films in variety of applications including low-loss, low-scatter optical coating for visible and near infrared optics [8-11] and electrical devices [12,13]. These applications have stimulated a considerable amount of activity in fabrication of dielectric films with high refractive index and low absorption. Compact thin films of TiO_2 on conducting glass are used in new types of solar cells: liquid and solid dye-sensitized photo-electrochemical solar cells [14,15]. These films are also of interest for the photo-oxidation of water [16], photo-catalytisis [17], electro chromic devices [18] and other uses [19]. Due to the need of cost competitive devices in these application areas, simple and inexpensive techniques are required for film deposition and preparation.

Different properties of thin films are strongly influenced by the nanostructure of films such as grain sizes, nano-strain, crystallographic orientation and other features. It is shown that nanostructure of thin films are strongly affected by film preparation procedures and deposition conditions. For example, the substrate temperature [20-22], angle of incidence [23-25], deposition rate [26,27], and film thickness [28] have important effects on the morphology and nanostructure of thin films. Some synthetic methods for TiO_2 nanostructures are:

1.1 Sol-gel method

In a typical sol-gel process, a colloidal suspension, or a sol, is formed from the hydrolysis and polymerization reactions of the precursors, which are usually inorganic metal salts or metal organic compounds such as metal alkoxides.

TiO_2 nanomaterials have been synthesized with the sol-gel method from hydrolysis of a titanium precursor. This process normally proceeds via an acid-catalyzed hydrolysis step of titanium(IV) alkoxide followed by condensation. The development of Ti-O-Ti chains is favored with low content of water, low hydrolysis rates, and excess titanium alkoxide in the reaction mixture. Three-dimensional polymeric skeletons with close packing result from the development of Ti -O- Ti chains. The formation of $Ti(OH)_4$ is favored with high hydrolysis rates for a medium amount of water. The presence of a large quantity of Ti-OH and insufficient development of three-dimensional polymeric skeletons lead to loosely packed first-order particles. Polymeric Ti -O- Ti chains are developed in the presence of a large excess of water. Closely packed first-order particles are yielded via a three-dimensionally developed gel skeleton. From the study on the growth kinetics of TiO_2 nanoparticles in aqueous solution using titanium tetraisopropoxide (TTIP) as precursor, it is found that the rate constant for coarsening increases with temperature due to the temperature dependence of the viscosity of the solution and the equilibrium solubility of TiO_2. Secondary particles are formed by epitaxial self-assembly of primary particles at longer times and higher temperatures, and the number of primary particles per secondary particle increases with time. The average TiO_2 nanoparticle radius increases linearly with time, in agreement with the Lifshitz-Slyozov-Wagner model for coarsening.

A series of thorough studies have been conducted by Sugimoto et al. using the sol-gel method on the formation of TiO_2 nanoparticles of different sizes and shapes by tuning the reaction parameters Typically, a stock solution of a 0.50 M Ti source is prepared by mixing TTIP with triethanolamine (TEOA) ([TTIP]/[TEOA] = 1:2), followed by addition of water. The stock solution is diluted with a shape controller solution and then aged at 100 °C for I day and at 140 °C for 3 days. The pH of the solution can be tuned by adding $HCl0_4$ or NaOH solution. Amines are used as the shape controllers of the TiO_2 nanomaterials and act as surfactants. These amines include TEOA, diethylenetriamine, ethylenediamine, trimethylenediamine, and triethylenetetramine.

By a combination of the sol-gel method and an anodic alumina membrane (AAM) template, TiO_2 nanorods have been successfully synthesized by dipping porous AAMs into a boiled TiO_2 sol followed by drying and heating processes. [38]

1.2 Micelle and inverse micelle methods

Aggregates of surfactant molecules dispersed in a liquid colloid are called micelles when the surfactant concentration exceeds the critical micelle concentration (CMC). The CMC is the concentration of surfactants in free solution in equilibrium with surfactants in aggregated form. In micelles, the hydrophobic hydrocarbon chains of the surfactants are oriented toward the interior of the micelle, and the hydrophilic groups of the surfactants are oriented toward the surrounding aqueous medium. The concentration of the lipid present in solution determines the self-organization of the molecules of surfactants and lipids. The lipids form a single layer on the liquid surface and are dispersed in solution below the CMC. The lipids organize in spherical micelles at the first CMC (CMC-I), into elongated pipes at the second CMC (CMC-II), and into stacked lamellae of pipes at the lamellar point (LM or CMC-III). The CMC depends on the chemical composition, mainly on the ratio of the head area and the tail length. Reverse micelles are formed in nonaqueous media, and the hydrophilic head groups are directed toward the core of the micelles while the hydrophobic groups are directed outward toward the nonaqueous media. There is no obvious CMC for reverse micelles, because the number of aggregates is usually small and they are not sensitive to the surfactant concentration. Micelles are often globular and roughly spherical in shape, but ellipsoids, cylinders, and bilayers are also possible. The shape of a micelle is a function of the molecular geometry of its surfactant molecules and solution conditions such as surfactant concentration, temperature, pH, and ionic strength. [38]

1.3 Sol method

The sol method here refers to the nonhydrolytic sol-gel processes and usually involves the reaction of titanium chloride with a variety of different oxygen donor molecules, e.g., a metal alkoxide or an organic ether.

$$TiX_4 + Ti(OR)_4 \rightarrow 2TiO_2 + 4RX \tag{1}$$

$$TiX_4 + 2ROR \rightarrow TiO_2 + 4RX \tag{2}$$

The condensation between Ti-CI and Ti-OR leads to the formation of Ti-O-Ti bridges. The alkoxide groups can be provided by titanium alkoxides or can be formed in situ by reaction of the titanium chloride with alcohols or ethers. In the method by Trentler and Colvin, a metal alkoxide was rapidly injected into the hot solution of titanium halide mixed with trioctylphosphine oxide (TOPO) in heptadecane at 300 °C under dry inert gas protection, and reactions were completed within 5 min. For a series of alkyl substituents including methyl, ethyl, isopropyl, and tert-butyl, the reaction rate dramatically increased with greater branching of R, while average particle sizes were relatively unaffected. Variation of X yielded a clear trend in average particle size, but without a discernible trend in reaction rate. Increased nucleophilicity (or size) of the halide resulted in smaller anatase nanocrystals. Average sizes ranged from 9.2 nm for TiF_4 to 3.8 nm for TiI_4. The amount of passivating agent (TO PO) influenced the chemistry. Reaction in pure TOPO was slower and resulted in smaller particles, while reactions without TOPO were much quicker and yielded mixtures of brookite, rutile, and anatase with average particle sizes greater than 10 nm. [38]

1.4 Hydrothermal method

Hydrothermal synthesis is normally conducted in steel pressure vessels called autoclaves with or without Teflon liners under controlled temperature and/or pressure with the reaction in aqueous solutions. The temperature can be elevated above the boiling point of water, reaching the pressure of vapor saturation. The temperature and the amount of solution added to the autoclave largely determine the internal pressure produced. It is a method that is widely used for the production of small particles in the ceramics industry. Many groups have used the hydrothermal method to prepare TiO_2 nanoparticles. [38]

1.5 Solvothermal method

The solvothermal method is almost identical to the hydrothermal method except that the solvent used here is nonaqueous. However, the temperature can be elevated much higher than that in hydrothermal method, since a variety of organic solvents with high boiling points can be chosen. The solvothermal method normally has better control than hydrothermal methods of the size and shape distributions and the crystallinity of the TiO_2 nanoparticles. The solvothermal method has been found to be a versatile method for the synthesis of a variety of nanoparticles with narrow size distribution and dispersity. The solvothermal method has been employed to synthesize TiO_2 nanoparticles and nanorods with/without the aid of surfactants. [38]

1.6 Chemical vapor deposition

Vapor deposition refers to any process in which materials in a vapor state are condensed to form a solid-phase material. These processes are normally used to form coatings to alter the mechanical, electrical, thermal, optical, corrosion resistance, and wear resistance properties of various substrates. They are also used to form free-standing bodies, films, and fibers and to infiltrate fabric to form composite materials. Recently, they have been widely explored to fabricate various nanomaterials. Vapor deposition processes usually take place within a vacuum chamber. If no chemical reaction occurs, this process is called physical vapor deposition (PVD); otherwise, it is called chemical vapor deposition (CVD). In CVD processes, thermal energy heats the: gases in the coating chamber and drives the deposition reaction. [38]

1.7 Physical vapor deposition

In PVD, materials are first evaporated and then condensed to form a solid material. The primary PVD methods include thermal deposition, ion plating, ion implantation, sputtering, laser vaporization, and laser surface alloying. TiO2 nanowire arrays have been fabricated by a simple PVD method or thermal deposition. [38]

1.8 Electrodeposition

Electrodeposition is commonly employed to produce a coating, usually metallic, on a surface by the action of reduction at the cathode. The substrate to be coated is used as cathode and immersed into a solution which contains a salt of the metal to be deposited. The metallic ions are attracted to the cathode and reduced to metallic form. With the use of the template of an AAM, TiO_2 nanowires can be obtained by electrodeposition. [38] Optical properties of Ti thin films, despite their importance in different technologies, are only reported by Johnson and Christy [29] and for Bulk Ti samples by Lynch et al [30] and Wall et al [31].Therefore, it is of interest to find out the relationship between different theories [32-34] given for optical parameters and the structural changes described by variation of film thickness [35-37]. Accordingly, it was decided to investigate the influence of film thickness on the optical and structural properties of Titanium oxide films produced at high temperature of 473 K.

2. Experimental details

Titanium oxide films of 10, 50, 100 and 200 nm thickness were deposited by evaporation of TiO_2 powder, on glass substrates at 473 K deposition temperature. The residual gas was composed mainly of H_2, H_2O, CO and CO_2 as detected by quad ro pole mass spectrometer. The substrate normal was at 8.5 degree to the direction of evaporated beam and the distance between the evaporation crucible and substrate was 54.5 cm. Substrates were glasses ($2\times2\times1$ cm^3) of minimum roughness.

Just before use, all glass substrates were ultrasonically cleaned in heated acetone, then ethanol. The near normal incidence reflectance spectra were obtained using a double beam spectrophotometer (carry 500) in the spectral range of 200-2500 nm corresponding to the energy range of 0.6-6.215 eV.

Nanostructures of these films were obtained using a Philips XRD X pert MPD Diffractometer (CuK_α radiation) with a step size of 0.03 and count time of 1s per step, while the surface physical morphology and roughness were obtained by means of AFM (Dual Scope™DS95-200/50)analysis

3. Results and discussion

3.1 Optical constants of TiO_2 fils as a function of film thickness

Kramers-Kronig relations were used to convert the measured reflectivity spectra of complex dielectric function, from which the optical conductivity absorption coefficient and other parameters were calculated.

Figures 1(a) and 1(b), show spectra of ε_1 and ε_2 for TiO_2/glass films of different thicknesses (10 nm to 200 nm) produced at 473 K deposition temperature, respectively. Johnson and Christys results [29] for thin Ti films (30 nm) and Lynch et al results [30] for bulk Ti samples are also included for comparison. Ti is a getter metal and a rare Ti film can not be prepared even in UHV conditions. The general trend of our results is similar to those of Johnson and Christy, and Lynch et al (Figure 1(a,b)). As it can be seen in Figure 1(a), all curves begin from negative values and reach to a maximum pick at about 0.9 eV. Generally by increasing film thickness, real part of dielectric constant increases (Figure 1(a)) and that is because of producing TiO_2 dielectric layers. As it can be seen in Figure 1(b), there is a peak at about 1.5 eV for all layers. By increasing the thickness, in the energy range of 1 eV up to 4 eV, general trend of ε_2 curves increases. For the rest of energy range (4 eV up to 6 eV), there is no general trend for ε_2 curves, and that is because of competition between increasing film thickness and surface and bulk diffusion of grains at high 473 K temperature.

3.2 Effective-media appraximation (EMA) approach

The correlation between the nanostructure of TiO_2 thin film and it's optical property achieved through using the effective-media approximation (EMA) method. The changes in fraction of voids can be attributed to the change of nanostructures in the evolution of film nanostructures by substrate temperature, film thickness and deposition rate.

The effect of voids on optical properties of thin films can be investigated by the Bruggman effective-media approximation method [33] or its version developed by aspens and coworkers [32]. As it can be seen in Figure (2), by increasing film thickness, in 1 eV up to 4 eV energy range, fraction of voids decreases. For the rest of analysis range (4 eV up to 6 eV), fraction of voids due to competition between increasing film thickness and surface and bulk diffusion of grains at high 473 K temperature, cross each other. So there is a good correlation between nanostructures and optical properties in our results.

3.3 Interband region

The energy bands for TiO_2 were calculated using potential in Shroudinger equation. The obtained experimental absorption coefficients (α=2Ek/hc) for TiO_2 films, were plotted in Figure (3). The plots have a good agreement with those of Janson and Cheristy and Lynch et al. In general by increasing film thickness, absorption coefficient also increases. By increasing the thickness, fraction of voids decreases, specially in the energy renge of

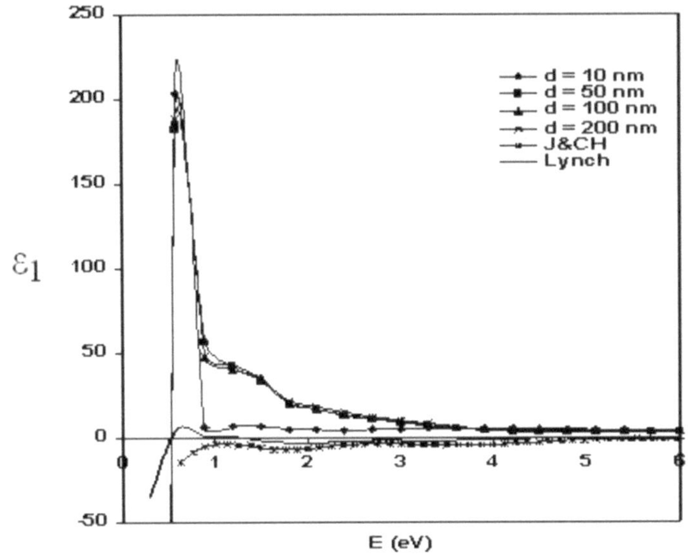

(a)

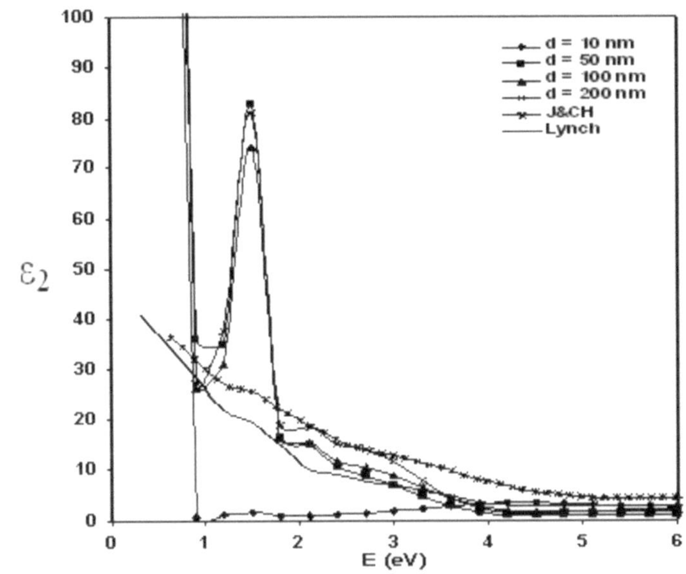

(b)

Fig. 1. The dielectric constants of deposited TiO_2/ glass films of Different thicknesses at 473 k (a) Real port (b) Imaginary Port.

1 eV – 4 eV, which is in agreement with fraction of voids (Figure 2). In the energy renge of 1 eV - 4 eV, transmittance decreases and absorbance increases. In the energy renge of 4 eV - 6 eV, there is no general trend for absorption coefficient curves and they cross each other. That is because of competition between surface and bulk diffusion in one hand and increasing thickness on the other hand. This result is also in agreement with EMA results. There was an inter band transition at energy value of about 4.5 eV.

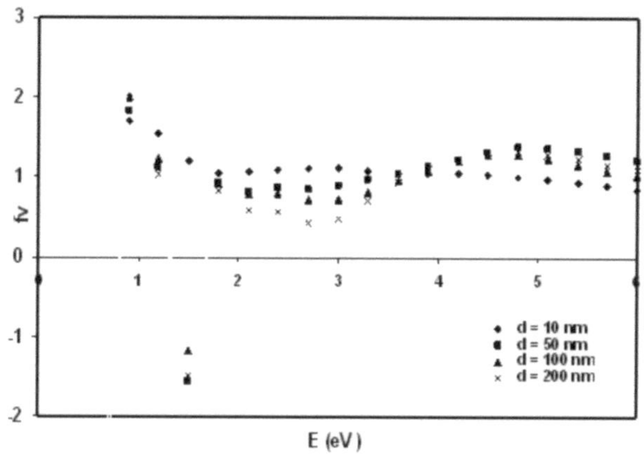

Fig. 2. Void fraction Vs. energy for different film thicknesses deposited TiO_2/glass films at 473 k.

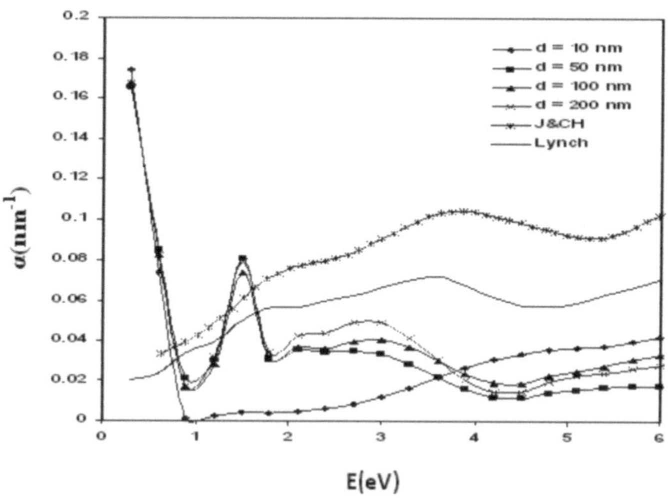

Fig. 3. Absorption coefficient vs. energy for different film thicknesses deposited TiO_2/glass films at 473 k.

Transmission spectroscopy was used to study optical properties of thin TiO_2 films. It was found that, low independent transmission absorbance measurements of a thin film, if the thickness is unknown, are sufficient for numerical inversion to determine the complex index of refraction of a film. Real part of refractive index (n) and imaginary part of refractive index (k) are shown in Figure 4 (a) and (b) respectively. As it can be seen in Figures 4(a) and 4(b), by increasing film thickness, general trend of the curves increases. 10 nm thickness is very thin so produced layer is uncompleted and we expect unusual behavior. By increasing thickness at high temperature (473 K), because of surface and bulk diffusion bigger grains form (will be discus in AFM analysis), that tends to increase refractive indexes (Figure 4(a)). Also by increasing thickness at high temperature (473 K), due to migration of grains, fraction of voids decreases, which results in, less transmittance as well as an increase in imaginary part of refractive index (k), see Figure 4 (b).

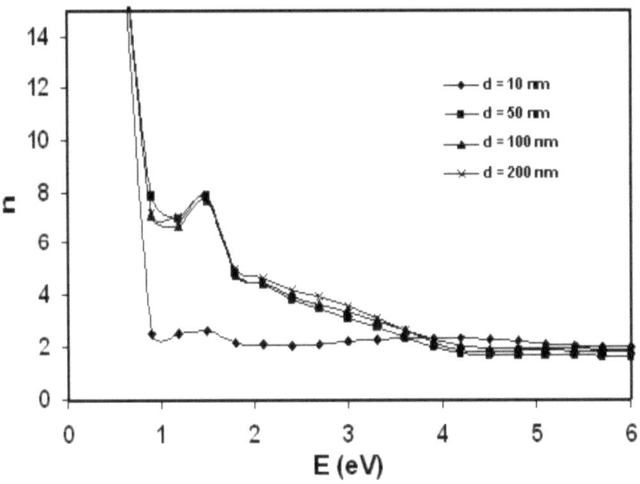

(a)

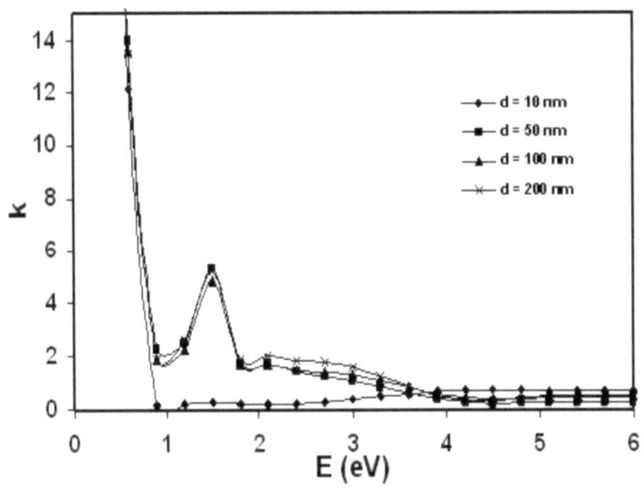

(b)

Fig. 4. Refractive index Vs. energy for different film thicknesses deposited TiO_2/glass films at 473 k,(a) Real part (b) Imaginary part.

3.4 AFM analaysis

Figure 5, shows the AFM images of TiO_2 / glass layers of different thicknesses at 473 k temperature. Figure 5(a), shows topography of TiO_2 / glass of 10 nm thickness. As it can be seen there are small grains on surface and the film surface is almost the same as substrate. By increasing thickness to 50 nm and in presence of 473 K temperature, surface diffusion happens and it forms bigger grains with small grains between them (Figure 5(b)). By increasing thickness to 100 nm in Figure 5 (c), most of the holes are covered with metallic grains and smoother layer produces and by increasing thickness to 200 nm, due to surface and bulk diffusion, domed grains appear (Figure 5(d)). By increasing thickness at high temperature, needle like and small grains change to big and domed grains.

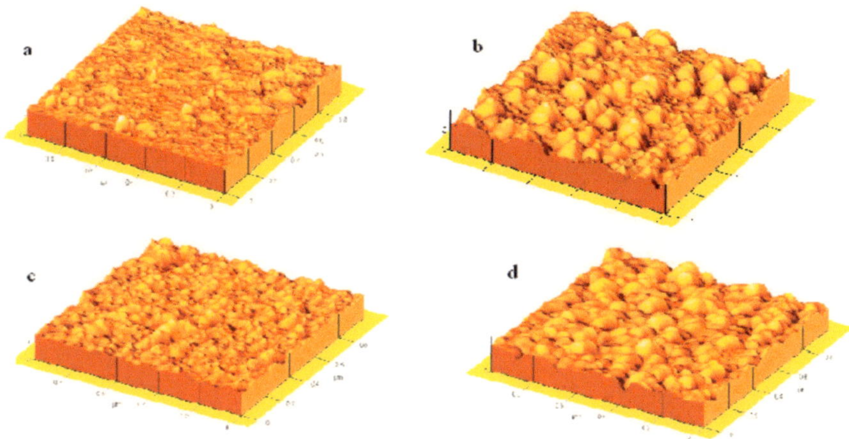

Fig. 5. The topography of TiO_2 / glass with (a)10 nm, (b)50 nm, (c)100 nm, (d)200 nm thickness at 473 k.

3.5 Roughness

Figure (6), shows the roughness curve of produced layers. As it can be seen by increasing thickness to 50 nm, roughness increases. At 100 nm and 200 nm thickness, due to surface and bulk diffusion and coalescence of grains, a decrease in fraction of voids happens (as discussed in optical properties) and roughness decreases. Generally by increasing thickness in high temperature almost roughness decreases.

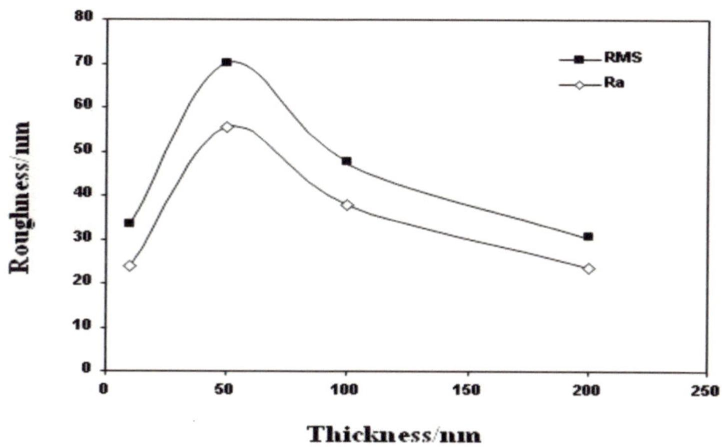

Fig. 6. The roughness curve of produced TiO_2 / glass layers at 473 k.

3.6 XRD analysis

Figure (7), shows the XRD patterns of produced layers. As it can be seen in Figure 7(a), for the layer of 10 nm thickness, no clear peak is appeared so the layer is amorphous. By increasing thickness to 50 nm, TiO_2 layer is getting crystallized and anatase A(200) crystallographic direction begin to grow (Figure 7 (b)). By increasing the thickness to 100 nm and 200 nm (Figure 7(c) and 7(d)), layers are crystallized and anatas A(004) crystallographic direction appear. By increasing thickness the pick becomes sharper. High temperature and thickness play an important role on the layers crystallization.

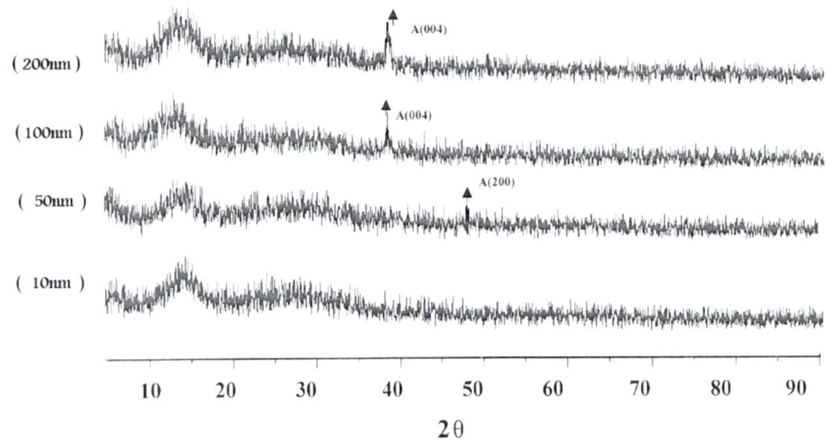

Fig. 7. The XRD pattern of TiO_2 / glass layers with different thicknesses at 473 k.

3.7 AFM Analaysis of other samples

Figure 8(a-d), shows AFM images of different thicknesses 20 nm, 70 nm, 200 nm and 250 nm respectively in presence of 600 K annealing temperature and Oxygen flow. As it can be seen heat and oxygen have different effects on different thicknesses. Morphology of these layers are completely different. For thinner layers (Figure 8(a) and 8(b)), Oxygen penetrate to

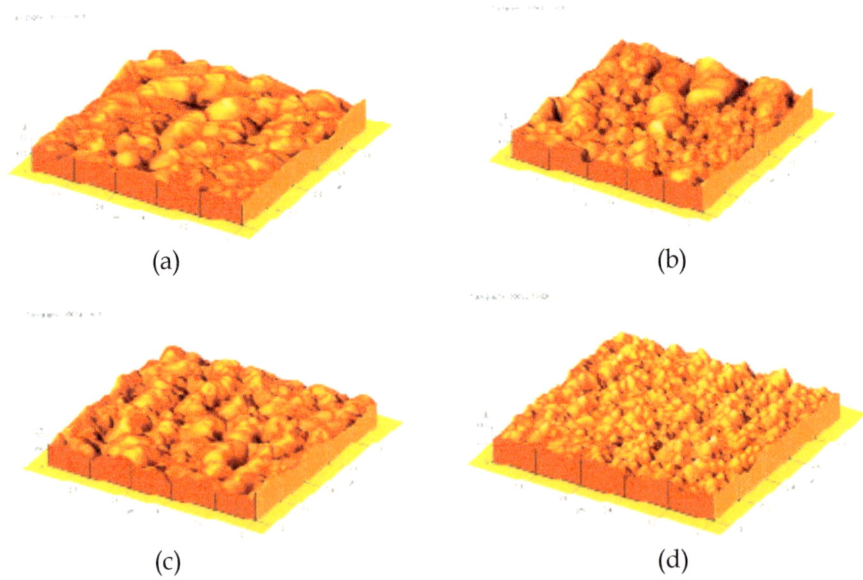

Fig. 8. The topography of TiO_2 / glass with (a)20 nm, (b)70 nm, (c)200 nm, (d)250 nm thickness at 600 k.

depth of the layers and by exerting high temperature, coalescence of grains happen and big domed grains appear. For thicker layers of 200 nm and 250 nm layers (Figure 8(c) and 8(d)), it seems that oxygen is penetrating to a special depth of surface and by exerting heat coalescence happens but grains are smaller in comparison with thinner layers. It was found that, film thickness play an important role on nanostructures of produced thin layers.

4. Conclusion

The relationship between nanostructure of TiO_2 thin films of different thicknesses (10 nm to 200 nm) produced at 473 K deposition temperature were studied. Optical properties, topography, roughness and crystallization of these films were investigated. Optical properties were investigated by studying the relationship between nanostructure and EMA, while the EMA results were dependent on dielectric constant of both film and bulk samples. The optical constants of the films were effected by the film thickness and high deposition temperature. By increasing film thickness, dielectric constants, absorption coefficient and real and imaginary parts of refractive indexes increased. Almost for all plots of 4 eV - 6 eV energy range, curves cross each other, that is because of competition between surface and bulk diffusion in one hand and increasing thickness on the other hand. The fraction of voids were obtained using the EMA method. The deviation from a general increasing or decreasing trend is due to the competition between surface and bulk diffusion of grains and film thickness. The results of absorption coefficient versus photon energy in inter band region for TiO_2 films were found to fall into a band, which is confined by Johnson and Christy's results (for thin film) at the top of band and those of Lynch et al (bulk Ti sample) at the bottom.

Topography of layers showed that, by increasing thickness at high temperature (473 K), surface and bulk diffusion happened and changed the shape of grains. Roughness curve showed that only for 50 nm thickness, roughness increased but for 100 nm and 200 nm due to surface and bulk diffusion and migration of the grains, roughness decreased.

XRD pattern of the layers showed that the 10 nm TiO_2 layer is amorphous and by increasing thickness, layers become crystallized. Crystallographic direction as A (200) for 50 nm and A(004) for 100 nm and 200 nm TiO_2 layers, appeared. By increasing thickness A(004) peak become sharper.

Finally the AFM images of another samples of 20 nm, 70 nm, 200 nm and 250 nm thicknesses deposited at 600 k were studied and it was found that, film thickness play an important role on nanostructures of produced thin layers.

5. References

[1] L. Kavan, M. Gratzel, Electrochim. *Acta*, 40 (1995) 643.
[2] S. D. Burnside, V. Shklover, C. Barbe, *Chem. Mater.*, 10 (1998) 2419.
[3] P. Sawunyyama, A. Yasumori,m K. Okada, *Mater. Res. Bull.*, 33 (1998) 795.
[4] S. Deki, Y. Aoi, *J. Mater. Res.* 13 (1998) 883.
[5] S. Yin, Y. Inoue, S. Uchida, Y. Fujishiro, T. Sato, *J. Mater. Res.*, 13 (1998) 844.

[6] E. Vigil, L. Saadoun, R. Rodriguez-Clemente, J. A. Ayllon, X. Domenech, *J.Mater. Sci. Lett.*, 18 (1999) 1067.

[7] H. Savaloni, K. Khojier, M. S. Alaee, *J. Mater. Sci.*, 42 (2007) 2603.

[8] K. A. Vorotilov, E. V. Orlova, V. I. Petrovsky, *Thin Solid Films*, 207 (1992) 180.

[9] M. G. Krishana, K. Narasimaha Rao, S. Mohan, *J. Appl. Phys.*, 73 (1983) 434.

[10] J. Rancourt, *User's Handbook: Optical Thin Film, McGrow-Hill*, New York, 1987.

[11] J. A. Dobrowski, in: W. Driscoll (Ed.), *Coating and Filters in Handbook of Optics, MacGrow-Hill*, New York, 1987.

[12] P. Babelon, A. S. Dequiedt, H. Mostesa-Sba, S. Bourgeois, P. Sibillot, M. Sacilotti, *Thin Solid Films*, 322 (1998) 63.

[13] Y. Leprince-Wang, K.-Y. Zhang, V. Nguyen, V. An, D. Souche, J. Rivory, *Thin Solid Films*, 307 (1997) 38.

[14] B. O'Regan, M. Gratzel, *Nature*, 353 (1991) 737.

[15] U. Bach, D. Lupo, P. Comte, J. E. Moser, F. Weissortel, J. Salbeck, H. Spreitzer,M. Gratzel, *Nature*, 395 (1998) 583.

[16] L. Kavan, M. Gratzel, Electrochim. *Acta*, 40 (1995) 643.

[17] R. L. Pozzo, M. A. Baltanas, A. E. Cassano, *Catalysis Today*, 39 (1997) 219.

[18] T.-S. Kang, D. Kim, K.-J. Kim, *J. Electrochem. Soc.*, 145 (1998) 1982.

[19] Y. Paz, Z. Luo, L. Rabenberg, A. Heller, *J. Mater. Res.*, 10 (1995) 2842.

[20] Ida, T.& Toraya, H., *J. Appl. Cryst.* 35 (2002) 58.

[21] M. Cernasky, *J. Appl. Cryst.* 16 (1983) 103.

[22] R. W. Cheary and A. Coelho, *J. Appl. Cryst.* 25 (1992) 109.

[23] D.Balzar and S.Popovic, *J. Appl. Cryst.* 29 (1996) 16.

[24] D.Balzar, *J. Appl. Cryst.* 25 (1992) 559.

[25] J. K. Yau and S. A. Howard, *J. Appl. Cryst.* 22 (1989) 244.

[26] S. A. Howard, and R. L. Snyder, *J. Appl. Cryst.* 22 (1989) 238.

[27] S. Enzo, G. Fagherazzi, A. Benedetti and S. Polizzi, *J. Appl. Cryst.* 22 (1989) 184.

[28] H. Savaloni, A. Taherizadeh, A. Zendehnam, *Physica B*, 349 (2004) 44.

[29] R. B. Johnson and R. W. Christy, *Physical Review B*, 9 (1974) 5056.

[30] D. W. Lynch, C. G. Olson, J. H. Weaver, *Physical Review B*, 11 (1975) 3617.

[31] W. E. Wall, M. W. Ribrasky and J. R. Stevenson, *J. App. Phys*, 51 (1980) 661.

[32] E. Aspnes, *Thin Solid Films*, 89 (1982) 249.

[33] D. A. G. Brauggeman, *Ann. Phys. (Leipzig)*, 24 (1935) 636.

[34] E. Aspnes, E. Kinsbron, and D. D. Bacon, *Phys. Rev.* B21 (1980) 3290.

[35] R. Messier and J. E. Yahoda, J. *Appl. Phys*, 58 (1985) 3739.

[36] P. B. Barna, M. Adamik, *Thin Solid Films*, 317 (1998) 27.

[37] Petrov, P. B. Barna, L. Hultman, J. E. Greene, *J. Vac.Sci. Technol*,S117-S128.

[38] Xiaobo Chen and Samuel S. Mao, *Chem. Rev.* 2007, 107, 2891-2959.

Section 3

Characterization of Mechanical Properties

7

Microtribological Behavior of Polymer-Nanoparticle Thin Film with AFM

Xue Feng Li[1], Shao Xian Peng[1] and Han Yan[2]
[1]*School of Chemical and Environmental Engineering, Hubei University of Technology, Wuhan*
[2]*School of Civil Engineering, Hubei University of Technology, Wuhan*
China

1. Introduction

1.1 Polymer-CNT nanocomposits

Carbon nanotubes (CNTs) have attracted significant scientific attention because of their remarkable mechanical properties and many potential applications (Balani et al., 2008; Vail et al., 2009). The tribological properties of CNTs as reinforcement agent in metal-based composite were studied by a few groups (Meng et al., 2009; Pei et al., 2008). The results indicated that metal-CNT composite coating was better wear resistance and self-lubricity than the metal coating. Ni-carbon nanotubes composite coating has better wear resistance and self-lubricity than the Ni coating. The poor solubility characteristics of CNTs have hindered it as an additive to liquid lubricants (Decher, 1997). If CNTs were dispersed in liquid lubricants by some physical or mechanical ways, CNTs still assembled and presented in the form of cluster. The soft chains of polymer molecules are reasonable to consider anti-wear for its flexibilities, great interests have been aroused in the thin polymer films on solid surface for their application in lubrication. Meanwhile the existence of rigid rod-like segments in the films is expected to be beneficial to bearing applied loads. Recently some chemists have adopted chemistry method for the preparation of soluble carbon nanotubes containing polymer (Riggs et al., 2000; Shaffer & Koziol, 2002). Attachment of surface polymer chain is expected to provide much more effective performance of nanocomposite materials, as dictated by the nature of the polymer. A lot of reports have described the attachment of short chains to CNTs to improve the performance of nanocomposite materials by esterification / amidation of surface acid groups like the preparation of poly(ethylene glycol)-carbon nanotubes (PEG-CNTs) described in the chapter (Chen et al., 1998; Li et al., 2011; Liu et al., 1998). However, very a few studies have been conducted in the grafting of CNTs by free radical polymerization, even though it is the commonest synthesis method and there are a lot of choices on the monomers and initiators. Exceptions include the attachment of prepolymerised poly(methyl methacrylate) and the formation of polystyrene grafted CNTs (Jia et al., 1999; Shaffer & Koziol, 2002). Here we report the synthesis of individual polyacrylamide-carbon nanotube (PAM-CNT) copolymer by an in-situ UV radiation initiated polymerization and water-soluble poly(vinyl alcohol)-carbon nanotube (PVA-CNT) copolymer by an in-situ radical initiated emulsion polymerization (Li et al.,

2004). Composite thin films were obtained by spin casting the copolymer aqueous solution onto freshly cleaved mica. The microtribological properties of the films were initial investigated using atomic force microscopy / friction force microscopy (AFM/FFM).

1.2 Diazoresin-nanoparticle self-assembly films

Another class of technologies known as self-assemble has very interesting applications in area of nano-composites and optoelectronic materials (Kwon et al., 2007). The self-assembly technique is exhibited to be a rapid and easy way to obtain molecular assemblies with precise control of composition which can be used in a wide variety of technologies ranging from microelectronic, corrosion protection and adhesion to electrochemistry, etc (Bai & Cheng, 2006; Kim et al., 2007). Tribological properties in microscales between two sliding solid surfaces are essential to supply super lubrication for micro-electro-mechanical systems (MEMS) (Cheng et al., 2006; Zhao et al., 2009). Therefore, the process of self-assembly films is considered critical technology for the realization of nano-scale devices (Bhushan et al., 2002; Medintz et al., 2005).

Since Cao's pioneering work, the covalently attached self-assembly films from diazoresin (DR) and different negatively charged materials have been prepared (Lu et al., 2004; Yang et al., 2005). This method, usually involving the alternate of two oppositely charged components on a substrate such as mica, silica wafer from aqueous solution, is environmentally friendly and easy in processing and does not require specific equipment. Nevertheless, studies on converting the physical force into covalent bonds to link the orderly carbon nanotubes / polymer thin films have seldom been reported. We describe a preparation method of poly (acrylic acid)-carbon nanotubes (PAA-CNTs), and a fabrication process of multilayer thin films constructed from PAA-CNTs and DR via a layer-by-layer technique, then the linkage bonds are converted under UV irradiation (Li et al., 2011). By this way, we also describe a preparation method of poly (acrylic acid)-titanium oxide (PAA-TiO_2) and poly (acrylic acid)-ferric hydroxide (PAA-$Fe(OH)_3$) sol, and a fabrication process of multilayer thin films constructed from PAA-TiO_2 and DR or PAA-$Fe(OH)_3$ and DR via a layer-by-layer technique (Li et al., 2009, 2011).

Then decomposition of the diazonium group could be detected after exposed under UV irradiation which leading to the conversion from ionic bond to covalent bond. In addition, to study surface morphology and friction of thin films on micro-scales, AFM and FFM are considered as excellent tools.

2. Experimental

2.1 Synthesis and microtribology of polymer-CNT nanocomposits

2.1.1 Synthesis PEG-CNT nanocomposites

Multi-wall CNTs were produced via the chemical vapor deposition method and purified using methods similar to those reported in the literatures (Hiura et al., 1995; Tsang et al., 1994). For example, CNT samples (300 mg) were suspended in an aqueous solution of hydrofluoric acid (20 wt.%, 60 ml) to prolonged sonication for 5 h and filtrated. The remaining solids were washed repeatedly with distilled water. The CNT samples thus were refluxed with an aqueous solution of nitric acid (22 wt.%, 60 ml) for 10 h. The mixture was centrifuged, and the remaining solids were washed repeatedly with distilled water many times, until the pH value of CNT solution approached 7, and then dried in a vacuum oven to eliminate impurities. CNT sample was treated with HNO_3 solution to attach the carboxylic acid groups on the CNT surface, followed by treatment in thionyl chloride to convert the carboxylic acid into acyl chlorides. These functionalized CNTs can be grafted PEG by esterification.

2.1.2 Preparation of PVA-CNT copolymer

A typical preparation process of PVA-CNTs is as follows: the mixture of 10 ml of vinyl acetate, 50 ml distilled water, 100 mg sodium dodecylsulfate, 0.1 ml emulsifier OP-10 and 50 mg potassium persulfate (initiator) was charged into a 250 ml reaction flask. The mixture was emulsified by continuous stirring for 1.5 h at 65℃. The 100 mg purified CNTs and 50 mg potassium persulfate were then added to the emulsion system. The emulsion was sonicated by an ultrasonic cleaner for 30 min, stirred for 2.5 h and maintained at 65℃. After that the emulsion was filtered to remove insoluble particles. After that 100 ml 13 wt.% of sodium chloride aqueous solution was added to precipitate the polymeric product and then dried under vacuum at 50℃ in a constant weight. The product was dissolved in methanol and saponified. The fractions were repeated re-suspension in water and centrifugation at 5000 rpm to sediment PVA-CNT products and remove all of the free PVA that was stably soluble in water. These samples of PVA-CNTs are, forming stable suspension in water.

2.1.3 Synthesis of PAM-CNT copolymer

Acrylamide was distilled under reduced pressure and sealed for preservation at low temperature. Typical PAM-CNT copolymer was prepared by UV radiation initiated polymerization. The mixtures of 5 g acrylamide and 30 ml distilled water containing 50 mg CNTs were added into a 100 ml quartz reaction flask. After ultrasonic dispersing for 30 min and adding 10 ml UV initiator solution including 30 mg diphenylketone, 30 mg benzoin and 0.1 ml triethanolamine, the mixtures were radiated and stirred for 1.5 h at room temperature and under the protection of N_2. The products were filtrated and dried in a vacuum oven at 50 ℃. Dried products were repeatedly extracted with water to obtain soluble fraction. The fraction was centrifuged at 5000 rpm to sediment the grafted CNT products. Repeated re-suspension in water and centrifugation was used to remove all of the free PAM that was stably soluble in water. These samples of PAM-CNTs are, forming stable suspension in water.

2.1.4 Characterization of nanocomposites

The PEG-CNT and PAM-CNT copolymer were characterized using FT-IR (Bruker Equindx 55). For transmission electron microscope (TEM, JEM 100CX) observations, PVA-CNT samples were dispersed in water, then scooped up onto a holey carbon micro-grid. The purified PAM-CNT copolymer was also characterized by UV-visible spectrometer (SHIMADZU UV-2250 UV-VISIBLE SPECTRIPHO METER) and fluorescence spectrometer (SHIMADZU DATA RECORDER DR-3).

2.1.5 Tribological and microtribological test of nanocomposites by AFM and FFM

The tribological measurements were carried out by an MQ-800 four-ball tribotester at a rotational speed of 1450 rpm and at a temperature of 20℃. The maximum non-seizure load (PB value) was obtained by GB3142-82, similar to ASTMD2783; Wear scar diameter (WSD) was measured under a test duration of 30 min; The stainless steel balls used in the tests were made of GCR15(AISIE52100) bearing steel with the 64-66 surface HRC hardness and 0.012 μm of surface roughness Ra. At the end of each test, the average WSD on the three lower balls was determined using optical microscopy to an accuracy of 0.01 mm. Since triethandamine is widely used in the metalworking fluid as a multifunctional additive, 2 wt.% of triethandamine aqueous solution as the base stock. And another water-based fluid with certain load-carrying and anti-wear properties, 0.5 wt.% OPZ (a type of water-soluble zinc alkoxyphosphate) fluid was chosen as the based stock (Duan, 1999). The different concentrations of PVA-CNTs prepared above were used as lubricant additive in the two kinds of base stock. WSD was measured for 30 min under a load of 200 N using the base stock without OPZ and 400 N using the base stock with OPZ respectively. The same concentration of the homopolymer reference additive was measured under the same condition.

The wear scar morphology was visualized with a JSM-35C scanning electron microscope (SEM) at a voltage of 25 kV. Specimens were wear scars on the steal balls in anti-wear tests which were performed at speed of 1450 rpm/min for 30 min, and were washed by alcohol and distilled water before inspection.

The Polymer-CNT aqueous solution (0.5 wt.%) was filtered through a 0.45 μm Teflon filter. Freshly cleaved micas were used as substrates. Thin films were obtained by spin casting (100 rpm, 1.5 min) the solution onto the mica substrates, and then dried in ambient temperature 2 days before used.

The measurements were performed with a Nanoscope a (Digital Instrument Inc.) equipped with a bioscopy G scanner (90μm) in contact mode and tapping mode. A commercial silicon cantilever (MikroMasch, Estonia, Russian) with triangular shape and force constant of K =0.12 $N \cdot m^{-1}$ were used. The images were obtained in air at 20 ℃ and relative humidity 50 % with 1 Hz scan rates.

Roughness images and root mean square (RMS) were obtained simultaneously when the scan angle was 0°. Friction images in trace and retrace scan directions were recorded when the scan angle was 90°. The applied load including the intrinsic adhesive force on thin films was calculated from set-point and "Force - Distance" curve .The applied load was changed and then the friction signals recorded, and so on, the friction-load figure was obtained (Li et al., 1999; Martínez-Martínez et al., 2009).

2.2 Fabrication and microtribology of diazoresin-nanoparticle self-assembly films

2.2.1 Preparation of PAA-CNTs

Acrylic acid (AA) was distilled under reduced pressure, dried with $CaCl_2$ overnight, and sealed for preservation at low temperature. Benzoyl peroxide (BPO) was recrystallized by chloroform. Other chemicals and solvent were used as received.

In a typical polymerization, 4 ml AA was dissolved in 25 ml 1, 4-dioxane, 60 mg BPO was added to the solution under stirring. After they have been deoxygenated by bubbling dry nitrogen gas for 10 min, the solution was heated to 85 ℃ and kept for 2.5 h under N_2. The 10 mg purified CNTs and 30 mg BPO was then added to the solution. The solution was ultrasonic dispersed for 30 min and maintained at 85 ℃ for 14 h with stirring under protection of N_2. The crude products dried in a vacuum oven were dissolved in 20 ml ethanol and precipitated by adding 200 ml petroleum ether, then filtered off. And the procedure was repeated twice. After the product was dried in a vacuum oven, a deep grey product was obtained. The products can be dissolved in polar solvents, such as water and methanol.

2.2.2 Preparation of PAA-TiO_2

Titanium dioxide (TiO_2) nanoparticles were prepared by tetrabutyl titanate hydrolysis as described in reference (Eremenko et al., 2001). Nano TiO_2 solid particles was 2.0 wt.% in the hydrosol. For TEM (HITACHI H-500, with a voltage of 80 kV) observations, samples were dispersed in water solution, and then scooped up onto holey carbon micro-grids. The distributions of TiO_2 particles in the hydrosol were determined by a particle size distribution analyzer (LB-550, HORIBA Inc.).

Potassium persulfate ($K_2S_2O_8$) was recrystallized by water. In a typical polymerization process of PAA-TiO_2, TiO_2 sol 2.50 g, acrylic acid 3 ml and deionized water 9 ml were mixed to make a solution under stirring and ultrasonic dispersing for 30 min. Then 6 mg $K_2S_2O_8$ was added under stirring. The solution was deoxygenated by bubbling dry nitrogen gas for 10 min, then heated to 85 centigrade and kept for 3 h with stirring under protection of N_2 (Chen et al., 2007; Liufu et al., 2005).

2.2.3 Preparation of PAA-$Fe(OH)_3$

$Fe(OH)_3$ sol was prepared by hydrolysis as follows: Under strong stir, 50 ml deionized water was slowly dripped into 1 ml fresh prepared ferric chloride aqueous solution with the concentrate of 0.6 mol/l, then 0.5 ml acetic acid aqueous solution was also slowly dripped under the strong stir, after that, stirred the solution in boiling water bath for 1h.

In a typical polymerization process, $Fe(OH)_3$ sol 2.0 g prepared just now, acrylic acid 3 ml and deionized water 9 ml were mixed to make a solution under ultrasonic dispersing for 30 min. Then 6 mg $K_2S_2O_8$ with 9 ml deionized water were slowly dripped into the solution under stirring. The solution was deoxygenated by bubbling dry nitrogen gas for 10 min, then heated to 85 ℃ and kept for 3 h with stirring under protection of N_2. PAA-$Fe(OH)_3$ was forming stable suspension in water.

2.2.4 Fabrication of self-assembly films

The diazoresin (Mn≈2500g/mol) aqueous solution 0.5 mg/ml was obtained from college of chemistry and molecular engineering, Peking University. PAA-nanoparticles was dissolved in the weak alkaline aqueous solutions (pH=8) with the concentrate of 0.5 mg/ml. Surface-negatively charged quartz glass was used as the substrate. Quartz glass was first cleaned by immersed in the mixed hydrogen peroxide and concentrated sulphuric acid solution with the ratio of 3 to 7 for 24 hours and then rinsed in deionized water. Cleaned quartz glass was immersed in an aqueous solution of 0.5 mg/ml DR for 5 min, rinsed with deionized water and then dried under a stream of air. After totally dried, the quartz glass was dipped into an aqueous solution of 0.5 mg/ml PAA-CNTs for 5 min again, followed by rinsing and drying. This procedure was repeated five times to yield a 5-bilayer self-assembled ultrathin multilayer film. All stages of growth were carried out at room temperature in the dark. The fabricated films were irradiated with 200 W medium pressure mercury lamp at a distance of

25 cm for 30 S. The self-assembly process was monitored after each fabrication cycle of the films by UV-vis spectra determinations (SHIMADZU UV-1700).

2.2.5 Study of diazoresin-nanoparticle self-assembly films by AFM and FFM

A commercial atomic force microscopy (SPA400, SEIKO Co.) with Si_3N_4 cantilever was used to conduct studies of surface morphology of the thin film. The surfaces of the films were studied by the AFM with tapping mode. The images were obtained in air at 25 ℃ and the relative humidity 60 %.

Friction images in trace and retrace scan directions were recorded when the scan angle was 90°. The applied load was changed and then the friction signals recorded to study the microscopic friction of the films with different layers.

3. Results and discussion

3.1 Characterization of the PEG-CNTs

In order to obtain the evidence of the bond conversion of CNTs, the FTIR spectra are shown in Fig.1. We can see the new peaks at 846 and 950 cm^{-1} (stretching vibration of the acyl chlorides group) after the reaction from CNTs-COOH to PEG-CNTs. The absorption at 1720 cm^{-1}, which is assigned to the carboxylic ester groups, increases contrasting with CNTs-COOH due to the conversion of the acyl chlorides to ester. The reaction between PEG and CNTs can be schematically represented as shown in scheme 1.

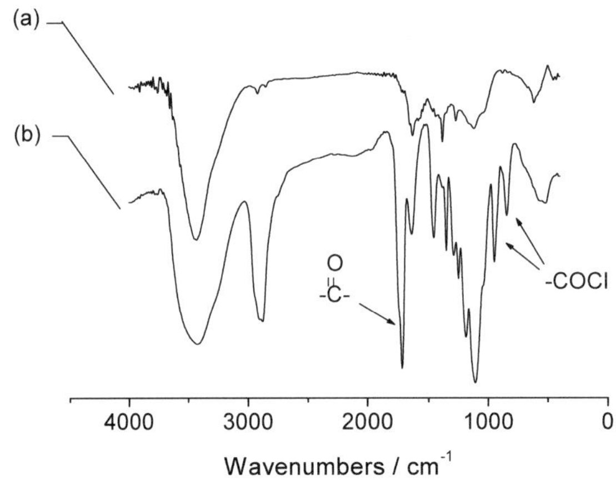

Fig. 1. FTIR spectra of (a) CNTs-COOH and (b) PEG-CNTs

3.2 Morphology of the PVA-CNT copolymer

Fig.2 displays typical TEM of the PVA-CNT copolymer. It is clear that PVA is coated on the surface of CNTs. The diameter of PVA-CNTs is much larger than that of CNTs, i.e. about 20-30 nm in diameter for CNTs and about 50-70 nm for PVA-CNTs were observed respectively. Meanwhile the much more PVA is assembled at part of curvature points of CNTs. Jia et al. also have observed similar phenomenon in the poly(methyl methacrylate)-CNT copolymer system (Jia et al., 1999). It might be proposed that CNTs had been broken in the process of

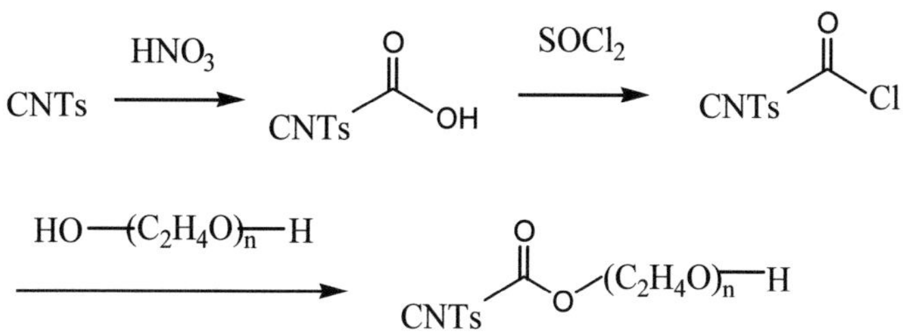

Scheme 1. Structure of CNTs conversion from carboxylic acid groups to carboxylic ester groups

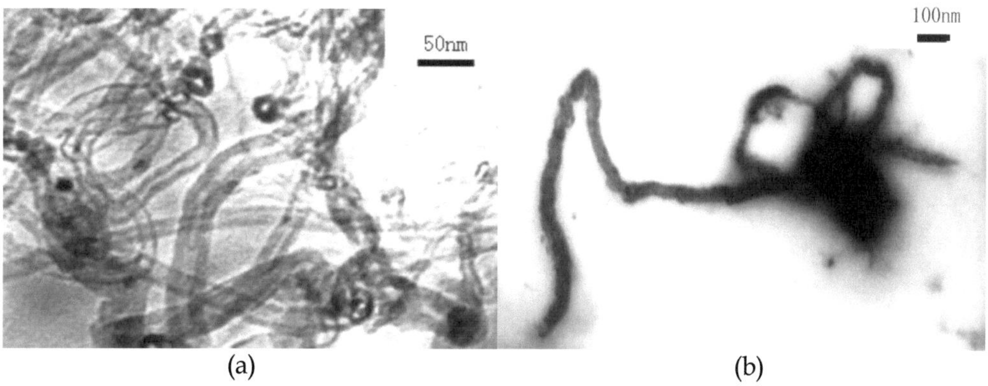

Fig. 2. Typical TEM images of (a) CNTs and (b) PVA-CNTs (Wrapping of PVA chains around CNT shell)

the reaction, because at these points structure consists of carbon atom pentagons, which is not as stable as a hexagon structure of carbon atoms. So it may be assumed that some C-C bonds at these points in CNTs can be more easily broken than other bonds, and then linked and wrapped by polymer.

3.3 Characterization of the PAM-CNT copolymer

To obtain the evidence of the bonding between the CNTs and the PAM, the PAM-CNT copolymer was characterized by FTIR, UV-visible absorbance spectra, fluorescence spectra and TEM.

FTIR spectra of the PAM-CNTs are shown in Fig.3. There is a peak at 1650 cm^{-1} (Fig 3b), which is attributed to the stretching vibration of the $-CONH_2$ group. In the spectra of CNTs in Fig3a, peaks at 1500-1700 and 3500 cm^{-1} are attributed to C=O and -OH functional groups from the acid treatment. The absorbencies of PAM-CNTs and PAM in the region of 1000 and 1500 cm^{-1} are different. The formation of these peaks may be proposed that deflections had occurred to the C-C bond (at about 1000 cm^{-1}) and the >C=O (at about 1500 cm^{-1}) on PAM-CNTs due to the existence of a chemical bond between the PAM and CNTs. The UV-visible absorption spectra and the fluorescence spectra for PAM-CNTs prepared above and PAM were measured (Fig.4). The peak at 195 nm split into two peaks on the UV-visible spectra of the PAM-CNTs. When the PAM-CNTs are excited at 347 nm, it emits fluorescence with a peak maximum at 430 nm, and increases in intensity in copolymer, but the fluorescence

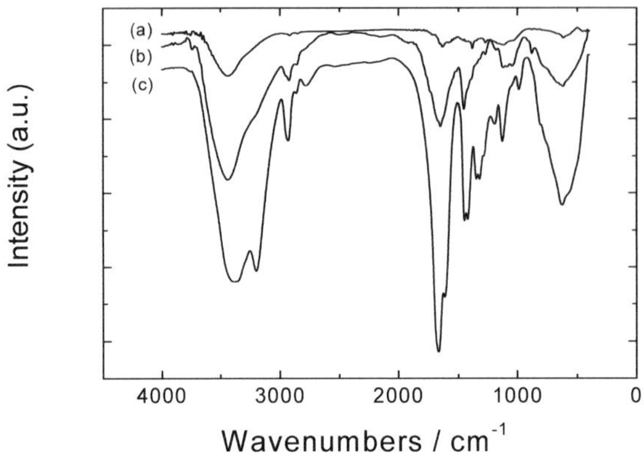

Fig. 3. FTIR spectra of (a) CNTs (b) PAM-CNTs (c) PAM

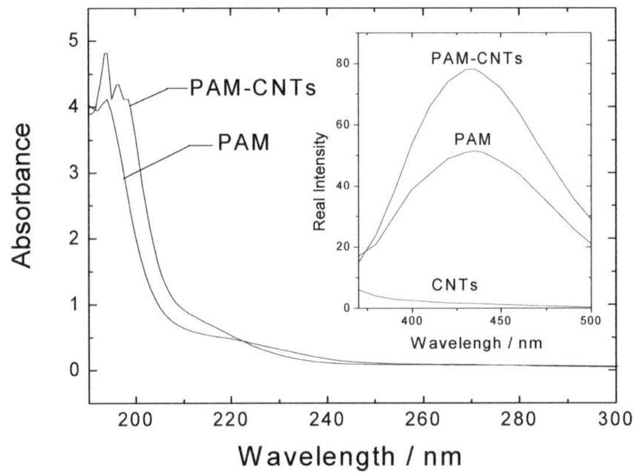

Fig. 4. UV-visible absorbance spectra with PAM and PAM-CNTs. Fluorescence spectra of CNTs, PAM and PAM-CNTs (Emission spectra under 347 nm irradiation)

spectral profile of the CNT sample is not fluorescent. Sun et al. also have observed that pristine CNTs are not luminescent and the functionalization facilitates the exhibition of luminescence from CNTs via the dispersion of the nanotubes. For fullerenes, another cage-like form of carbon, multiple functionalized derivatives also are more fluorescent (El-Hami & Matsushige, 2003). CNTs are long, slender fullerenes where the walls of the tubes are hexagonal carbon (graphite structure) and often capped at each end. Mechanistically, the luminescence from CNTs was explained in terms of two possibilities: the existence of extensive conjugated electronic structures and the excitation energy trapping associated with defects in the nanotubes (Riggs et al., 2003). The solubilization of CNTs via linking PAM might contain extended π-electronic structures that are isolated as a result of nanotubes surface modification. It may be supposed that the bonds in CNTs were opened in the reaction, and CNTs can be linked with PAM by their opened π-bonds too.

Fig.5 displays typical TEM images of the PAM-CNT and CNT samples. It is clear that PAM is coated on the surface of CNTs. The diameter of PAM-CNTs is much larger than that of CNTs, i.e. about 20-30 nm in diameter for CNTs and about 100-150 nm for PAM-CNTs were observed respectively. Meanwhile the CNTs in the PAM-CNTs are much shorter than the pristine CNTs samples.

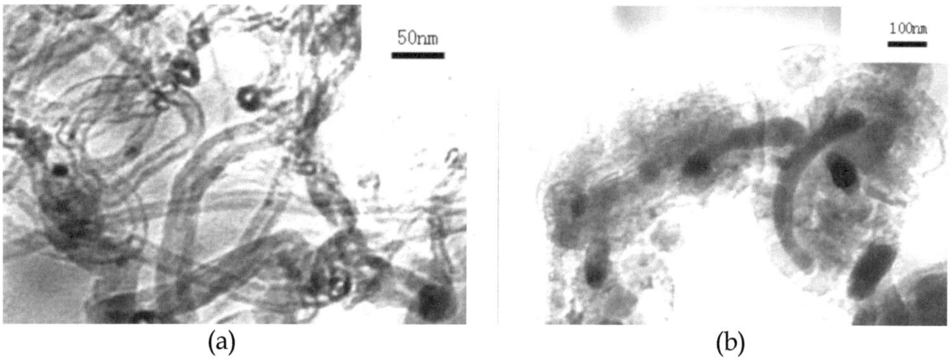

Fig. 5. Typical TEM images of (a) CNTs and (b) PAM-CNTs (Wrapping of PAM chains around CNTs shell)

3.4 Tribological and microtribological behavior of polymer-CNT nanocomposits

3.4.1 Effect of the PEG-CNT

The PB value represents the load-carrying capacity of the lubricant. The PB values of the two kinds of base stock described above containing different concentrations of PEG-CNTs are shown in Fig.6. With the increasing of the PEG-CNTs in the fluid, the load carrying capacity of the fluid is increased to a concentration of 0.5 wt.%, and then decreased. The WSD represents the anti-wear capacity of the lubricant. It is seen that the addition of the PEG-CNTs can decrease the WSD of the base stock. When the PEG-CNT content reach 0.5 wt.%, the WSD is minimum to 0.68 mm and the PB is maximum to 340 N. As a result, the obvious effects of PEG-CNTs added in the water base fluid are discovered. This infers that PEG-CNTs have pretty good load carrying and anti-wear performance in water fluid and its properties of lubrication is not proportional to its content. Since a higher concentration of PEG-CNTs makes the solution less mechanically stable, the decrease at 0.8 wt.% of PEG-CNT concentration could properly be due to the part of the PEG-CNT sediment in the tribological experiment at 1450 rpm rotating speed.

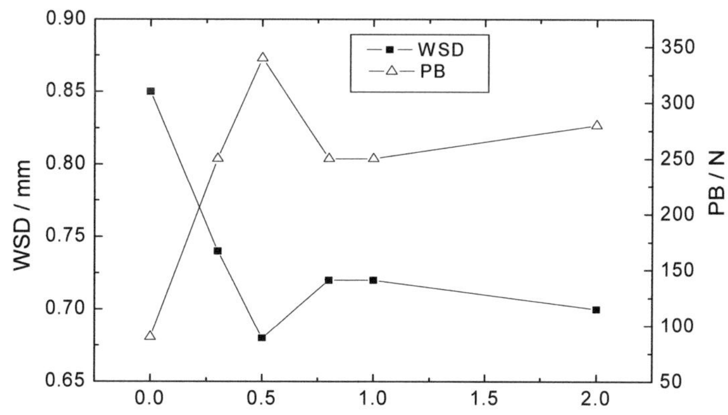

Fig. 6. Effect of the PEG-CNT content on maximum non-seized load (PB) and wear scar diameter (WSD, 30 min under a load of 200 N)

The dependence of WSD on load is shown in Fig.7. Under testing loads, the WSDs of the two kinds of base stock with 0.5 wt.% PEG-CNTs are smaller than those with 0.5 wt.% PEG, especially, at the load of 400 N, the difference in WSD between the PEG-CNTs and PEG is larger, which may be attributed to the lower load-carrying capacity of the PEG. At same time it means that the presence of CNTs can strengthen the wear resistance of the water base stock.

Dependence of WSD on rubbing time is shown in Fig.7. At the testing time, the WSD of the two kinds of base stock with 0.5 wt.% PEG-CNTs is smaller than that with 0.5 wt.% PEG. Especial after 15 min friction, the difference in WSD becomes larger, which indicates further that the presence of CNTs can strengthen the anti-wear performance of the base stock.

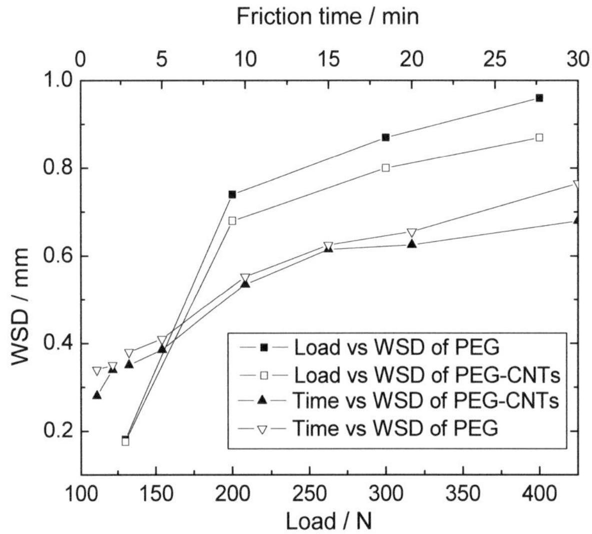

Fig. 7. Effect of load (time was 30min) and friction time (load was 200 N) on WSD (Base stock with 0.5 wt.% sample, 200 N)

In order to further study the effect of CNTs on wear and also to disclose the lubricating mechanism of CNTs, the worn surface in four-ball machine testing, which was obtained under a load of 200 N and a testing time of 30 min, was observed by optical microscopy. The wear scars of the base stock with 0.5 wt.% PEG-CNT nanocomposites and that with 0.5 wt.% PEG are shown in Fig. 8. They indicate that the wear scar obtained with the CNT nanocomposites additive is obviously smaller and exhibits mild scratches. Compared with the worn surfaces, it can be seen that the nanocomposite PEG-CNTs is relatively smoother than PEG. In other words, the PEG-CNT nanocomposites can improve microcosmic wear condition, Lei et al. have observed similar phenomenon in their fullerene-styrene sulfuric acid nanocomposite system (Lei et al., 2002). It is supposed that CNTs penetrate into the interface during friction process, and they have the possibility to cause microcosmic rolling effect between two rubbing surfaces as nanometer tiny tube. So the anti-wear abilities of the base fluid can be improved. Since CNTs have very high load-carrying capacity and the CNT nanocomposites is nanometer tiny structure, which can penetrate into rubbing surfaces and deposit there, it is reasonable to speculate that the polymer-CNTs maybe more effective than its corresponding homopolymer to support and isolate two relative motion surfaces, and therefore, the anti-wear performance of the base stock was improved.

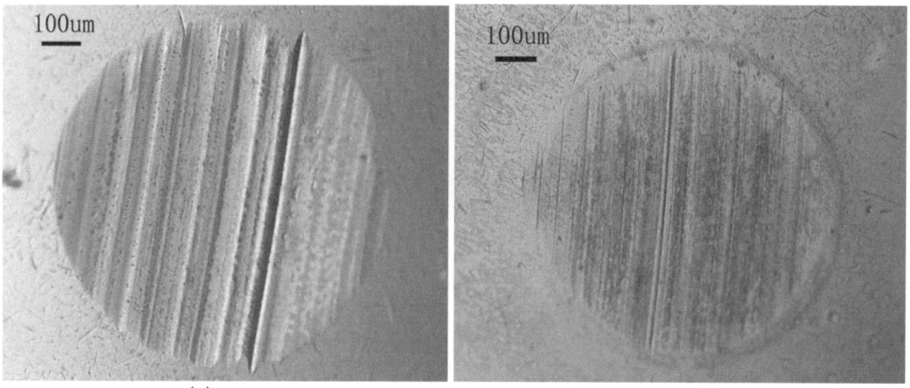

(a) (b)

Fig. 8. The morphology of worn surface (under load of 200 N and testing time of 30 min). (a) lubricated with 0.5 wt. % PEG; (b) lubricated with 0.5 wt. % PEG-CNTs

To study surface morphology and microtribological properties of thin film on a micro nanoscale, AFM/FFM is considered an excellent tool (Theoclitou et al., 1998). The surface roughnesses of PEG and PEG-CNT thin films on mica were visualized using AFM in tapping mode, shown in Fig.9. The data of RMS of PEG and PEG-CNTs are 2.244 and 12.865 nm respectively. The RMS revealed the both films are flat. The CNTs are dispersed well in the thin film and no apparent aggregation can be seen in the figures. Islands with diameter of 80-150 nm (Fig.10b) are distributed uniformly. And then it is observable that PEG film is rather flat and uniform compared to the PEG-CNT nanocomposite film. We presume that the intrinsic molecular chains are locally disordered in the CNTs-containing nanocomposites, and the existence of interpenetrating structure between the polymer phase and CNT phase.

We assume that the soft PEG chains as a matrix have reasonable anti-wear properties. Meanwhile the existence of rod and rigid segments of CNTs dispersed in thin film is expected to be beneficial for bearing the applied loads. The investigation of the relation between friction force and load could help the interpretation of AFM/FFM images at the molecular level. Fig.10 shows the overlap of two different friction forces versus load curves obtained from PEG-CNT and PEG films respectively. It is observed that both films exhibit stable and lower friction force signals below the 100 nN load due to polymer soft chains and flat surface. However, after load of 120 nN, the difference in friction force signals between the two films is larger, which may be attributed to the lower load carrying capacity of the PEG. The slopes of linear fit of PEG-CNTs and PEG in the friction signals versus load are 1.23 and 2.03 respectively, which represents the friction coefficient of the films. The friction coefficient decreased significantly as the CNTs addition. Obviously, the PEG-CNT nanocomposite film is better in bearing load and anti-wear than that of PEG film. The stability of the film can be attributed to the additional load-bearing ability afforded by CNTs chemically bonded on the nanocomposite chains.

The microtribological properties of the PEG-CNT films measured here are consistent with our perspective of CNTs-containing nanocomposites as potentially solid lubricant films, and that should encourage the idea that polymer-CNT films can be lower in friction and wear at the micro-scale.

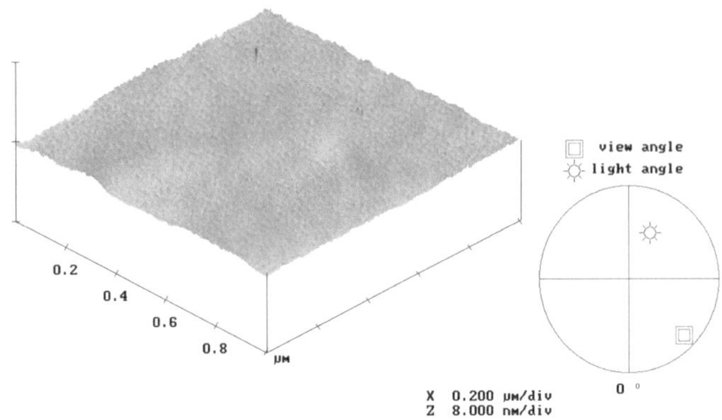

(a)

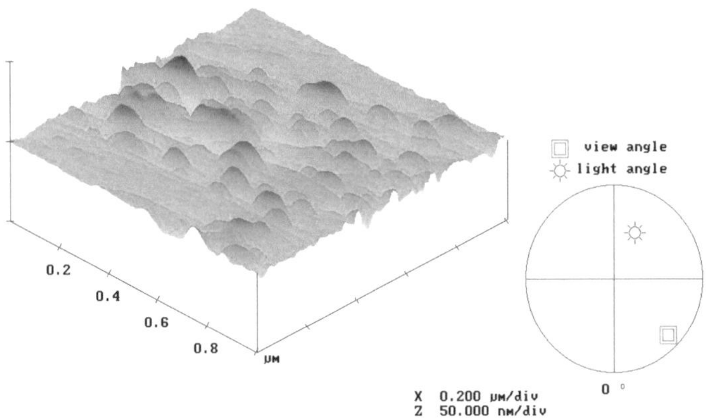

(b)

Fig. 9. Tapping mode AFM images of the spin casting films. Topographic picture of 1μm×1μm. (a) PEG film (b) PEG-CNT film

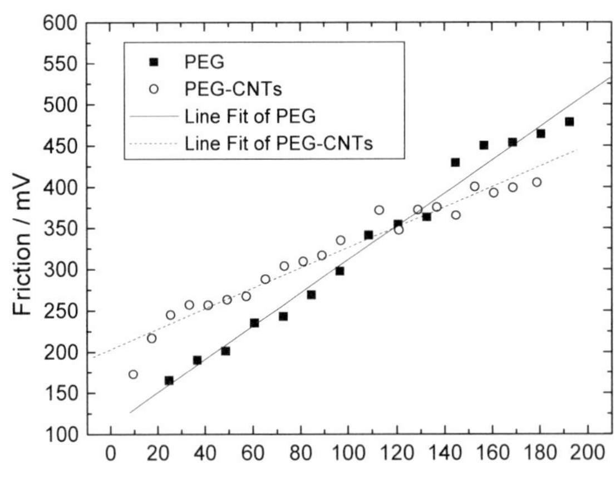

Fig. 10. Diagrams of frictional force signal versus load for PEG-CNT and PEG spin casting films

3.4.2 Effect of the PVA-CNT

The PB values of the two kinds of base stock described above containing different concentrations of PVA-CNTs are shown in Fig.11a. With the increasing of the PVA-CNTs in the fluid, the load carrying capacity of the fluid is increased to a concentration of 0.25 wt.%, and then decreased. The WSD represents the anti-wear capacity of the lubricant. The WSD data are given in Fig.11b. It is seen that the addition of the PVA-CNTs can decrease the WSD of the base stock. When the PVA-CNT content reach 0.25 wt.%, the WSD is minimum to 0.379 mm and the PB is maximum to 610 N. As a result, the obvious effects of PVA-CNTs added in the water base fluid are discovered. This infers that PVA-CNTs have pretty good load carrying and anti-wear performance in water fluid and its properties of lubrication is not proportional to its content.

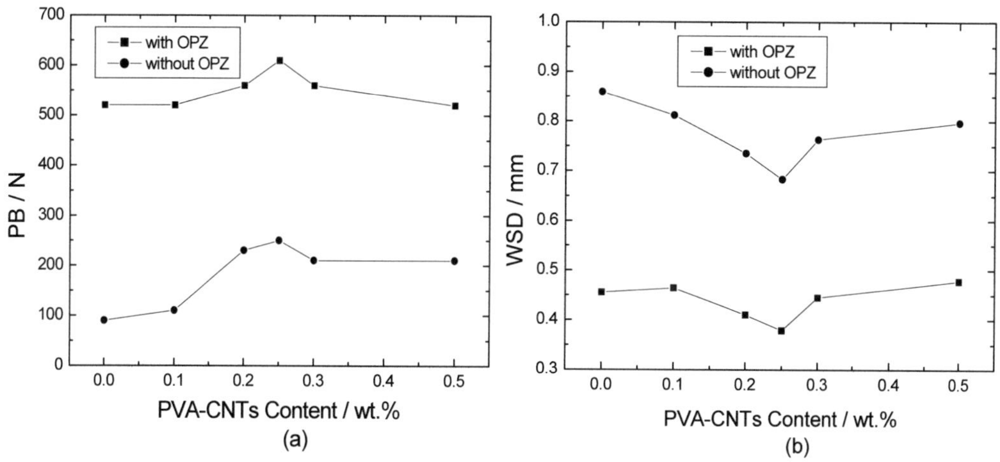

Fig. 11. Effect of the PVA-CNTs content on (a) maximum non-seized load (PB) and (b) wear scar diameter (WSD)

The dependence of WSD on load is shown in Fig.12. Under testing loads, the WSD of the two kinds of base stock with 0.25 wt.% PVA-CNTs are smaller than those with 0.25 wt.% PVA, especially, at the load of 400 N, the difference in WSD between the PVA-CNTs and PVA is larger. It means that the presence of CNTs can strengthen the wear resistance of the water base stock.

Dependence of WSD on friction time is shown in Fig.13. At the testing time, the WSD of the two kinds of base stock with 0.25 wt.% PVA-CNTs is smaller than that with 0.25 wt.% PVA. Especial after 15 min friction, the difference in WSD becomes larger, which indicates further that the presence of CNTs can strengthen the anti-wear performance of the base stock.

The worn surface of the base stock with 0.25 wt.% PVA-CNT copolymer and that with 0.25 wt.% PVA are shown in Fig.14 and 15, respectively. They indicat that the wear scar obtained with the CNT copolymer additive is obviously smaller and exhibits mild scratches. Compared with the corresponding partly enlarged micrographs of the worn surfaces, it can be seen from Fig.15 (b) that the worn surface of the copolymer PVA-CNTs is relatively smoother than that in Fig.14 (b). In other words, the PVA-CNT copolymer can improve microcosmic wear condition.

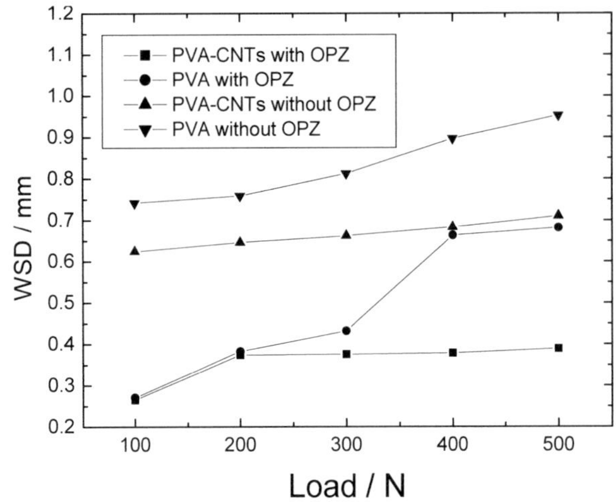

Fig. 12. Effect of load on wear scar diameter (Base stock with 0.25 wt.% sample)

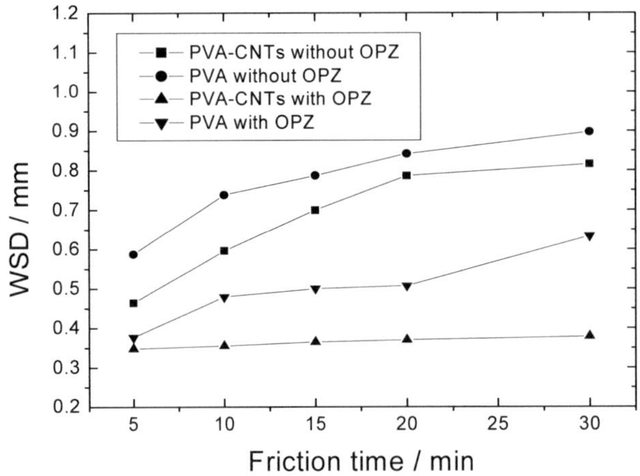

Fig. 13. Effect of friction time on wear scar diameter (Base stock with 0.25 wt.% sample)

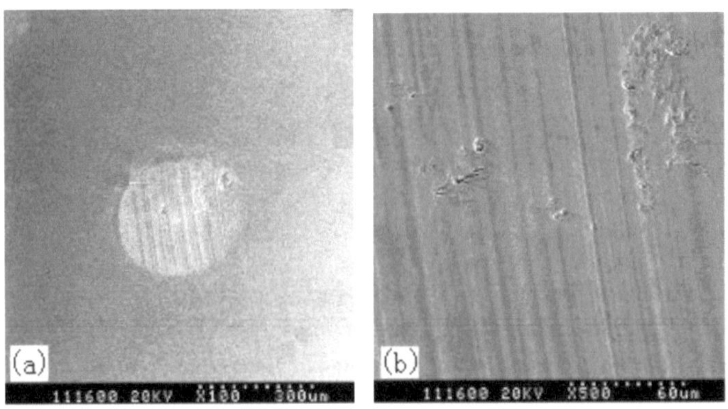

Fig. 14. SEM morphology of worn steel surface lubricated with the base stock (0.5 wt.% OPZ) containing 0.25 wt.% PVA (four-ball, 1450 rpm, 30 min, 400 N) (a) (×100) (b) (×500)

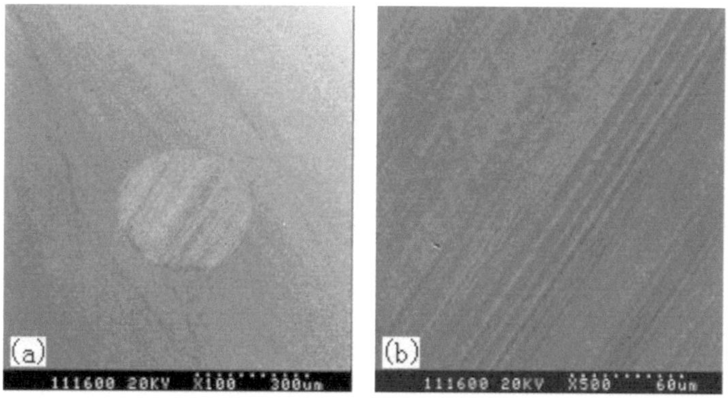

Fig. 15. SEM morphology of worn steel surface lubricated with the base stock (0.5 wt.% OPZ) containing 0.25 wt.% PVA-CNTs (four-ball, 1450 rpm, 30 min, 400 N) (a) (×100) (b) (×500)

The surface roughness of PVA and PVA-CNT composites thin films on micas was visualized by AFM in tapping mode, shown in Fig.16. The data of RMS of PVA and PVA-CNTs are 1.24 and 2.91 nm respectively, which revealed the PVA and PVA-CNT films are flat. The CNTs are dispersed uniform in PVA-CNT thin film and no apparent aggregation can be seen in the figures. And then the PVA film has a more flat and closely surface.

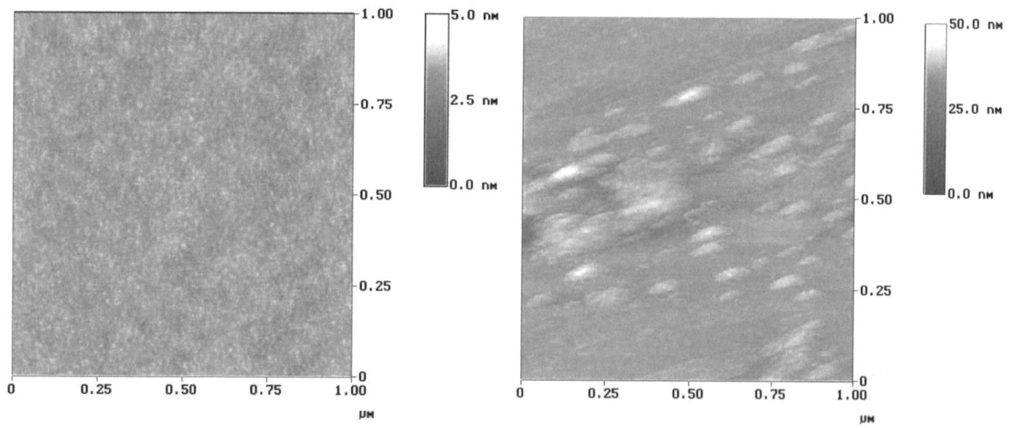

(a) (b)

Fig. 16. Tapping mode AFM images of (a) PVA and (b) PVA-CNTs spin casting films

The friction signals of PVA-CNT and PVA films could be measured at different applied loads for the same area in the FFM map. Fig.17 shows the overlap of two different friction signals versus load curves obtained from PVA-CNT and PVA films respectively. It is observed that both films exhibit stable and lower friction signals below the 110 nN load due to polymer soft chains and flat surface. However, after load of 130 nN, the difference in friction signals between the two films is larger, which may be attributed to the lower load carrying capacity of the PVA. The slopes of linear fit of PVA-CNTs and PVA in the friction signals versus load are 1.18 and 1.82 respectively, which represents the friction coefficient of the films. The friction coefficient decreased significantly as the CNTs addition. Obviously, the PVA-CNT film is better in bearing load and anti-wear than that of PVA.

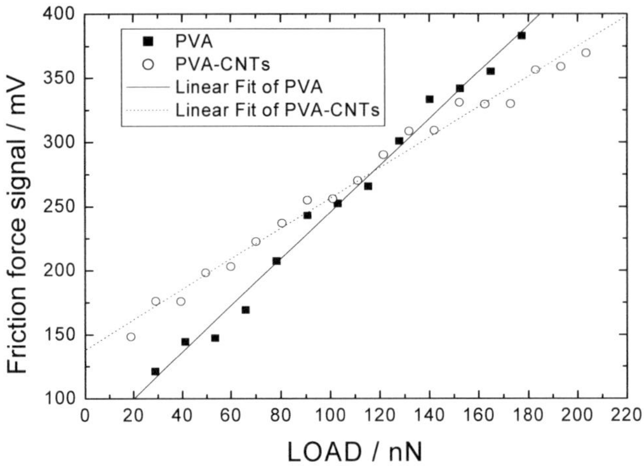

Fig. 17. Diagrams of frictional force signal versus load for PVA-CNTs and PVA spin casting films

3.4.3 Effect of the PAM-CNT

The surface roughnesses of PAM, PAM-CNT and PAM/CNT composites thin films on micas were visualized by AFM in tapping mode, shown in Fig.18. The data of RMS of PAM, PAM-CNTs and PAM/CNTs are 1.10, 2.18 and 12.44 nm respectively. The RMS revealed the PAM and PAM-CNT films are flat. The CNTs are dispersed uniform in PAM-CNT thin film and no apparent aggregation can be seen in the figures. And then the PAM film has a more flat and closely surface. In the films of physical blending samples of PAM/CNTs there are typical two separate phases and uneven surface. It also indicates that CNT surface modification disperses nanotubes uniform in polymer matrix, on the contrary, CNTs aggregation easily appears in the physical blending samples.

We assume that the soft PAM chains as a matrix have reasonable anti-wear properties. Meanwhile the existence of rigid rod-like segments of CNTs dispersed in thin films is expected to be beneficial to bearing the applied loads.

The friction signals of the blending samples films were unstable because of its uneven surface. So it was difficult to measure the microtribological properties of the films, which were coincident with their morphologies. The friction signals of PAM-CNT and PAM films could be measured at different applied loads for the same area in the FFM map. Fig.19 shows the overlap of two different friction signals versus load curves obtained from PAM-CNT and PAM films respectively. It is observed that both films exhibit stable and lower friction signals below the 130 nN load due to polymer soft chains and flat surface. However, after load of 140 nN, the difference in friction signals between the two films is larger, which may be attributed to the lower load carrying capacity of the PAM. The slopes of linear fit of PAM-CNTs and PAM in the friction signals versus load are 1.21 and 3.52 respectively, which represents the friction coefficient of the films. The friction coefficient decreased significantly as the CNTs addition. Obviously, the PAM-CNT film is better in bearing load and anti-wear than that of PAM. The stability of the film can be attributed to the additional load-bearing ability afforded by CNTs chemically bonded on the copolymer chains.

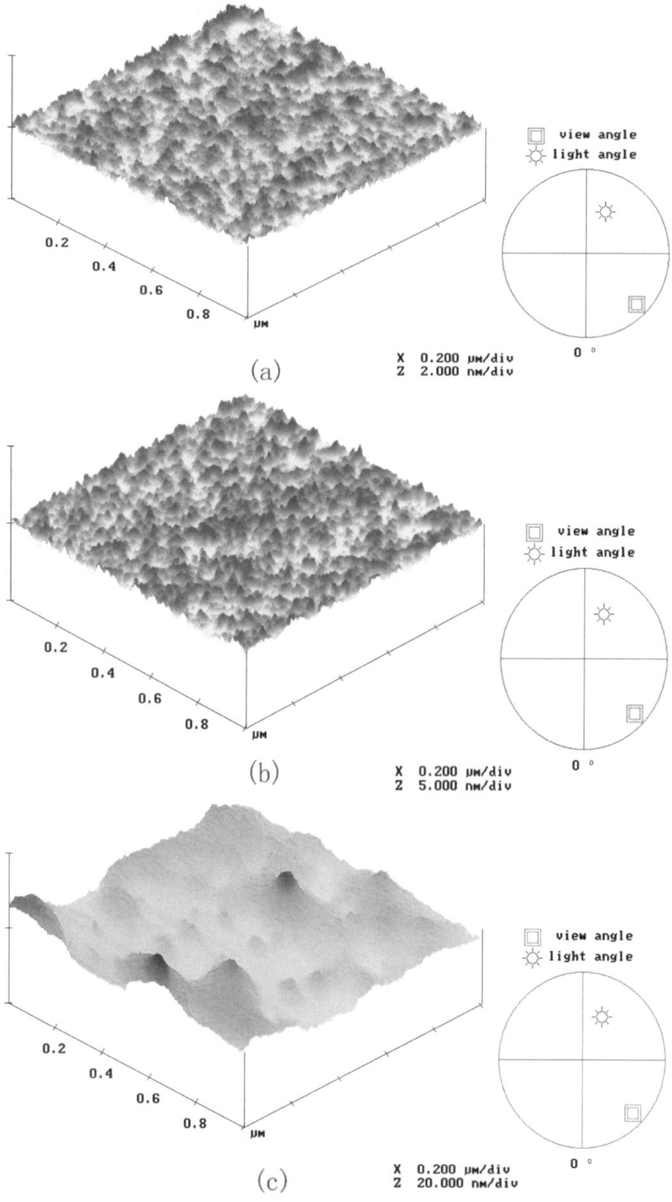

Fig. 18. Tapping mode AFM images of spin casting films. Three dimensional representation of topographic picture of 1μm×1μm. (a) PAM film (b) PAM-CNTs film (c) Physical blending sample of PAM/CNTs film

3.5 Mechanism of diazoresin-nanoparticle self-assembly films

As a cationic polyelectrolyte, DR can be easily deposited on the quartz glass surface. Then PAA-CNT deposits on the DR layer to form a DR and PAA-CNT bilayer on both sides of the quartz glass in each fabrication cycle. The absorbance of DR on quartz glass after each cyclic deposition was recorded via a UV-vis scanning spectrometer to monitor the self-assembly process (Fig. 20). The peak at 383 nm is assigned to the absorption of the diazonium group of DR and increase linearly with increasing bilayer number. It can be seen that the absorbance increases by ca. 0.03 every two bilayers indicating smooth step-by-step deposition.

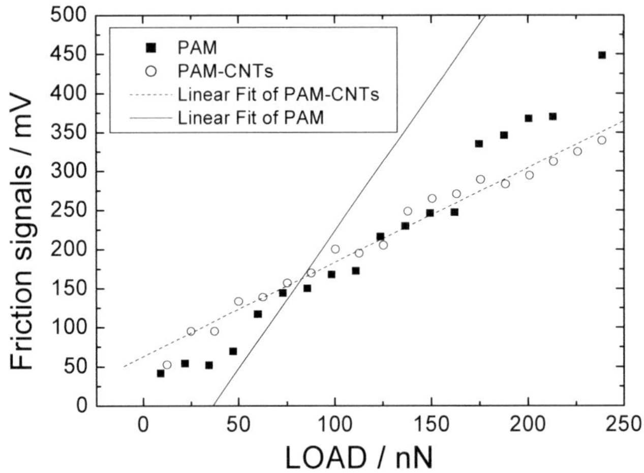

Fig. 19. Diagrams of friction signals versus load for PAM-CNTs and PAM thin films

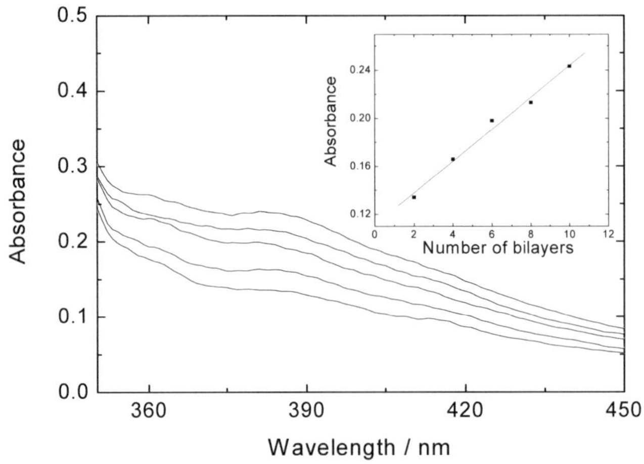

Fig. 20. The UV-vis spectra of the multilayer PAA-CNTs/DR films at 383 nm with difference numbers of bilayer, bilayer number (bottom to top): 1, 2, 3, 4, 5

The five bilayers film was then irradiated with UV light and the resulting absorbance determined. The UV-vis spectra before and after irradiation and decomposition of the diazonium groups of the films are shown in Fig. 21, respectively. The absorbance at 383 nm (diazonium group absorption) decreased with irradiation, which indicates that the ionic bonds of DR and PAA-CNTs convert partially to ester bonds. The diazonium group of the DR and PAA-CNT film decomposes easily under UV irradiation because it is sensitive towards UV light. Before irradiation, the multilayer film is formed via electrostatic attraction

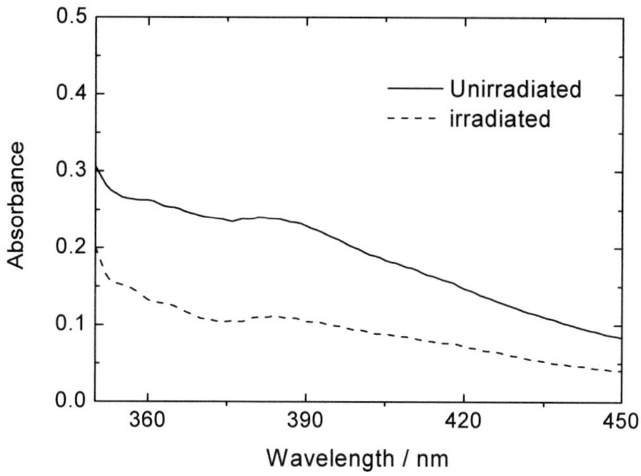

Fig. 21. UV-vis spectra of a 5-bilayer PAA-CNTs/DR film before and after UV irradiated (unirradiated and irradiated)

between the diazonium groups and carboxyl groups. Under UV irradiation, the diazonium group decomposes leading to phenyl cations, which combine with the carboxyl groups and produce covalent linkages. The conversion of the ionic bonds to covalent bonds is shown in scheme 2. This was confirmed by the FTIR studies elsewhere (Luo et al., 2001).

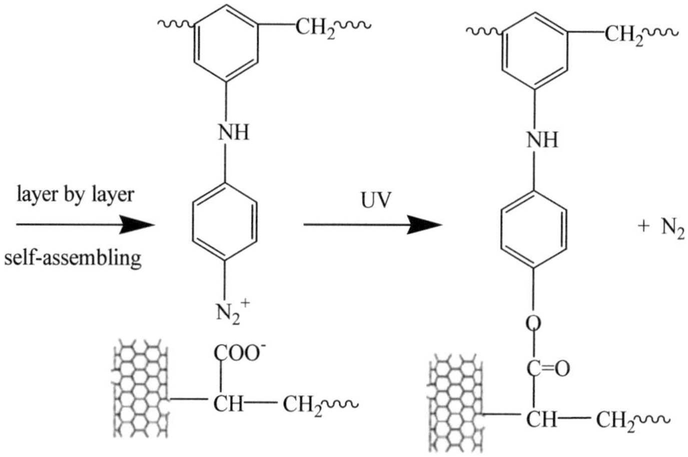

Scheme 2. The bond conversion taken place in PAA-CNTs / DR multilayer film

Under UV irradiation the films consisting of both cationic and anionic groups exhibit an extremely high tendency to aggregate by the ionic attractive force. The conversion of ionic bonds into covalent bonds makes the multilayer self-assembled films more stably packed. This is schematically represented in Scheme 3. The unirradiated films keep the original ionic structure and are much more unstable in polar solvents, such as dimethylformamide (DMF), and can be washed away. However, the irradiated films do not dissolve in DMF because of formation of the covalently crosslinked structure. The results show that the stability of the films towards polar solvents increases significantly after UV irradiation.

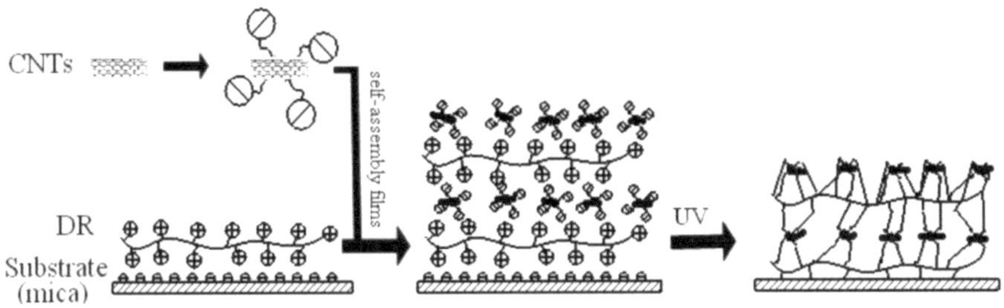

Scheme 3. Schematic diagram of the conversion of the linkage bonds from ionic to covalent in multilayer films of PAA-CNTs/DR after irradiation

Product of the TiO_2 sol was prepared by tetrabutyl titanate hydrolysis in acidic medium. TiO_2 particles were uniform and their sizes were about 50-70nm in Fig. 22. These sizes were consistent with the results of the particle size distribution analyzer (Fig. 23), and had the average diameters of 64 nm.

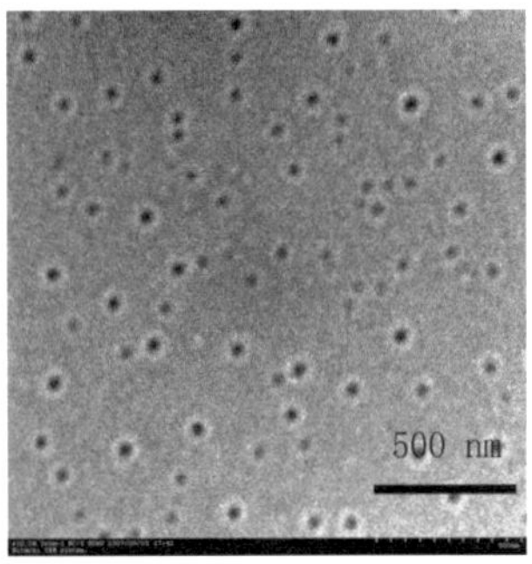

Fig. 22. TEM image of TiO_2

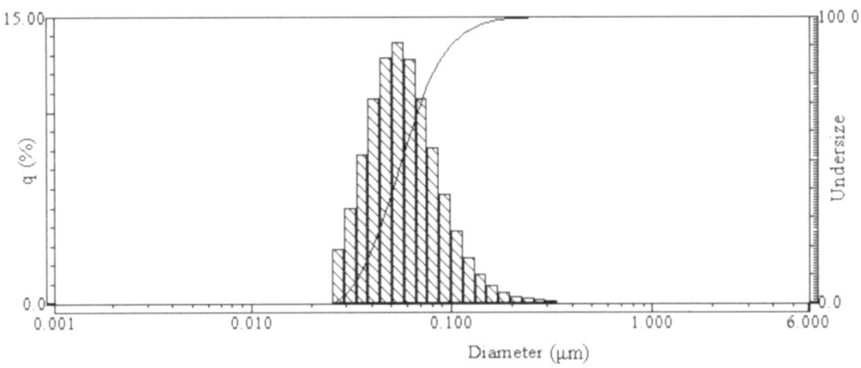

Fig. 23. Distribution of diameter of TiO_2 sol

PAA-TiO_2 and DR deposit on both sides of the quartz glass in each fabrication cycle. The absorbance of DR on quartz glass after each cyclic deposition was recorded via a UV-vis scanning spectrometer to monitor the self-assembly process (Fig. 24). It can be seen that the absorbance increases by ca. 0.03 every two bilayers indicating smooth step-by-step deposition.

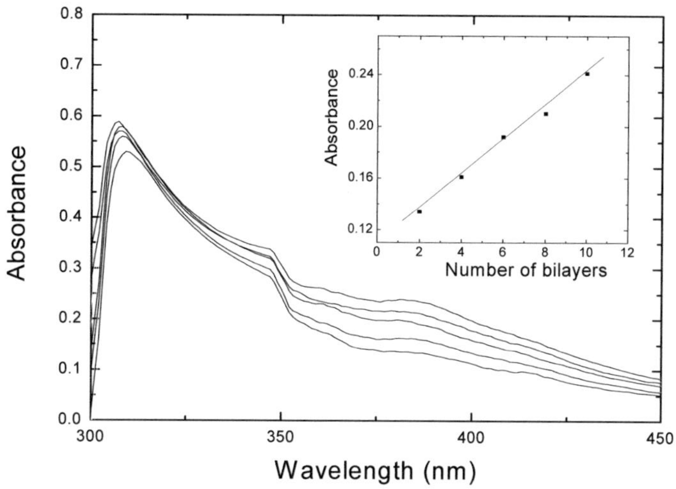

Fig. 24. The UV-vis spectra of the multilayer PAA-TiO_2/DR films at 383 nm with different numbers of bilayer. Bilayer number (bottom to top): 1, 2, 3, 4 and 5. Insert: Relationship between the absorbance at 383nm and the bilayer number

DR as a cationic polyelectrolyte, and PAA-$Fe(OH)_3$ solution as an anion polyelectrolyte deposits on the quartz glass in turn to form DR and PAA-$Fe(OH)_3$ bilayer films through electrostatic attraction. The absorbance of DR on the quartz glass after each cyclic deposition was recorded via a UV-vis scanning spectrometer to monitor the self-assembly process (Fig. 25). It can be seen that the peak at 383 nm which is assigned to the absorption of the diazonium group of DR increases linearly by ca. 0.06 every two bilayers. The linear relationship indicates smooth step-by-step deposition.

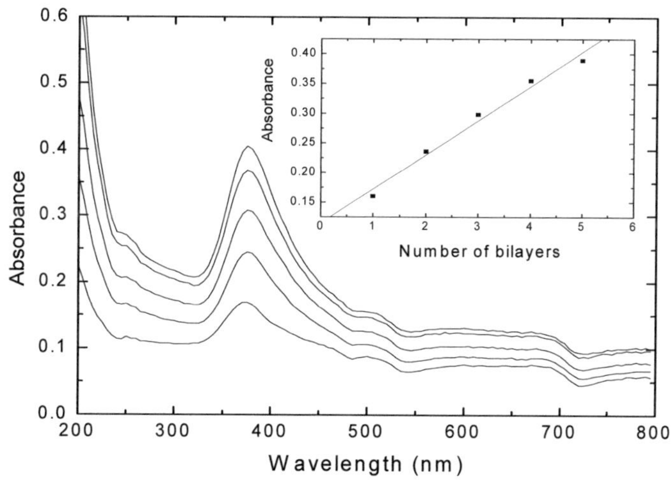

Fig. 25. UV-vis spectra of the multilayer PAA-$Fe(OH)_3$/DR films at 383 nm with difference numbers of bilayer. Bilayer number (bottom to top): 1, 2, 3, 4, 5

The UV-vis spectra of the 5-bilayer films before and after irradiation and decomposition of the diazonium groups of the films are exhibited in Fig. 26. The absorbance at 383 nm

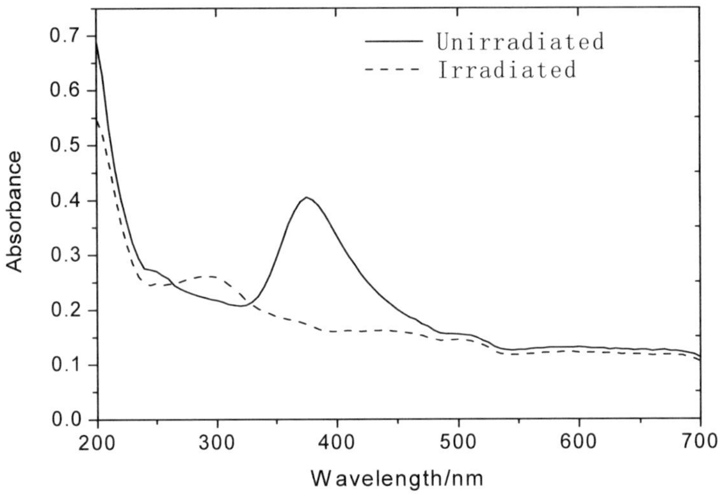

Fig. 26. UV-vis spectra of a 5-bilayer PAA-$Fe(OH)_3$/DR film before and after UV irradiated (unirradiated and irradiated)

(diazonium group absorption) decreased with irradiation, which indicates that the ionic bonds of DR and PAA-Nanoparticle converted partially to ester bonds.

3.6 Study of diazoresin-nanoparticle self-assembly films by AFM

To study surface morphology of thin films on a micro nano-scals, atomic force microscopy (AFM) is considered as an excellent tool (Sriram et al., 2009). The surface roughnesses of one layer DR film, five bilayers films of PAA-CNTs/DR and PAA-TiO_2/DR on micas were visualized using AFM in tapping mode, as shown in Fig. 27 (a), (b) and (c). The images reveal that the films are rather flat and uniform through the step-by-step technique. The CNTs and the particles of TiO_2 are dispersed well in the thin film and no apparent aggregation can be seen in Fig. 27.

AFM images with difference numbers of bilayer of PAA-$Fe(OH)_3$/DR self-assembly films (Fig.28) show surface morphology on micro-scales. The data of surface roughness of 1, 3 and 5 bilayer are 3.82, 5.47 and 8.73nm respectively. The images reveal that the films are rather flat and uniform through the step-by-step technique. The particles of $Fe(OH)_3$ are dispersed well in the thin film and no apparent aggregation can be seen.

The investigation of the relation between friction signals and load could help the interpretation of AFM/FFM images at the molecular level. A typical approach to characterize the friction properties of thin films is to investigate the friction force in the trace-retrace friction image. Fig.29 shows the friction coefficients of different number of bilayers were rather low which nearly equals to value of mica (0.0471). It is observed that the films exhibit stable and low friction signals due to polymer soft chains and flat surface. The stability of the films can be attributed to the additional load-bearing ability afforded by nanoparticles in the polymer composites. Therefore, the microtribological properties of PAA-$Fe(OH)_3$/DR self-assembly films measured here are consistent with our perspective of nanoparticles-containing polymer as potentially solid lubricant films.

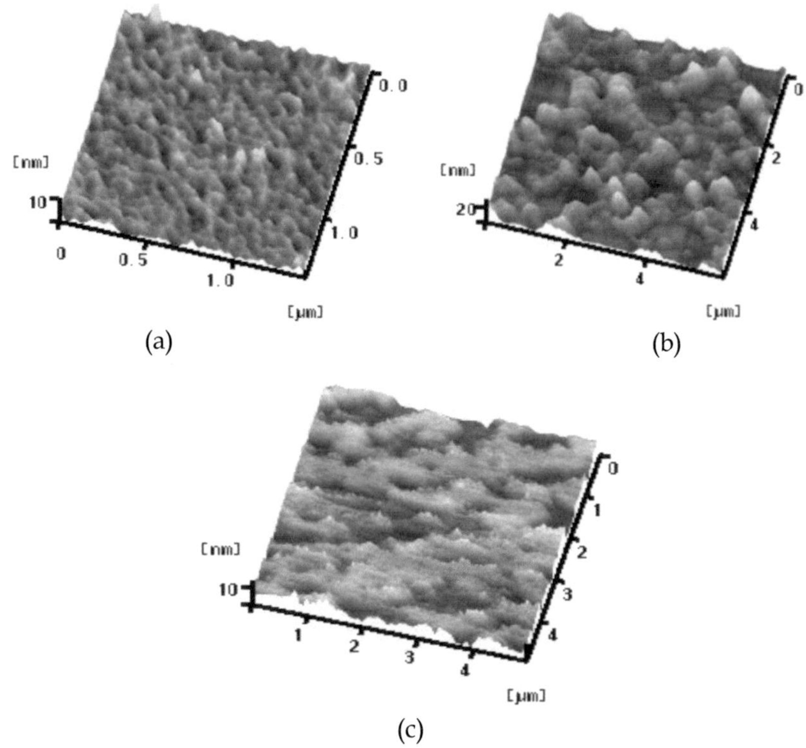

Fig. 27. AFM images of (a) 1 layer DR film, (b) 5-bilayer PAA-CNTs/DR and (c) PAA-TiO_2/DR film on mica surface

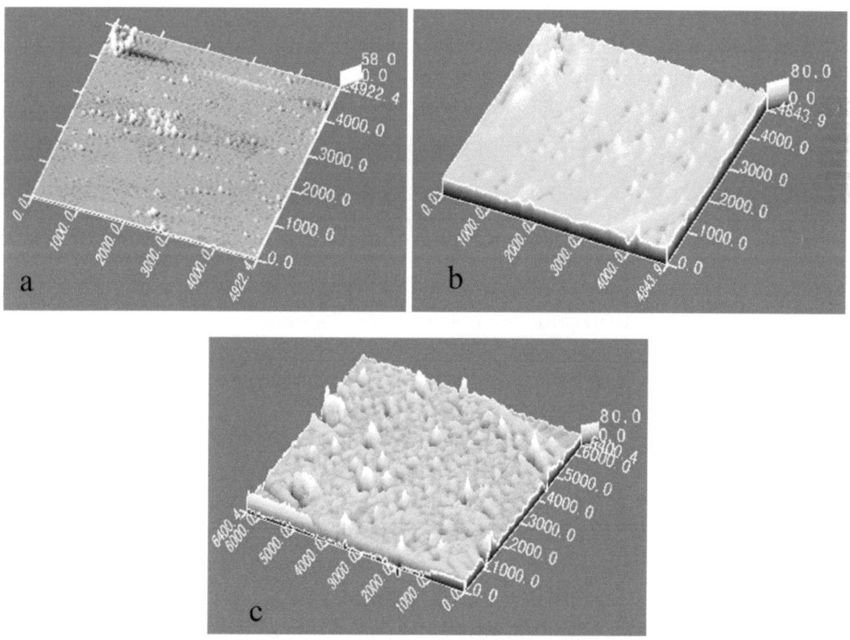

Fig. 28. AFM images of films with difference numbers of bilayer, bilayer number (a to c):1, 3, 5

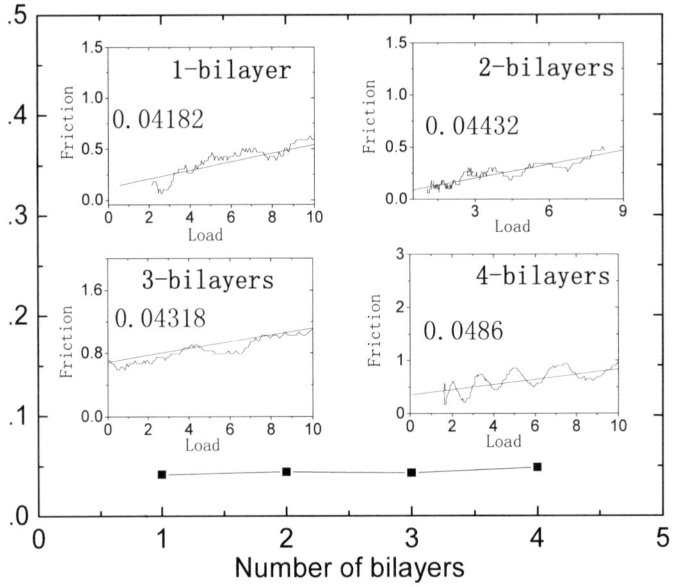

Fig. 29. Relationship between the friction coefficients and the bilayer number

4. Conclusion

We have succeeded in fabricated polymer-CNT nanocomposites. As a lubricant additive in water-based fluid, it can improve the wear resistance, load-carrying capacity and anti-wear ability of base stock. And then the lubrication mechanism of polymer-CNTs was deduced. The spin-casting nanocomposite thin films on mica were characterized using AFM and their height profiles were obtained. The films surface morphology is relatively flat. The difference of friction force on applied tiny load between polymer-CNT and polymer thin films are noted and discussed in terms of the role CNT segments. This supports the prediction that the load-bearing property of CNTs-containing nanocomposites thin films is enhanced.

We have succeeded in preparing PAA-nanoparticle polyelectrolytes. The multilayer thin films were fabricated on quartz glass with diazoresin as cationic polyelectrolyte and PAA-nanoparticle as anionic polyelectrolyte via the self-assembly technique, and then verified by UV-vis spectrum. The analysis of the surfaces of the self-assembly films indicate that the films are relatively flat, uniform and low friction signals.

5. Acknowledgment

The authors would like to acknowledge the support of Nature & Science Foundation of China (51143005) and Science Development Foundation of Department of Housing and Urban-Rural Development of Hubei ([2009]260).

6. References

Bai, T. and Cheng, X. (2006). Investigation of the Tribological Behavior of 3-Mercaptopropyl Trimethoxysilane Deposited on Silicon. *Wear,* 261(7-8): 730-737. ISSN: 0043-1648

Balani, K., Harimkar, S.P., Keshri, A., Chen, Y., Dahotre, N.B. and Agarwal, A. (2008). Multiscale Wear of Plasma-sprayed Carbon-nanotube-reinforced Aluminum Oxide Nanocomposite Coating. *Acta. Mater.*, 56: 5984-5994. ISSN: 1359-6454.

Bhushan, B., Israelachvili, J.N. and Landman, U. (2002). Nanotribology: Friction, Wear and Lubrication at the Atomic Scale. *Nature,* 374: 607-616. ISSN: 0028-0836.

Chen, J., Mamon, M.A., Hu, H., Chen, Y.S., Rao, A.M., Eklund, P.C. and Haddon, R.C. (1998). Solution Properties of Single-Walled Carbon Nanotubes. *Science,* 282: 95-98. ISSN: 0036-8075.

Cheng, X., Bai, T., Wu, J. and Wang, L. (2006). Characterization and Tribological Investigation of Self-assembled Lanthanum-Based Thin Films on Glass Substrates. *Wear,* 260(7-8): 745-750. ISSN: 0043-1648.

Chen, H.J., Jian, P.C., Chen, J.H., Wang L., and Chiu, W.Y. (2007) Nanosized-hybrid Colloids of Poly(acrylic acid)/Titania Prepared via In Situ Sol-Gel Reaction. *Ceram. Int.,* 33: 643-653. ISSN: 0272-8842.

Decher, G. (1997). Fuzzy Nanoassemblies: Toward Layered Polymeric Multicomposites. *Science,* 277: 1232-1237. ISSN: 0036-8075.

Duan, B.A. (1999). Study on Colloidal PSt-a New Type of Water-Based Lubrication Additive. *Wear,* 236: 235-239. ISSN: 0043-1648.

El-Hami, K. and Matsushige, K. (2003). Covering Single Walled Carbon Nanotubes by the Poly(VDF-co-TrFE) Copolymer. Chem. *Phys. Lett.,* 368: 168-177. ISSN: 0009-2614.

Eremenko, B.V., Bezuglaya, T.N., Savitskaya, A.N., Malysheva, M.L., Kozlov I.S. and Bogodist, L.G. (2001). Stability of Aqueous Dispersions of the Hydrated Titanium Dioxide Prepared by Titanium Tetrachloride Hydrolysis. *Colloid. J.,* 63(2): 173-178. ISSN: 1061-933X

Hiura, H., Ebbesen, T.W. and Tanigaki, K. (1995). Opening and Purification of Carbon Nanotubes in High Yields. *Adv. Mater.,* 7: 275-276. ISSN: 0935-9648.

Jia, Z., Wang, Z., Xu, C., Liang, J., Wei, B., Wu, D. and Zhu, S. (1999). Study on Poly(methyl methacrylate)/Carbon Nanotube Composites. *Mat. Sci. Eng.,* A271: 395-400. ISSN: 0921-5093.

Kim, S.H., Asay, D.B. and Dugger, M.T. (2007). Nanotribology and MEMS. *Nano today,* 2(5): 22-29. ISSN: 1748-0132.

Kwon, S.G., Piao, Y. and Park, J. (2007). Kinetics of Monodisperse Iron Oxide Nanocrystal Formation by "Heating-Up" Process. *J. Am. Chem. Soc.,* 129 (41): 12571-12584. ISSN: 0002-7863.

Li, J.W., Wang, C., Shang, G.Y., Xu, Q.M., Lin, Z., Guan, J.J. and Bai, C.L. (1999). Friction Coefficients Derived From Apparent Height variations in contact mode atomic force microscopy. *Langmuir,* 15: 7662-7669. ISSN: 0743-7463.

Lei, H., Guan, W. and Luo, J. (2002). Tribological Behavior of Fullerene – Styrene Sulfonic Acid Copolymer as Water-based Lubricant Additive. *Wear,* 252: 345-350. ISSN: 0043-1648.

Li, X.F., Guan, W., Yan, H. and Huang, L. (2004). Fabrication and Atomic Force Microscopy/Friction Force Microscopy (AFM/FFM) Studies of Polyacrylamide Carbon Nanotubes Copolymer Thin Films. *Materials Chemistry and Physics,* 88: 53-58. ISSN: 0254-0584

Li, X.F and Peng, S. (2009). Self-assembly Thin Films of Poly (acrylic acid)-Titanium Oxide. *The International Society for Optical Engineering,* 7493: 8-10. ISSN: 1022-6680

Li, X.F. and Peng, S. (2011). Poly(acrylic acid)-Ferric Hydroxide Photosensitive Self-assembly Film. *Materials Science Forum,* 663-665: 252-255. ISSN: 0255-5476

Li, X.F., Yan. H. and Peng, S. (2011). Tribological Behavior of Poly(ethylene Glycol) - Carbon Nanotubes. *Advanced Materials Research,* 217-218: 688-691.

Li, X.F, Yan, H. and Peng, S. (2011). Study of Poly(Acrylic Acid)–Carbon Nanotube Self-Assembly Films. *Designed Monomers and Polymers,* 14: 347-352. ISSN: 1385-772X

Liu, J., Rinzler, A.G., Dai, H.J., Hafner, J.H., Bradley, R.K., Boul, P.J., Lu, A., Iverson, T., Shelimov, K., Huffman, C.B., Rodriguez-Macias, F., Shon, Y.S., Lee, T.R., Colbert, D.T. and Smalley, R.E. (1998). Fullerene Pipes. *Science,* 280: 1253-1256. ISSN: 0036-8075.

Liufu, S.C., Xiao, H.N. and Li, Y.P. (2005). Adsorption of Poly(acrylic acid) onto the Surface of Titanium Dioxide and the Colloidal Stability of Aqueous Suspension. *J. Colloid. Interf. Sci.,* 281: 155-163. ISSN: 0021-9797.

Lu, C.G., Wei, F., Wu, N.Z., Zhao, X.S., Luo, C.Q. and Cao, W.X. (2004). Micropatterned Self-Assembled Film Based on Temperature-Responsive Poly (N-Isopropylacrylamide-co-Acrylic Acid), *J. Colloid. Interf. Sci.,* 277: 172-175. ISSN: 0021-9797.

Luo, H., Chen, J., Luo, G., Chen, Y. and Cao, W. (2001). Self-Assembly Films From Diazoresin and Carboxy-Containing Polyelectrolytes, *J. Mater. Chem.,* 11: 419-422. ISSN: 0959-9428.

Martínez-Martínez, D., Kolodziejczyk, L., Sánchez-López J.C. and Fernández, A. (2009). Tribological Carbon-Based Coatings: An AFM and LFM Study. *Surf. Sci.,* 603: 973-979. ISSN: 0039-6028.

Meng, H., Sui, G.X., Xie, G.Y. and Yang, R. (2009). Friction and Wear Behavior of Carbon Nanotubes Reinforced Polyamide 6 Composites Under Dry Sliding and Water Lubricated Condition. *Compos. Sci. Technol.,* 69: 606-611. ISSN: 0266-3538.

Medintz, I.L, Uyeda, H.T., Goldman, E.R. and Mattoussi, H. (2005). Quantum Dot Bioconjugates for Imaging, Labelling and Sensing. *Nat. Mater.,* 4: 435-436. ISSN: 1476-1122.

Pei, X., Hu, L., Liu, W. and Hao, J. (2008). Synthesis of Water-Soluble Carbon Nanotubes via Surface Initiated Redox Polymerization and Their Tribological Properties as Water-Based Lubricant Additive. *Eur. Polym. J.,* 44: 2458-2464. ISSN: 0014-3057.

Riggs, J.E., Guo, Z., Carroll, D. and Sun, Y.P. (2000). Strong Luminescence of Solubilized Carbon Nanotubes. *J. Am. Chem. Soc.,* 122: 5879-5880. ISSN: 0002-7863.

Shaffer, M.S.P. and Koziol, K. (2002). Polystyrene Grafted Multi-Walled Carbon Nanotubes, *Chem. Commun.,* 18: 2074-2075. ISSN: 1359-7345.

Sriram, S., Bhaskaran, M., Short, K.T., Matthews, G.I. and Holland, A. S. (2009). Thin Film Piezoelectric Response Characterisation Using Atomic Force Microscopy with Standard Contact Mode Imaging, *Micron,* 40: 109-113. ISSN: 0968-4328.

Theoclitou, M., Rayment, T. and Abell, C. (1998). *Nature,* 391: 566-568. ISSN: 0028-0836.

Tsang, S.C., Chen, Y.K., Harris, P.J.F. and Green, M.L.H. (1994). A Simple Chemical Method of Opening and Filling Carbon Nanotubes. *Nature,* 372: 159-162. ISSN: 0028-0836.

Vail, J.R., Burris, D.L. and Sawyer, W.G. (2009). Multifunctionality of Single-Walled Carbon Nanotube-Polytetrafluoroethylene Nanocomposites. *Wear,* 267: 619-624. ISSN: 0043-1648.

Yang, L.L., Yang, Z.H. and Cao, W.X. (2005).Stable Thin Films and Hollow Spheres Composing Chiral Polyaniline Composites. *J. Colloid. Interf. Sci.,* 292: 503-508. ISSN: 0021-9797.

Zhao, W.J., Zhu, M. Mo, Y.F. and Bai, M.W. (2009). Effect of Anion on Micro/nano-Tribological Properties of Ultra-thin Imidazolium Ionic Liquid Films on Silicon Wafer. Colloid. *Surface A.,* 332: 78-83. ISSN: 0927-7757.

8

Nanomechanical Evaluation of Ultrathin Lubricant Films on Magnetic Disks by Atomic Force Microscopy

Shojiro Miyake[1] and Mei Wang[2]
[1]Department of Innovative System Engineering, Nippon Institute of Technology, Saitama
[2]Department of Research and Development, OSG Corporation, Aichi
Japan

1. Introduction

With the development of information technology, marked progress has been made in hard disks. To realize high-density memory, it is necessary to decrease the flying height between the head and disk. However, there is the possibility of head-crash in high-density magnetic systems because the flying height of the most recent hard disks is less than 10 nm. Therefore, an extremely thin protective layer against wear and corrosion must now be applied to the magnetic head-disk interface (Miyake et al., 2002).

Sputtered carbon films with thicknesses as low as 10 nm are currently used as protective layers. Because the use of protective layers may lead to magnetic loss, their thickness must be decreased to 1–5 nm, corresponding to approximately ten layers of atoms to decrease the effective gap at the interface between the magnetic head and disk. However, the thinner the protective layer, the more difficult the maintenance of its tribological durability. If a magnetic protective layer with a graded composition could be formed, it would effectively improve the tribological characteristics of discontinuous protective films. Although sputtered carbonaceous films are currently employed as magnetic disk protective layers (Kaneko et al., 1990; Miyake, 1994), the possibility of realizing carbon nitride C_3N_4 (Cutinogco et al., 1996) has attracted interest, because the volumetric modulus of elasticity of C_3N_4 has been theoretically predicted to exceed that of diamond (Liu & Cohen, 1989). Amorphous nitrogen containing carbon films have been shown to have superior mechanical properties (Miyake et al., 1994, 1997).

Diamond-like carbon (DLC) films are currently employed as magnetic disk protective layers (Miyake et al., 1992, 1993; Miyake, 1994). A reduction in the effective clearance at the magnetic head-disk interface requires a reduction in the thickness of the protective film (Miyake, et al., 2005). That is, when considering a film of the above thickness, the corpuscular characteristics of atoms subjected to friction and wear should be taken into account. Nowadays, electron cyclotron resonance chemical vapor deposition (ECR-CVD) (Yamamoto et al., 2000) is used to deposit thin films, and filtered cathodic vacuum arc (FCVA) tetrahedral Amorphous Carbon (ta-C) (Hyodo et al., 2001; Robertson, 2008; Yamamoto et al., 2003) thin films are expected to be applied to magnetic disks.

On the other hand, the tribological properties of lubricant and protective DLC systems play a very important role in the reliability of hard disks. A number of studies have been carried out to clarify the characteristics of lubricant and DLC systems (Saitoh & Miyake, 2003; Saperstein & Lin, 1990; Tani, 1999). Previous works on clarifying the optimal quantities of lubricant on a hard disk surface by scanning probe microscopy (SPM) have employed the force modulation method (Saitoh & Miyake, 2003). On the surface of a hard disk, a low-lying part is defined as a valley and a hilltop part is defined as a hill. The hill and valley parts represent the surface roughness of the hard disk. The valley contains large quantities of lubricant, because tan δ for the valley is larger than that for the hill (Saitoh & Miyake, 2003).

Perfluoropolyether (PFPE) is employed as a lubricant in magnetic recording disk drives to improve tribological properties. An ultrathin PFPE lubricant film is particularly effective in decreasing the wear of the protective carbon layer due to sliding at the head-disk interface. The wear durability of the medium depends strongly on the retention and replenishment of the lubricant on the protective carbon surface. Both the retention and replenishment of lubricant are dependent on the interaction between the lubricant molecules and the carbon surface. A fundamental understanding of the lubricant-surface interactions in terms of their relationship to surface mobility is therefore required.

In this chapter, we firstly reported the method for evaluating the tribological properties of extremely thin protective nitrogen-containing carbon (C-N) films deposited on magnetic hard disks. The extremely thin C-N films with thicknesses of 3, 6 and 9 nm were prepared as a protective layer on CoCrTa magnetic disks by a complex treatment (Miyake et al., 2006). The bonding structure and composition of the films were studied by X-ray photoelectron spectroscopy (XPS). The nanohardness and elastic modulus of the films were measured by performing nanoindentation tests using atomic force microscopy (AFM). Wear tests were carried out to investigate the wear-resistance properties of the films.

Secondly, we reported the method for evaluating the nanometer-scale mechanical properties of extremely thin DLC films. The extremely thin DLC films were deposited by the FCVA and plasma chemical vapor deposition (p-CVD) methods (Miyake et al., 2009). Nanoindentation hardness and nanowear resistance were evaluated by AFM. The nanoindentation hardnesses of 100-nm-thick DLC films deposited by FCVA and p-CVD were 57 and 25 GPa, respectively. The nanowear test by AFM clarified the mechanical properties of the extremely thin DLC films.

Third, we reported the method for evaluating the microtribological properties of heat-treated hard disk evaluated by force modulation. The durability of PFPE lubricant and heat-treated perpendicular hard disks was evaluated. With regard to the heat treated perpendicular hard disks, their microtribological properties such as hardness, storage modulus, loss modulus and tan δ were evaluated by SPM. In a quasi-static nanoindentation hardness test, it was observed that nanoindentation hardness in the valleys was higher than that on the hills, and the hardness of the hard disk was increased by heat treatment. Consistent with the results of quasi-static nanoindentation in the dynamic nanoindentation test, the storage modulus and loss modulus were increased by heat treatment. Moreover, tan δ also increased as clarified by the evaluation of viscoelastic properties. On the other hand, regarding the microwear properties, the wear depth of the heat-treated disk and its wear volume were decreased by heat treatment, corresponding to the result of quasi-static nanoindentation.

2. Methods for evaluating nanomechanical properties of extremely thin films

2.1 Evaluation of nanoindentation hardness

To evaluate nanometer-scale deformation properties, the nanohardness and elastic modulus of all films were measured using a Hysitron Triboscope® nanomechanical testing system (Doerner et al., 1986; Farhat et al., 1997; Miyake, 2003). A total of ten indentations were performed for each sample and the mean values and standard deviation of the hardness and elastic modulus were computed; the results are provided later in the chapter. The diamond triangular pyramid (Berkovich-type) indenter that was used had an approximate tip radius of 50 nm and an indent angle of 120°. The hardnesses of the films were determined using the relation $H = P/A_{max}$, where A_{max} is the area of contact, $A_{max} = k \cdot h_p^2$ (the constant k for the Berkovich indenter was 24.5), h_p is the plastic penetration depth and P is the indenter load (Bhushan, 1995; Farhat et al., 1997; Miyake, 2003). The indentation was performed under a load of 100 μN. As shown in Fig. 1, from the load-displacement curves and the tangent of the unloading curves, the depth of plastic deformation can be evaluated. The deformation energy during indentation was analyzed. The total deformation energy including storage and dissipated energies was calculated as the integral of the loading curve. The storage energy was calculated as the integral of the unloading curve and the dissipated energy was evaluated as the total energy minus the storage energy (Farhat et al., 1997; Miyake, 2003). The modulus of dissipation was calculated as the dissipated energy divided by the total energy.

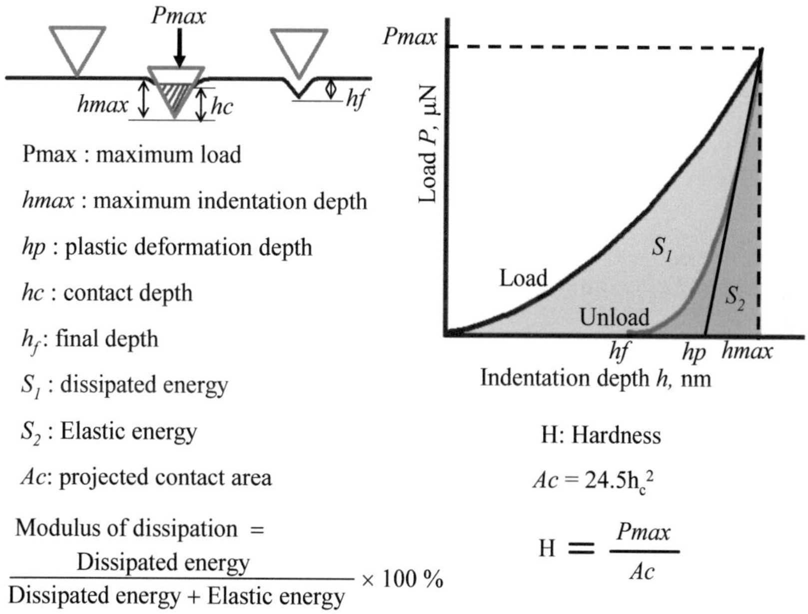

Fig. 1. Method of evaluating of nanoindentation hardness.

2.2 Evaluation of nanometer-scale wear resistance

The nanometer-scale wear resistance of the films was investigated by AFM using a diamond tip (Miyake et al., 1991). The dependence of the wear resistance of the C-N films on the number of scanning cycles, the dependence of the wear resistance of the DLC films on the film thickness, and the dependence of the wear volume of the PFPE lubricant on the applied load were evaluated, where the wear depths of the extremely thin CN films were evaluated by applying a load of 15 μN for 1, 5 and 10 scanning cycles, the nanowear depths of the 0.8-, 1- and 2-, 5- and 100-nm-thick ECR-CVD-DLC and FCVA-DLC films were evaluated by applying a load of 10 μN for 20 sliding cycles of tip scanning and the wear volumes of the PFPE lubricant films were evaluated by scanning a wear area of 500 × 500 nm^2 using a Berkovich type diamond tip with a radius of curvature of 200 nm under loads of 10-80 μN. As shown in Fig. 2, a triangular pyramid Berkovich diamond indenter was used with loads of 10 μN for the DLC films, 15 μN for the C-N films and 10-80 μN for the PFPE lubricant films in the wear tests, and the radius of the indenter tip was approximately 200 nm. During the experiment, the wear test area was 500 nm×500 nm and the observed areas were 1000 nm × 1000 nm and/or 2000 nm × 2000 nm, which were larger than the sliding area (Miyake et al., 1994, 2006).

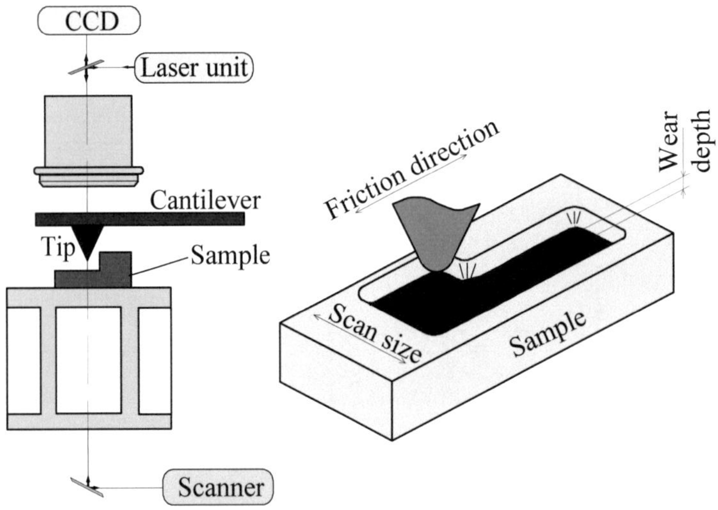

Fig. 2. Microwear test by atomic force microscopy (AFM).

2.3 Evaluation of viscoelastic properties

Viscoelastic properties were evaluated by the force modulation method of SPM as shown in Fig. 3 (Asif et al. 1999). This apparatus was added to nanoindentation system along with a lock-in amplifier. The Berkovich-type diamond indenter was vibrated in the vertical direction in the test. Phase lag and displacement were evaluated from the response of the tip indenter, which contained a transducer controller. Viscoelastic properties such as storage modulus, loss modulus and tan δ were analyzed by a computer. In the test the load was increased from 10 μN to 50 μN, the frequency was 300 Hz and the load amplitude was 5.0 μN. Table 1 shows the test conditions for the viscoelastic evaluation. To examine the existence of PFPE, test points were located on both a hill and a valley, as shown in Fig. 4. The hill and valley parts represent the surface roughness of a hard disk, the surfaces of hard disks were observed before the evaluation of viscoelastic properties to measure the amplitudes of the hill and valley regions. To determine the motion of PFPE on the wear test area, viscoelastic properties were evaluated after the microwear test.

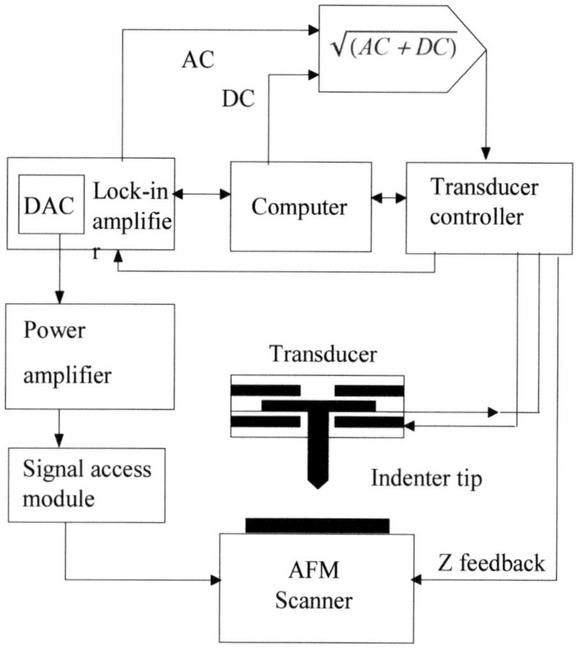

Fig. 3. Schematic illustration of nanoindentation system with lock-in amplifier.

Specimen	Heat treated and non-treated perpendicular type hard disk
Test temprature	20-25 centigrade degree
Humidity	30 -40%
Tip	Berkobitch diamond r = 200nm
Load	10 - 50 μN
Load amplitude	5.0 μN
Frequency	300 Hz

Table 1. The evaluation conditions for the viscoelastic properties.

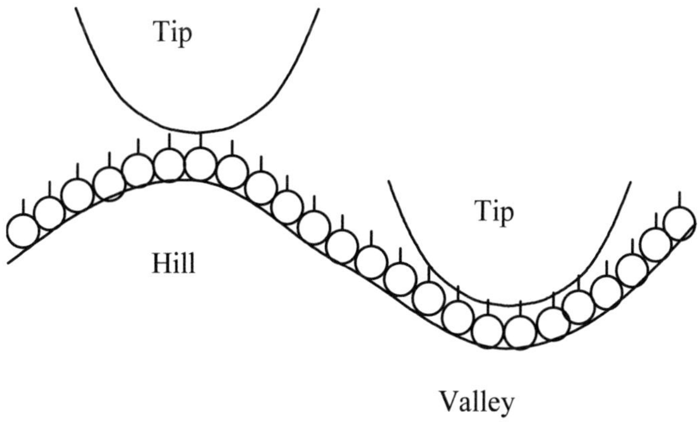

Fig. 4. Schematic illustration of evaluation of viscoelastic properties of hill and valley.

3. Nanoscale mechanical and tribological properties of extremely thin films

3.1 Mechanical properties of extremely thin C-N films

3.1.1 XPS evaluation of modified surface of C-N films

CoCrTa magnetic disks were used as substrates for the fabrication of extremely thin C-N films. The extremely thin C-N films were deposited on the CoCrTa magnetic disks in the sputtering chamber of radio frequency (RF) reactive sputtering equipment, which was able to supply an RF power of 13.56 MHz to both a target and a substrate. Two sets of C-N thin films were prepared, one set by RF reactive sputtering and the other set by a complex treatment (Miyake et al., 2006). The complex treatment, which involves plasma irradiation (plasma-irradiated) and the deposition of C-N film, where the substrate was pretreated by plasma irradiation and then treated by the deposition of a C–N film by RF reactive sputtering (Miyake et al., 2006) was carried out to form complex C–N films. The depositions of C-N and complex C-N films were performed by supplying N_2 or N_2+He (helium) under a pressure of 7×10^{-2} Torr and RF electric powers of 25 W on the substrate side and 300 W on the target side to deposit a nitrogen-containing carbon film with high wear resistance as a protective film (Cutinogco et al., 1996; Liu & Cohe, 1990; Miyake et al., 1994; White et al, 1996). The thickness of the C-N films was determined by controlling the deposition time. The addition of He to the plasma atmosphere was to enhance the excitation process (Miyake & Wang, 2004; Sugimoto & Miyake, 1989).

Both XPS and Auger electron spectroscopy (AES) were used to examine the elemental composition and bonding of the film. Figure 5 shows XPS spectra of carbon (C) and C-N films. A binding energy (BE) of 284.6 eV for the C 1s photoelectron peak in the XPS spectrum of the carbon film was obtained (Fig. 5(a)). A conspicuous shoulder in the XPS spectra on the high-energy side was observed in the XPS measurement of the C-N film (Fig. 5(b)), indicating the existence of C-N bonds (Miyake et al., 2006).

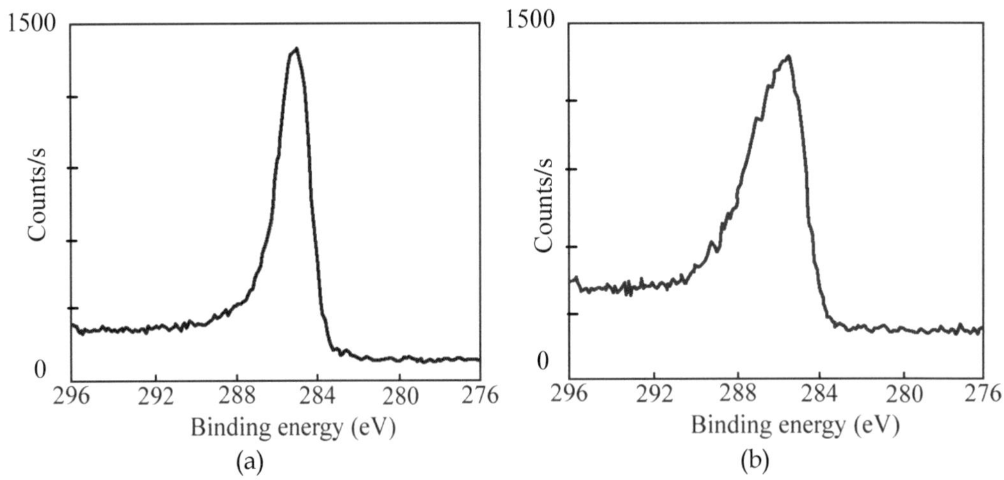

(a) XPS of carbon film
(b) XPS of C-N film deposited with complex treatment

Fig. 5. XPS spectra of carbon film and C-N film deposited by complex treatment.

3.1.2 Nanoindentation hardness of C-N films

The indentation curves for the untreated sample and the C-N and complex C-N films deposited in an atmosphere of N_2 or N_2+He were obtained at a maximum applied load of 100 μN. In these tests, nanoindentation was performed perpendicular to the plane of the coatings and each curve represents an average of ten measurements taken for each specimen. As shown in Figs. 6(a) and 6(b), the maximum depths (h_m) of the 3-, 6- and 9-nm-thick complex C-N films deposited in N_2+He were 5.2, 5.3 and 5.0 nm with residual depths of 0.1, 0.2 and 0.1 nm, respectively. In the case of 3-, 6- and 9-nm-thick C-N films, the maximum depths were 8.5, 8.1 and 7.1 nm and the residual depths were 3.8, 2.9 and 3.3 nm, respectively, indicating that more plastic deformation occurred in these films.

The hardness of the C-N and complex C-N films at the maximum applied load of 100 μN is plotted in Figs. 7(a) and 7(b). The hardness of the complex C-N films is greater than that of the C-N films. Moreover, the complex C-N films deposited in N_2+He exhibit the greatest hardness. The average hardness values of the complex 3-, 6- and 9-nm-thick C-N films deposited in N_2 are 62.6, 31.9 and 34.1 GPa, whereas those of the 3-, 6- and 9-nm-thick complex C-N films deposited in N_2+He were 67.8, 63.8 and 63.8 GPa, respectively.

Figures 8(a) and 8(b) show the deformation energies of the C-N and complex C-N films. The deformation energies of the films deposited in N_2+He were lower than those of the films deposited in N_2. Furthermore, the dissipated energy of the complex C-N films significantly decreased with decreasing film thickness. It is considered that the elastic-plastic property of the disks was improved by the complex treatment involving He addition, which explains why the residual depth of the 3-nm-thick complex film was very small, the surface was restored to the previous position to its previous position after unloading.

Figure 9 shows the modulus of dissipation as a function of nanoindentation hardness. As shown in Fig. 9(a), the modulus of dissipation of the untreated sample was 90.7% with a hardness of 15.6 GPa. The moduli of dissipation of the 3-, 6- and 9-nm-thick complex C-N films deposited in N_2 were 25.0%, 39.9% and 52.8%, with hardnesses of 62.6, 31.9 and 34.1 GPa, respectively. However, as shown in Fig. 9(b), the moduli of dissipation of the 3-, 6- and 9-nm-thick complex C-N films deposited in N_2 + He were only 0%, 8.3% and 9% with hardnesses of 67.8, 63.8 and 63.8 GPa, respectively. The 3-nm-thick complex C-N film had a small modulus of dissipation and a high hardness compared with the other films.

The method most widely used to determine the hardness of materials is the quasi-static indentation method. Since indentation hardness is essentially a measure of the plastic deformation properties of materials and depends only to a secondary extent on their elastic properties, the elastic-plastic deformation behavior of the films during indentation can be observed to study the nanomechanical properties of a topmost coating as thin as the monolayer of a substrate. In this study we considered the measured indentation hardness as a composite hardness that is affected by the type of substrate used. Although this composite hardness is derived from the impact with the substrate and cannot be regarded as an absolute hardness value for films, it can be used to investigate the plastic deformation of a topmost coating and a substrate by measuring comparative composite hardness values. For this reason, we analyzed the elastic-plastic deformation energies of the films to further understand their deformation properties. Although at higher indentation depths, the composite hardness of a film and a substrate changes with indentation depth, it is thought

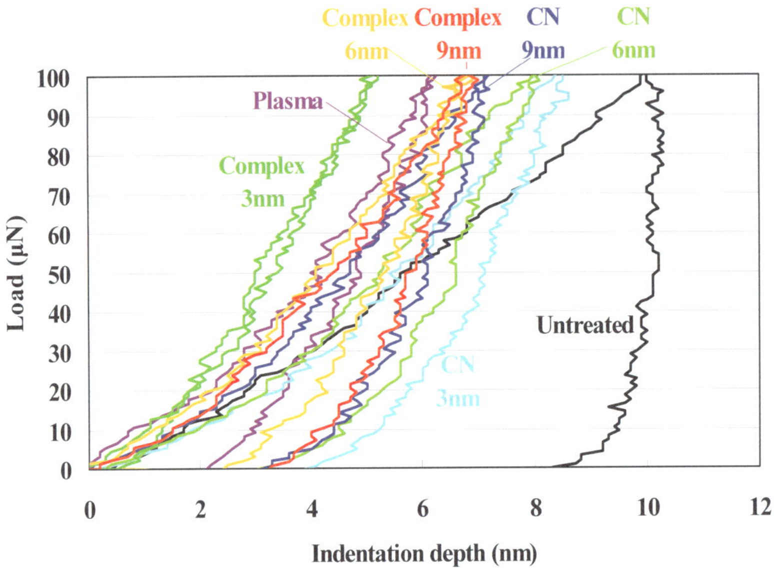

(a) Nanoindentation curves of C-N films deposited in N_2

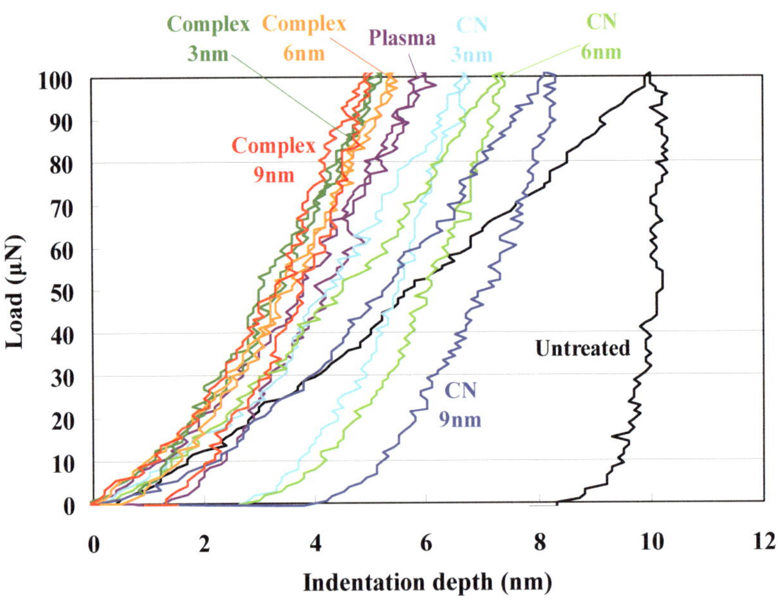

(b) Nanoindentation curves of C-N films deposited in N_2+He

Fig. 6. Load and displacement curves of untreated, plasma-irradiated, C-N and complex C-N films deposited in N_2 and N_2+He at 100 μN peak load.

that composite hardness is mainly determined by the topmost film on the substrate. Therefore, the results of the indentation tests also show that the complex C-N films had higher indentation hardness than the other films as well as good elastic-plastic deformation properties.

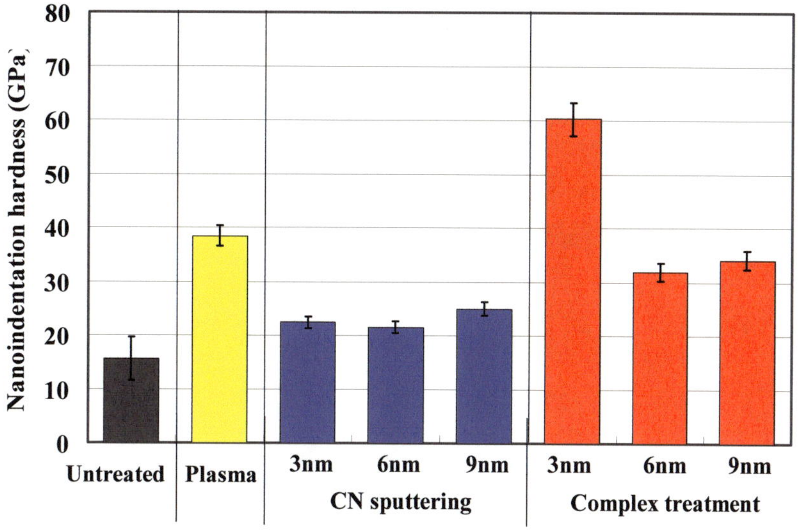

(a) Nanoindentation hardness of untreated and C-N films deposited in N_2

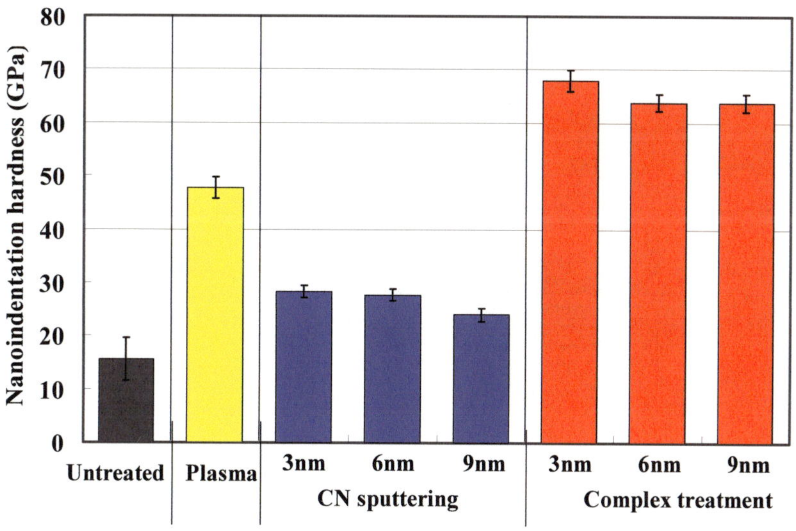

(b) Nanoindentation hardness of untreated and C-N films deposited in N_2+He

Fig. 7. Nanoindentation hardness of untreated, plasma-irradiated, C-N and complex C-N films deposited in N_2 and N_2+He.

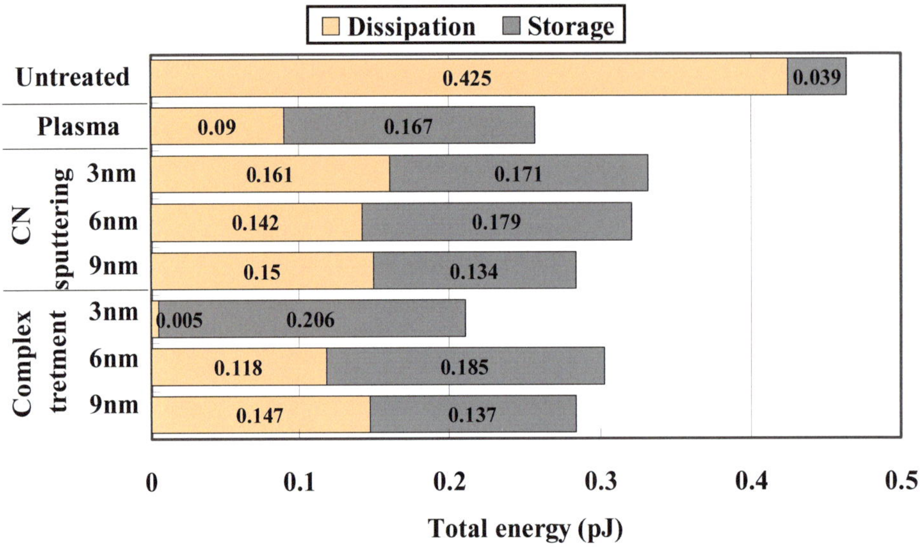

(a) C-N films deposited in N_2

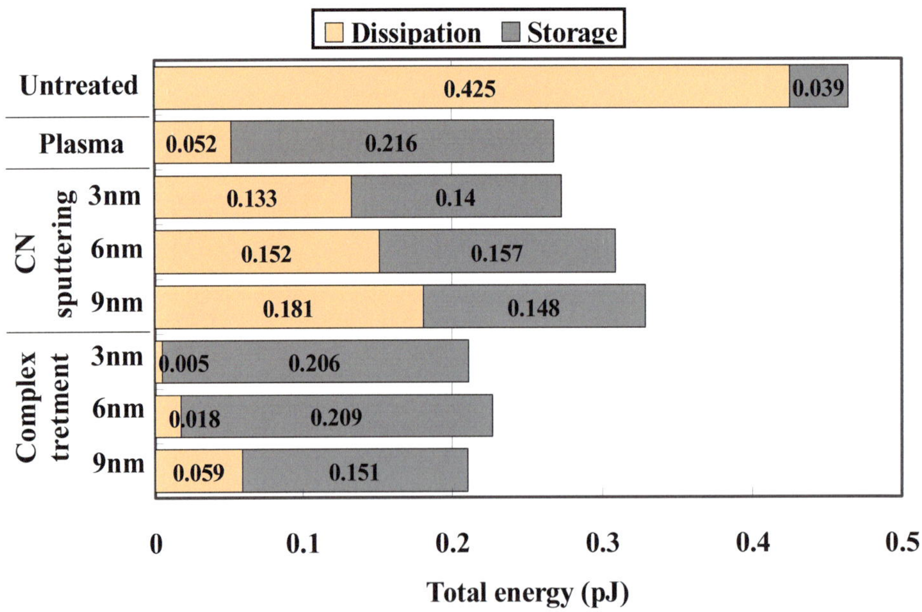

(b) C-N films deposited in N_2 and N_2+He.

Fig. 8. Measured dissipation and storage energies of untreated, plasma-irradiated, C-N and complex C-N films deposited in N_2 and N_2+He.

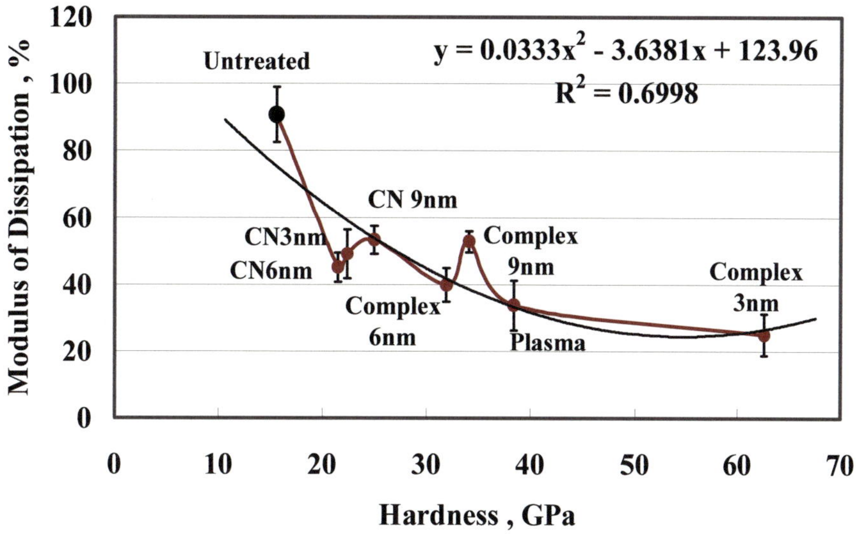

(a) C-N films deposited in N_2

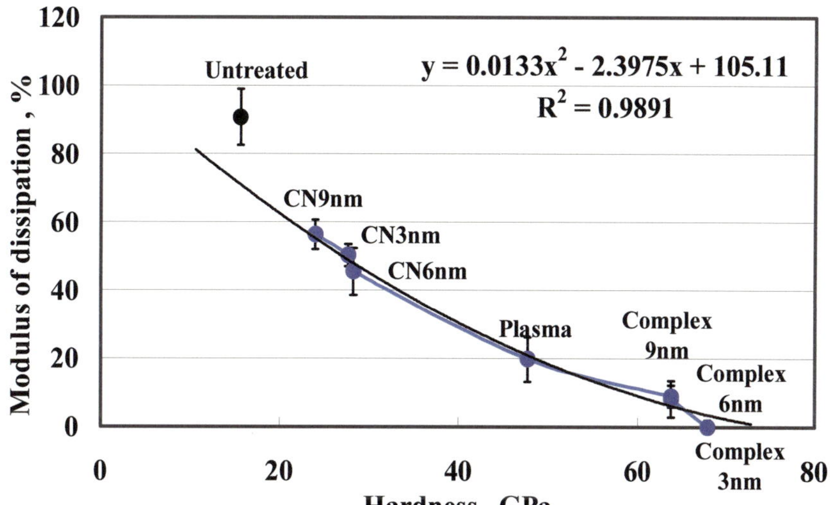

(b) C-N films deposited in N_2 and N_2+He.

Fig. 9. Modulus of dissipation vs indentation hardness of untreated, plasma-irradiated, C-N and complex C-N films deposited in N_2 and N_2+He.

3.1.3 Microwear test for C-N films

Figures 10 and 11 show the wear depth dependence of the films on the number of scanning cycles. Wear tests were performed by applying a load of 15 µN for 1, 5 and 10 scanning cycles. The results indicate that the wear resistance of the disks was improved by the C-N and complex C-N film coatings. In addition, the wear resistance of the films deposited in N_2+He was greater than that of the films deposited in N_2. The wear of the 3-nm-thick complex C-N film deposited in N_2+He was negligible compared with that of the other films.

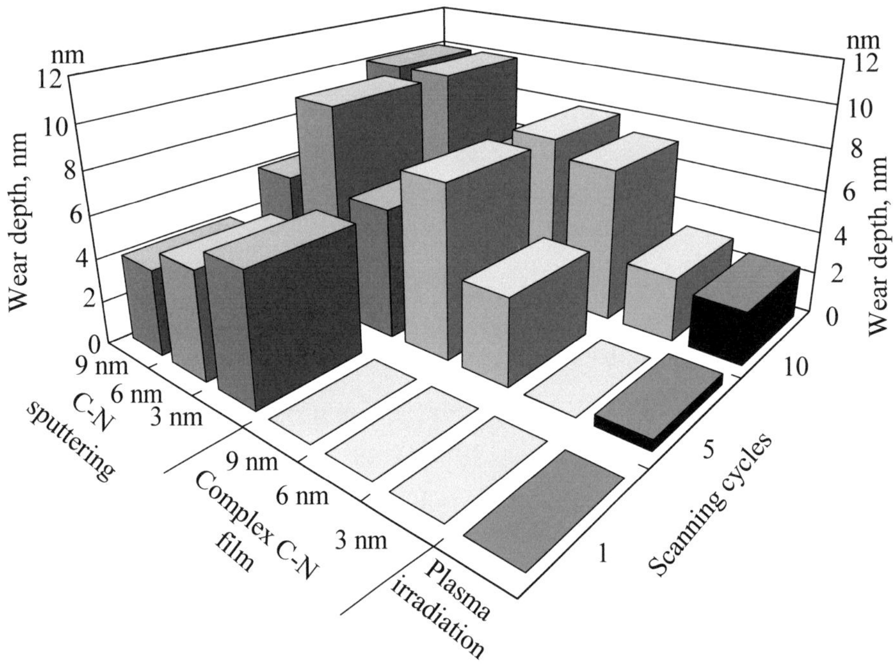

Fig. 10. Microwear dependence on number of scanning cycles for plasma-irradiated, C-N and complex C-N films deposited in N_2.

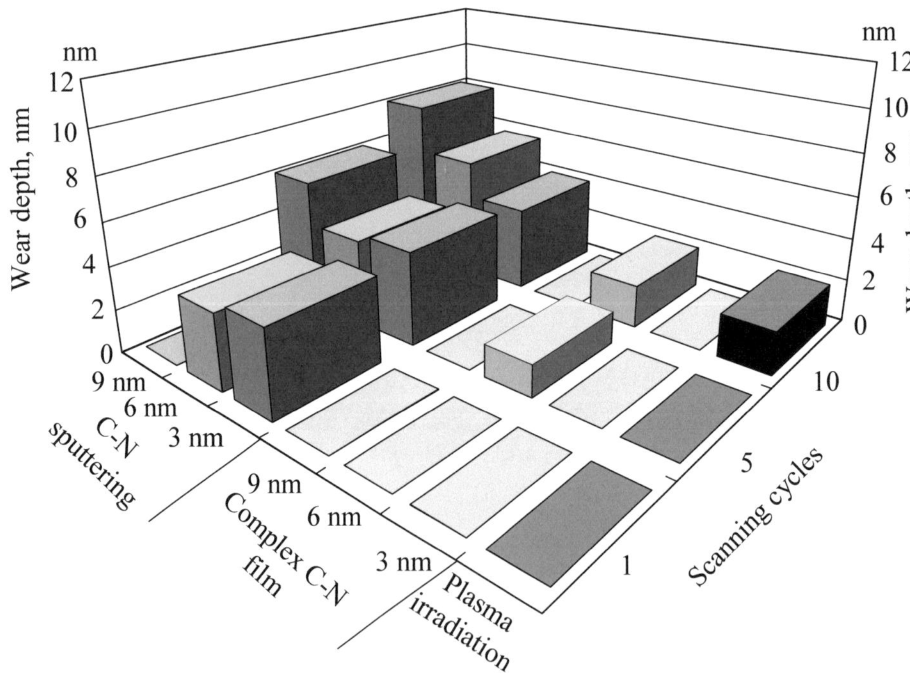

Fig. 11. Microwear dependence on number of scanning cycles for plasma-irradiated, C-N and complex C-N films deposited in N_2+He.

Figures 12 and 13 show cross-sectional profiles of the wear marks generated on the plasma-irradiated substrate and the substrates coated with 3-nm-thick C-N and complex C-N films. The normal force used for the image traces was 2 nN and the load used for the wear test was 10 μN. The results show that wear was 2.5 nm deep for the plasma-irradiated sample, 6.0 nm for the C-N film and 3.0 nm for the complex C-N film deposited in N_2. In the case of the films deposited in N_2+He, the wear was significantly lower, the wear was 2.0 nm deep for the plasma-irradiated sample, 4 nm for the C-N film and 0 nm for the complex C-N film. In addition to high indentation hardness, the C-N film exhibited excellent wear resistance owing to the plasma irradiation treatment in the presence of He. It is considered that the extremely thin C-N film had superior wear resistance because of its graded composition.

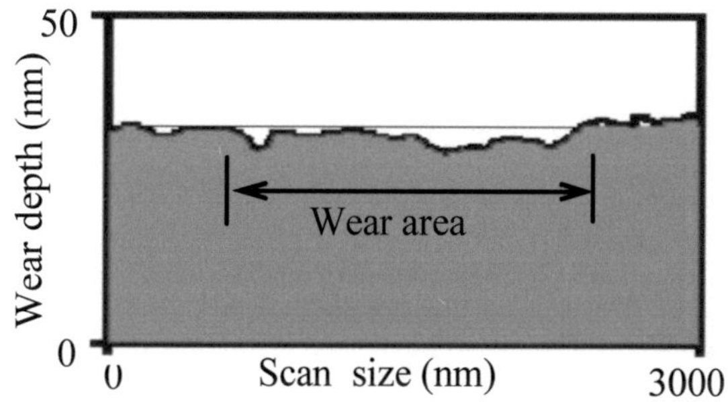

(a) Nitrogen plasma irradiation (wear depth 2.5 nm)

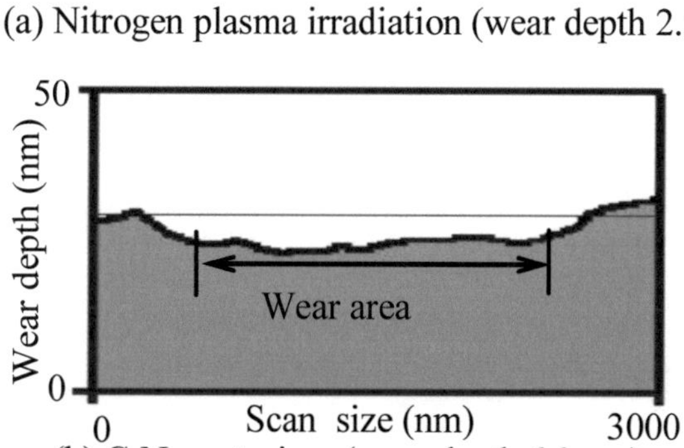

(b) C-N sputtering (wear depth 6.0 nm)

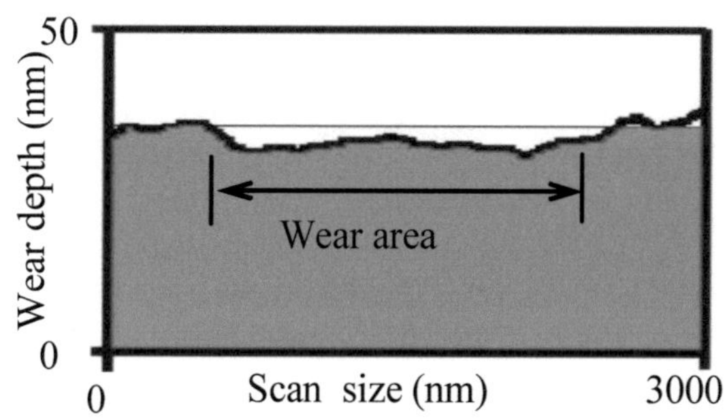

(c) Complex C-N film (wear depth 3.0 nm)

Fig. 12. Cross-sectional profiles of microwear of plasma-irradiated, 3 nm C-N and complex C-N films deposited in N_2.

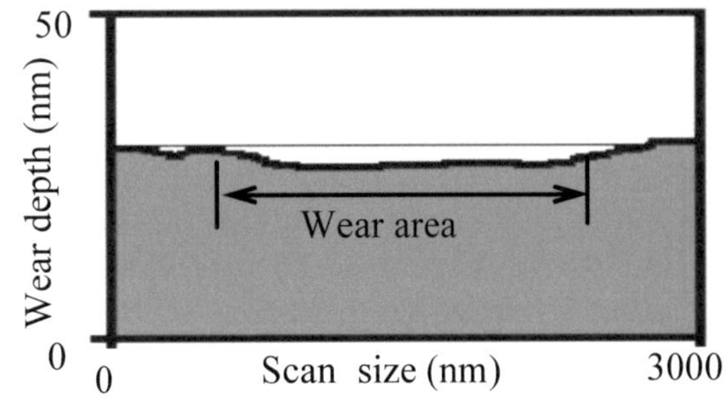

(a) Nitrogen plasma irradiation (wear depth 2.0 nm)

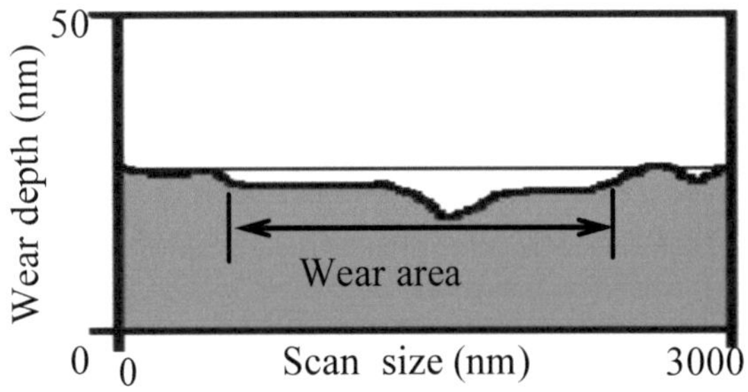

(b) C-N sputtering (wear depth 4.0 nm)

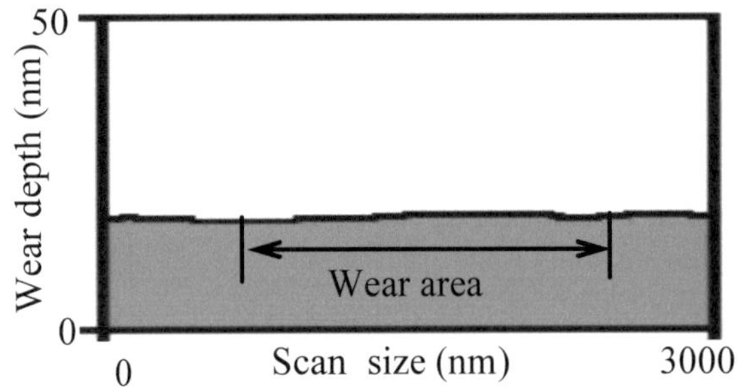

(c) Complex C-N film (wear depth 0 nm)

Fig. 13. Cross-sectional profiles of microwear of plasma-irradiated, 3 nm C-N and complex C-N films deposited in N_2+He.

3.2 Nanometer-scale mechanical properties of extremely thin protective DLC films

3.2.1 Surface analysis of DLC films

Extremely thin protective DLC films were deposited on Si (100) wafers by FCVA (ta-C) and ECR-CVD (Hyodo et al., 2001; Miyake et al., 2009; Robertson, 2008; Yamamoto et al., 2000, 2003). The target thicknesses of films were 0.8, 1.0, 2.0, 5.0, and 100 nm, where the film thickness was adjusted by changing the deposition time. The properties of these DLC films such as their structure, composition and actual thickness were evaluated using Raman spectroscopy, transmission electron microscopy (TEM) and AES (Miyake et al., 2009).

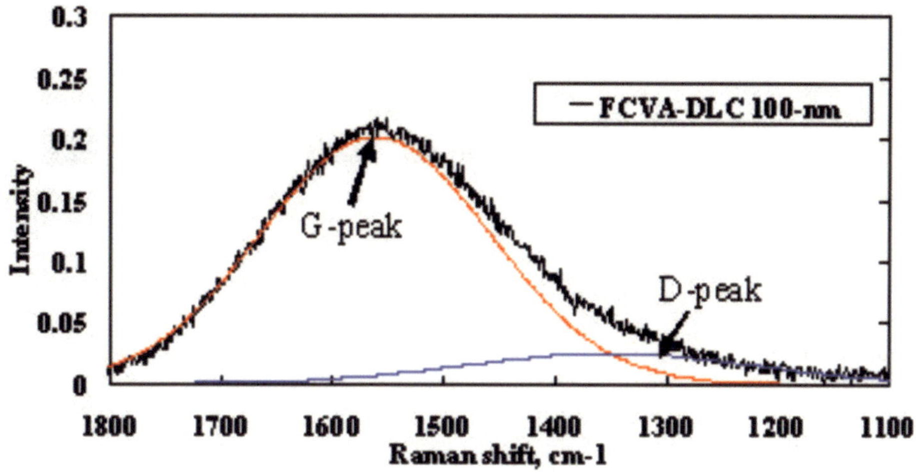

(a) FCVA-DLC 100nm

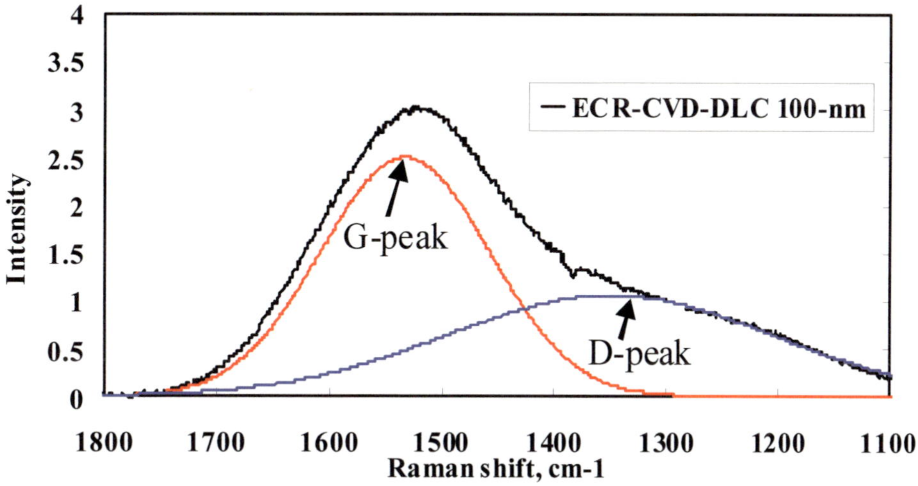

(b) ECR-CVD-DLC 100 nm

Fig. 14. Raman spectra of (a) 100-nm-thick FCVA-DLC and (b) ECR-CVD-DLC films.

Figure 14 shows the Raman spectra of DLC films with a thickness of 100 nm prepared by FCVA and ECR-CVD. The spectra show that the FCVA-DLC (ta-C) film contained a large number of sp^3 bonds than the ECR-CVD (a-CH) film (Ferrari, 2002; Lemoinea et al., 2007; Miyake et al., 2009).

From the cross-sectional TEM images and AES depth profiles of FCVA and ECR-CVD-DLC films of various thicknesses, the actual thicknesses of the extremely thin films were evaluated. The AES spectra of the top surfaces of the 1-nm-thick DLC films are shown in Fig. 15. For these films, according to the AES analysis of the surface, C is the main component with small amounts of O and Si, and no significant differences between the FCVA-DLC and ECR-CVD-DLC films were observed.

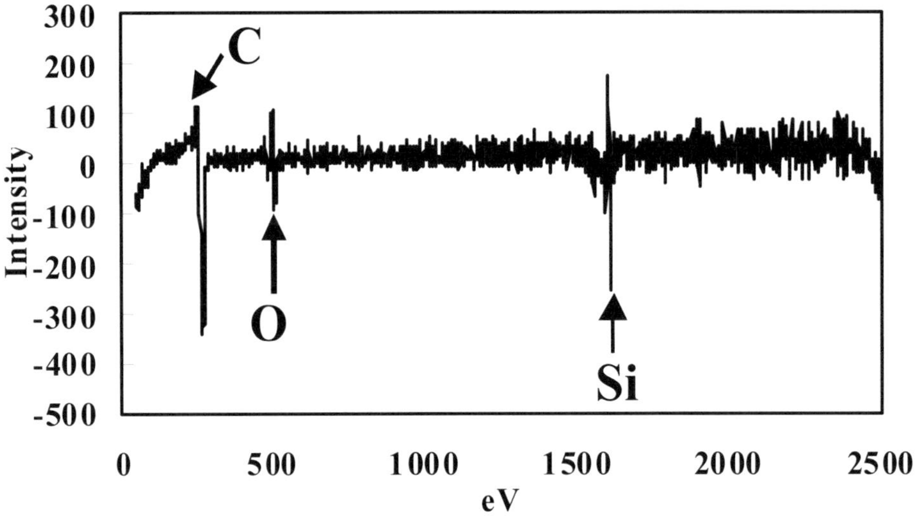

(a) FCVA-DLC 1 nm

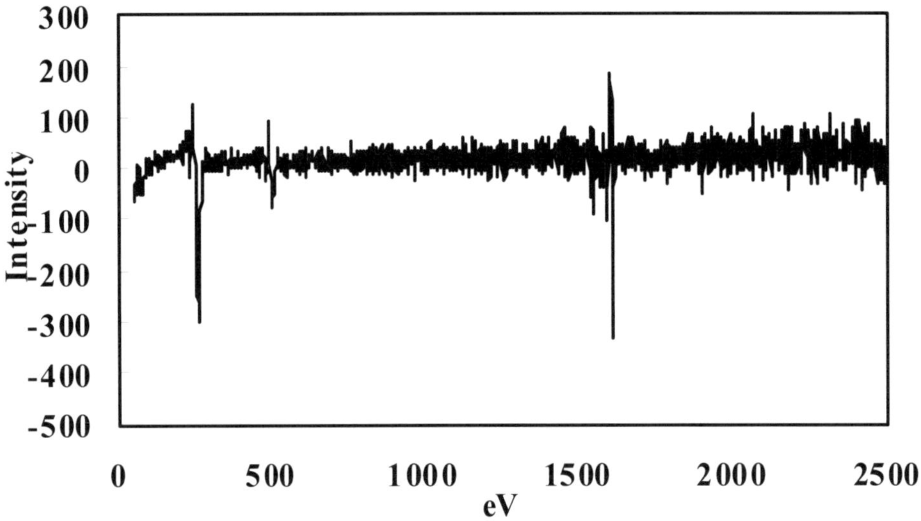

(b) ECR-CVD-DLC 1-nm

Fig. 15. AES spectra of top surfaces of 1-nm-thick FCVA and ECR-CVD DLC films.

The actual thicknesses of the 100-nm-thick DLC films were obtained by TEM and AES. The depth profiles were shown in Fig. 16.The FCVA-DLC and ECR-CVD-DLC films had actual thicknesses of 107 and 97 nm and their densities were 3.3 and 1.9 g/cm^3, evaluated by Rutherford back scattering (RBS), respectively (Miyake et al., 2009).

3.2.2 Nanoindentation properties of DLC films

The nanoindentation curves of FCVA-DLC and ECR-CVD-DLC films with a thickness of 100 nm are shown in Fig. 17. The indentation depth of the ECR-CVD-DLC film was greater than

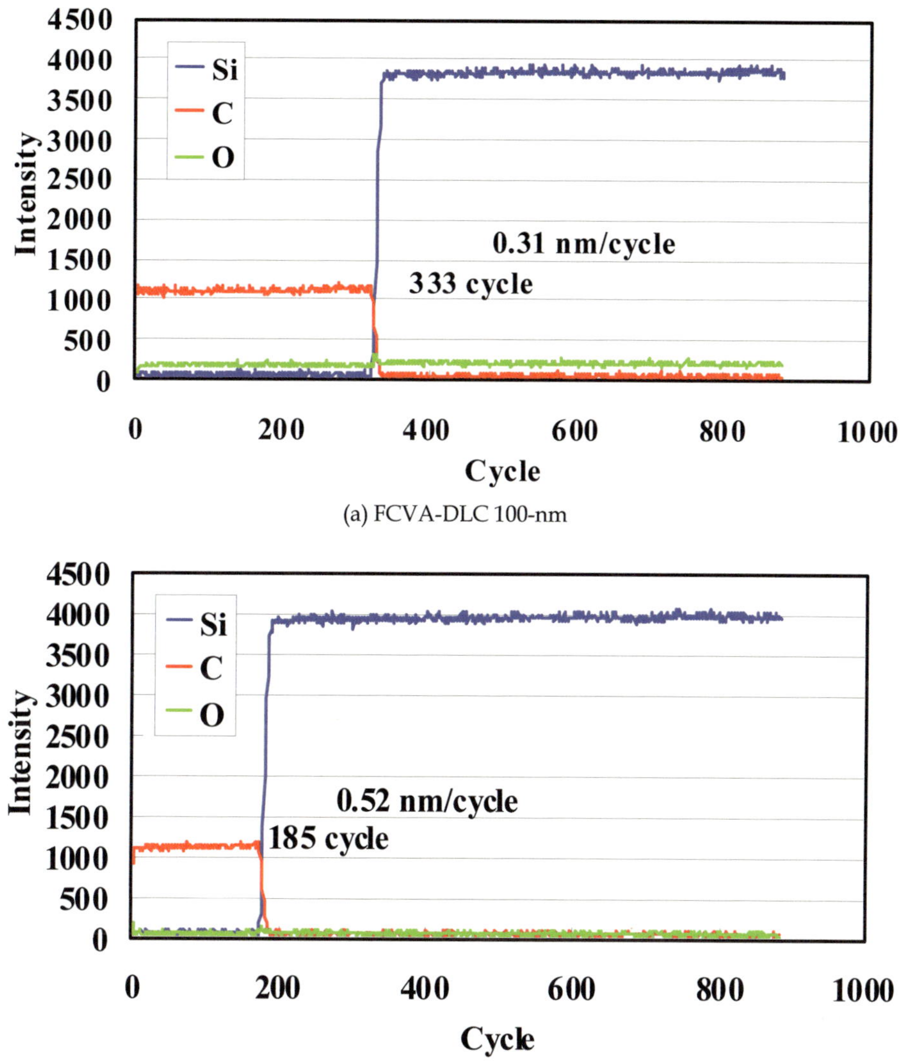

(a) FCVA-DLC 100-nm

(b) ECR-CVD-DLC 100-nm

Fig. 16. AES depth profiles of 100-nm-thick DLC films.

that of the FCVA-DLC film. The FCVA-DLC film mainly exhibited elastic deformation. However, the ECR-CVD-DLC film exhibited greater energy dissipation at the same indentation load (Farhat et al., 1997; Miyake, 2003). The nanoindentation hardnesses of the

DLC films deposited by FCVA and ECR-CVD were 57 and 25 GPa, respectively, at a 40 μN load. This high hardness of the FCVA-DLC film is similar to the reported hardness of 59 GPa for a 100-nm-thick FCVA-DLC film deposited under similar conditions (Shi et al., 1996). The density of our FCVA-DLC film was as high as 3.3 g/cm^3, nearly equal to the previously reported density of 3.37 g/cm^3 (Shi et al., 1996). Under our deposition conditions, the actual hardness of the films, regardless of the effect of the substrate, can be obtained. The FCVA-DLC film exhibited excellent resistance to nanometre-scale plastic deformation.

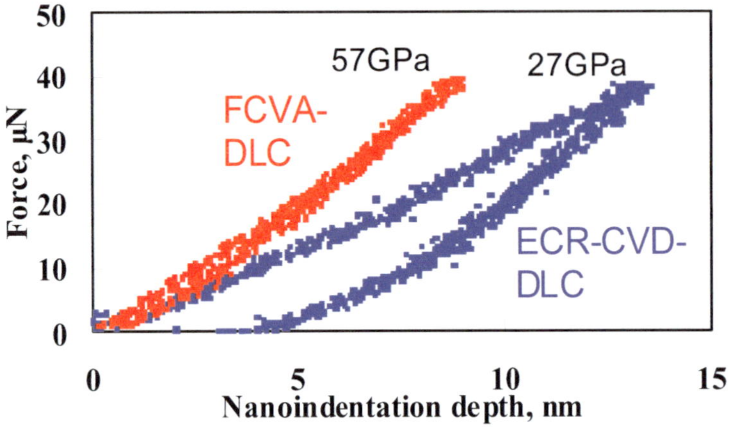

Fig. 17. Nanoindentation curves of 100-nm-thick DLC films.

The nanoindentation curves of the films with a target thickness of 5 nm are shown in Fig. 18. In the nanoindentation loading curves of both DLC films, points of inflection (arrows in Fig. 18) can be observed when the indentation depth exceeds the film thickness. The points of inflection correspond to the onset of nanometer-scale plastic deformation of the thin films due to damage to the surface. Hardly any difference between the nanoindentation curves of the extremely thin DLC films deposited by FCVA and ECR-CVD was detected. It was difficult to evaluate the hardness of the DLC films with a thickness of a few nm by this nanoindentation test owing to the effect of the substrate (Lemoine et al., 2004).

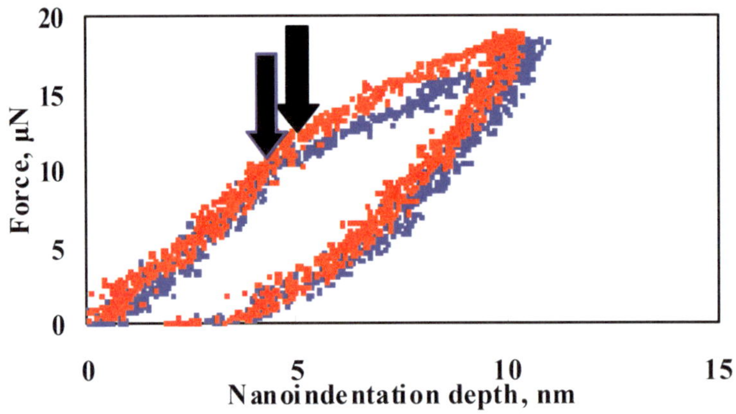

Fig. 18. Nanoindentation curves of various 5-nm-thick DLC films.

3.2.3 Nanowear properties of DLC films

Figure 19(a) shows the method used to evaluate nanowear and Fig. 19(b) shows cross-sectional profiles of the Si(100) substrate. Wear of nearly 3 nm depth and protuberances with a height of 1 nm were observed. The nanowear and cross-sectional profiles of 2- and 1-nm-thick DLC films are shown in Figs. 20 and 21, respectively. Wear of approximately 3-5 nm was observed in the ECR-CVD-DLC film. In contrast, no wear of the FCVA-DLC film can be observed in Figs. 20(a) and 21(a).

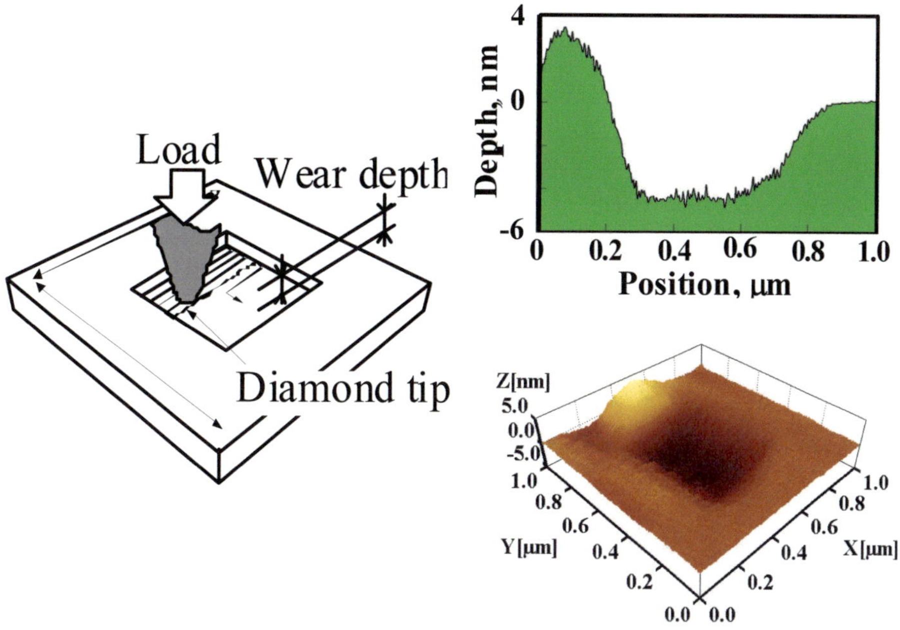

(a) A nanowear test (b) Wear

Fig. 19. Nanowear test (a) and wear profiles of Si substrate (b).

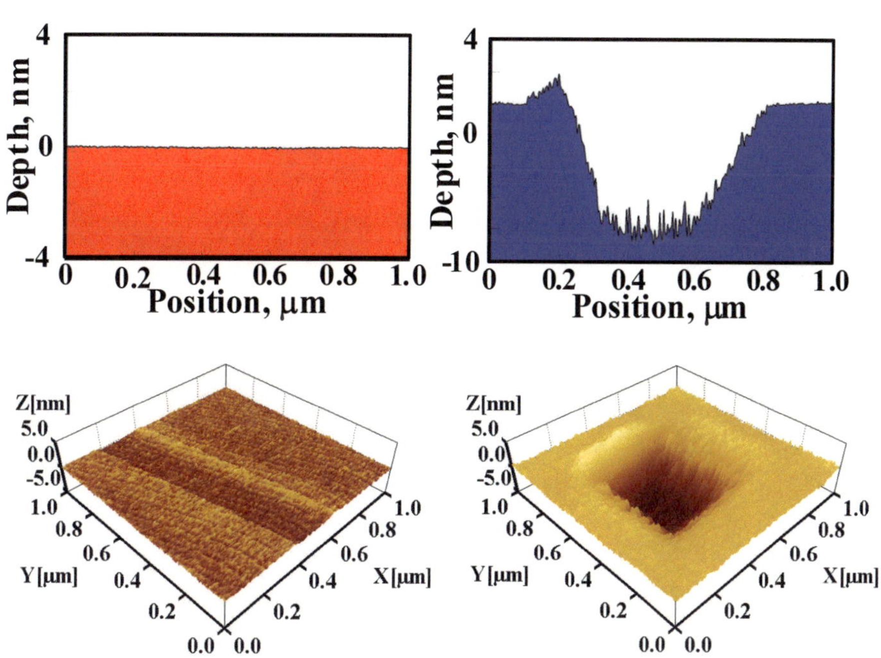

(a) FCVA-DLC (b) ECR-CVD-DLC

Fig. 20. Wear profiles of 2-nm-thick DLC films.

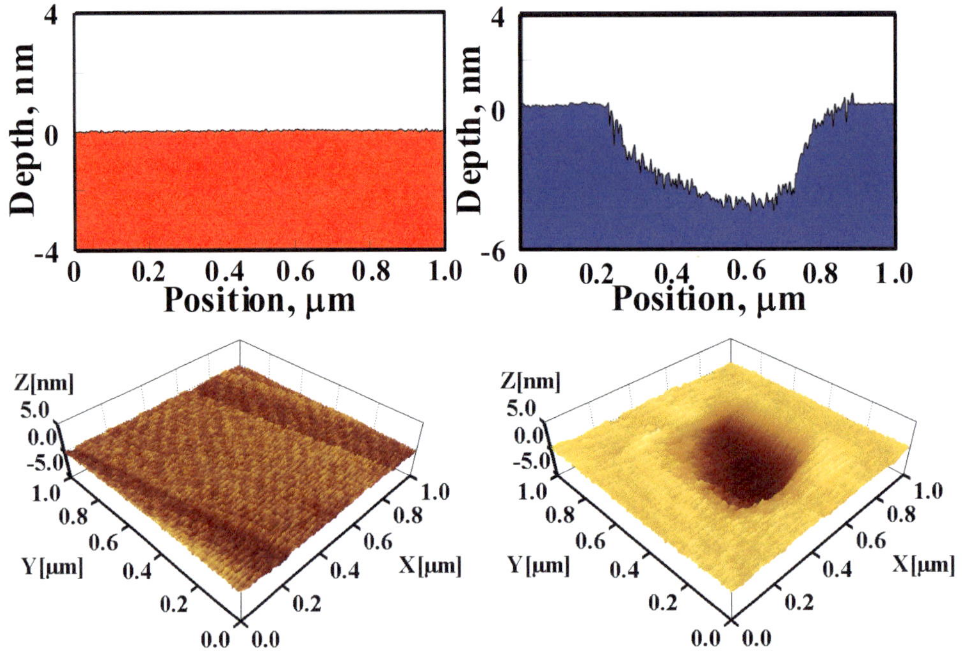

(a) FCVA-DLC (b) ECR-CVD-DLC

Fig. 21. Wear profiles of 1-nm-thick DLC films.

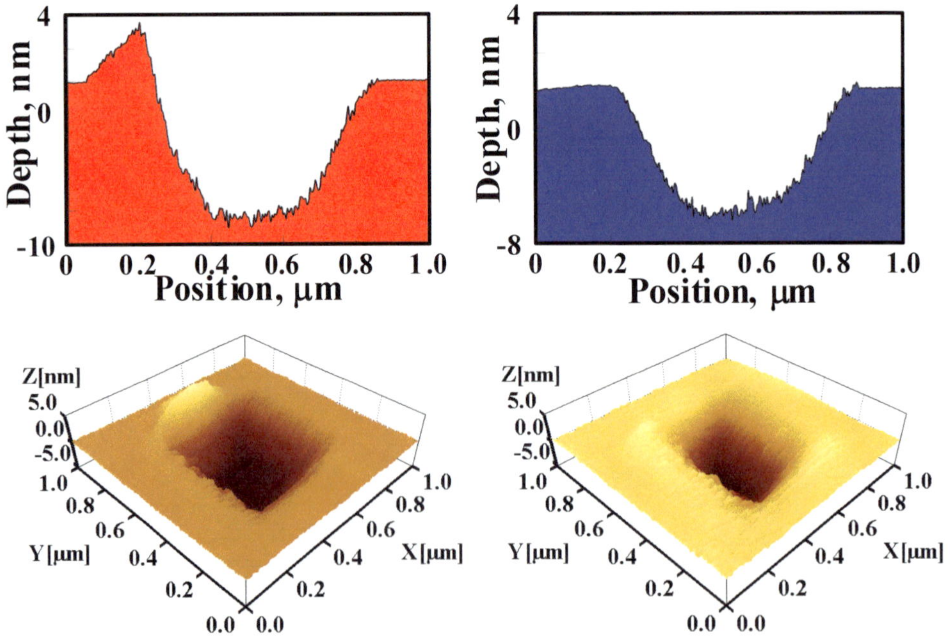

(a) FCVA-DLC (b) ECR-CVD-DLC

Fig. 22. Wear profiles of 0.8-nm-thick DLC films

These DLC films are extremely thin. Therefore, atomic-scale defects in the film affect the nanowear properties. The density of the FCVA-DLC film was higher than that of the ECR-CVD-DLC film. The ECR-CVD-DLC film contained a hydrogen terminated carbon network. Therefore, plastic deformation was more easily caused by friction stress in the ECR-CVD-DLC film than in the FCVA-DLC film. However, as shown in Fig. 22, the wear depth of the 0.8-nm-thick FCVA-DLC film increased rapidly with the number of sliding cycles and is similar to those of the ECR-CVD-DLC film and Si surface. The 0.8-nm-thick FCVA-DLC film did not exhibit superior nanowear resistance.

The nanowear depth dependence on the number of sliding cycles for various DLC films is shown in Fig. 23. The wear depths of 1- and 2-nm-thick FCVA-DLC films were extremely low, less than 1 nm even after 20 sliding cycles of diamond tip scanning, as shown in Fig. 23(a). In contrast, the wear depths of 0.8-, 1- and 2-nm-thick ECR-CVD-DLC films were nearly 1 nm after one sliding cycle and exceeded the their film thickness after a few sliding cycles, as shown in Fig. 23(b). These results reveal the variation of wear resistance among these extremely thin DLC films and the superior and excellent wear resistance of FCVA-DLC films compared with ECR-CVD films.

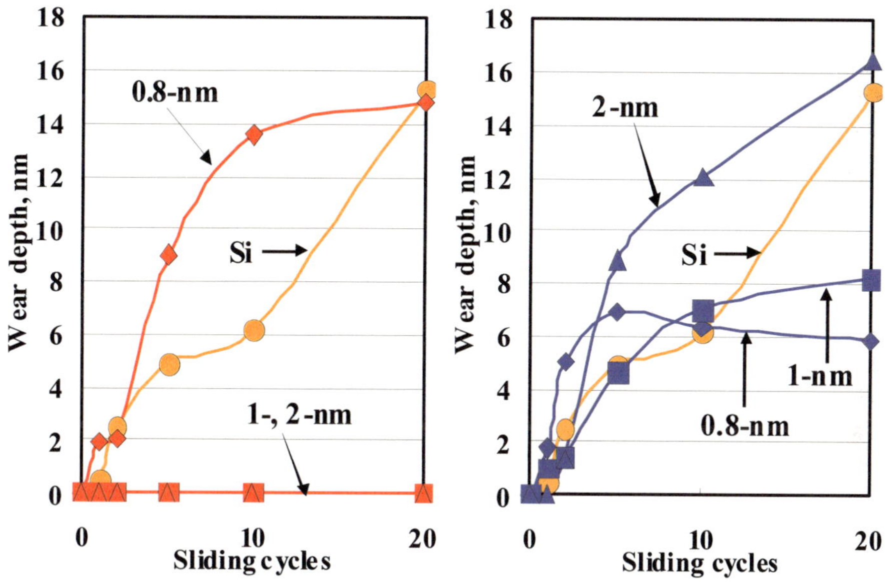

Fig. 23. Nanowear depth dependence on number of sliding cycles.

3.3 Micro-tribological properties of heat treated PFPE lubricant

3.3.1 Heat treatment of PFPF lubricant

Perpendicular hard disks with and without heat treatment were used as specimens in our study. PFPE lubricant was coated on these disks. Heat treatment was performed for 15 min at 100°. The lubricant thickness of the untreated hard disk was 1.1 nm and that of the disk with heat treatment was 0.95 nm. The roughness of the disks both with and without heat treatment was found to be 0.4 nm Ra.

3.3.2 Resonance of transducer system

To investigate the viscoelastic properties of the hard disks, it is necessary to know the resonance of the tip indenter system. Figure 24 shows the amplitude and phase lag of the indenter tip when the tip is free of contact. The diamond indenter tip was vibrated at 1-300 Hz. The resonance frequency was found to be approximately 130-140 Hz. Physical parameters such as the spring constant, mass and damping coefficient of the transducer system were evaluated based on the relationship between amplitude and vertical motion of the indenter tip, as shown in Fig. 24. The spring constant was found to be 209 N/m, the mass of the system was 262 mg and the damping coefficient was 0.049 kg/s.

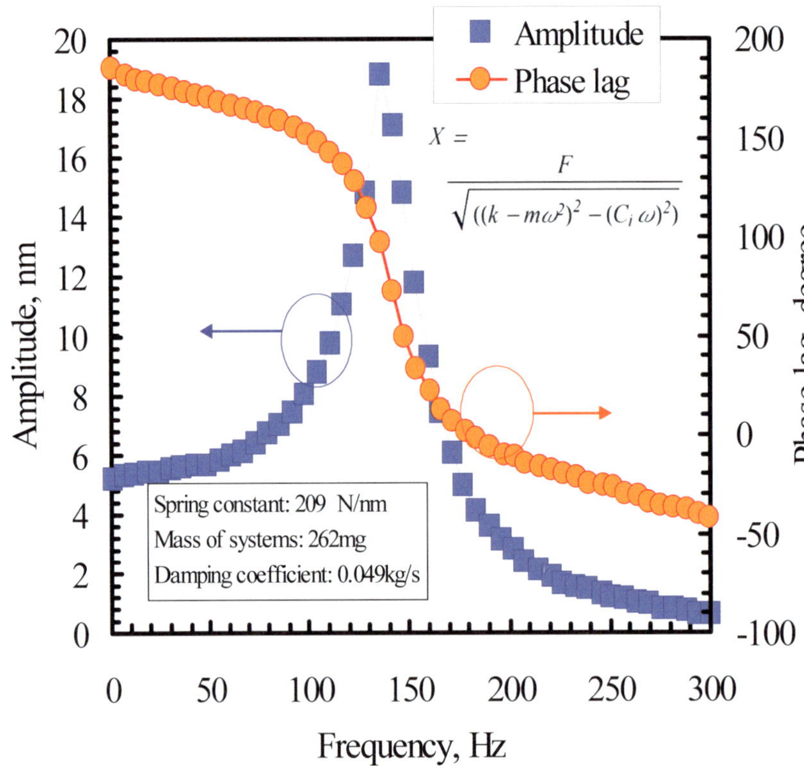

Fig. 24. Relationship between amplitude and vertical motion of indenter tip.

3.3.3 Nanoindentation hardness of PFPE coated disks

Figure 25 shows quasi-static nanoindentation curves obtained from the hardness test with a 20 N load. As shown in Fig. 25(a), the maximum indentation depth in the hill of the untreated disk was approximately 6.0 nm and that of the heat-treated disk was 4.0 nm. The maximum indentation depth of the hill was greater than that of the valley, as observed in Figs. 25(a) and 25(b). As shown in Fig. 25(c), for both disks, the hardness in the valley was higher than that on the hill. As a result, the nanoindentation hardness of the hill of the heat-treated hard disk was 8.0 GPa and that of the valley was 14.0 GPa. It is clear that the hardness of the hard disk was increased by heat treatment.

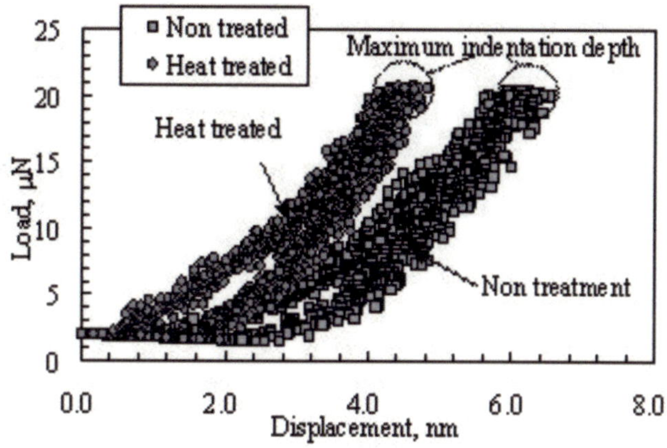

(a) Nano-indentation curve of hill part

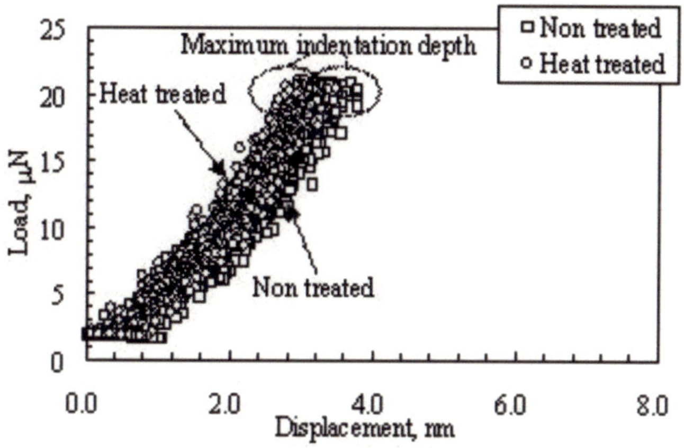

(b) Nano-indentation curve of valley part

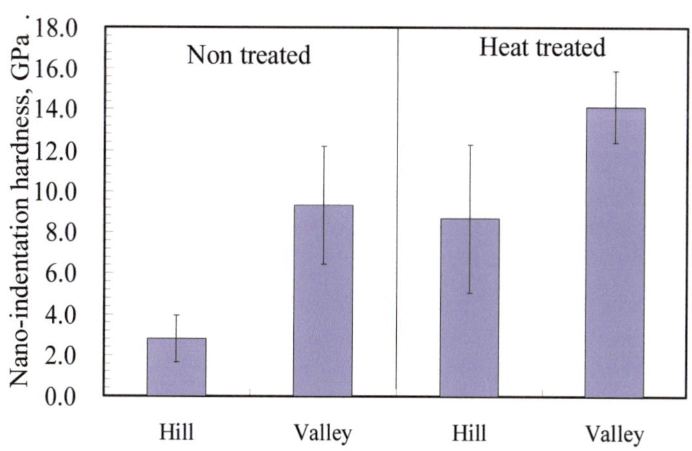

(c) Average of nanoindentation hardness

Fig. 25. Variation in nanoindentation hardness for untreated and heat-treated disks.

3.3.4 Viscoelastic properties of PFPE lubricant

Figure 26 shows the viscoelastic properties of untreated and heat-treated disks. The storage modulus of the hill of the untreated disk surface was larger than that of the valley at all loads. The storage moduli of the valleys were more variable than those of the hills as shown in Fig. 26(a). On the other hand, for the heat-treated disk, the storage modulus of the valley part was larger than that of the hill part. The storage modulus of both the hill and valley increased after heat treatment. These results are consistent with the results for the nanoindentation hardness. The loss modulus of the valley part of the untreated disk surface was larger than that of the hill part, as shown in Fig. 26(b). Moreover, the loss modulus of the heat-treated disk exhibited the same tendency as that of the untreated disk surface. As shown in Fig. 26(c), tan δ for the valley part of the untreated disk surface was larger than that for the hill part. Tan δ for the hill part and valley part of the heat-treated disk surface has the same behavior as that for the untreated disk surface. These results are related to the quantity of lubricant in the valley part.

3.3.5 Viscoelastic properties of PFPE lubricant after microwear test

To determine the motion of PFPE by sliding testing, the viscoelastic properties were evaluated in the hill and valley parts by a microwear test. Surface and cross-sectional profiles of the heat-treated and untreated disks under a load of 30 μN are shown in Fig. 27. The wear depth of the untreated disk surface was about 3.5 nm, while that of the heat-treated disk was about 3.0 nm. The microwear properties depended on the load. The same tendency was observed for the wear volume. Figure 28 shows the results of the microwear test of untreated and heat-treated disk surfaces. The wear depth of the heat-treated disk was decreased by heat treatment for all applied loads. This finding is consistent with the results of the quasi-static nanoindentation tests.

Figure 29 shows the storage moduli of the wear and nonwear parts. As shown in Fig. 29(a), in the hill part of the untreated disk surface, the storage modulus of the wear part was different from that of the nonwear part. However, in the hill part of the heat-treated disk, the storage modulus of the wear part was markedly different from that of the nonwear part. As shown in Fig. 29(b), the changes in the storage modulus of the wear part of the untreated surface disk were greater in the valley part than in the hill part. In the case of the heat-treated disk, the storage modulus of the wear part exhibited much greater fluctuations in both the valley part and the hill part than those of the nonwear part.

The loss moduli of the wear and nonwear parts for the heat-treated and untreated disks are displayed in Fig. 30. As shown in Fig. 30(a), for the untreated disk and the heat-treated disk surfaces, the loss moduli of the wear part in the hill are similar to those of the nonwear parts. Similarly, for both the heat-treated and untreated disk surfaces, the loss moduli of the wear part in the valley were different from those of the nonwear part.

Figure 31 shows tan δ for the wear part of the heat-treated and untreated disk surfaces. In the hill part of the untreated disk surface, at loads of 10–90 μN, tan δ for the wear part and nonwear part were the same. However, for the heat-treated disk, at higher loads of 70–90 μN, tan δ for the wear part was smaller than that for the nonwear part, and tan δ in the valley part exhibited the same tendency in the hill part. It is considered that PFPE lubricant was present on the wear part. On the other hand, it is thought that tan δ for the heat-treated disk surface was lower than that for the untreated disk surface because the PFPE lubricant was solidified and removed during the sliding test.

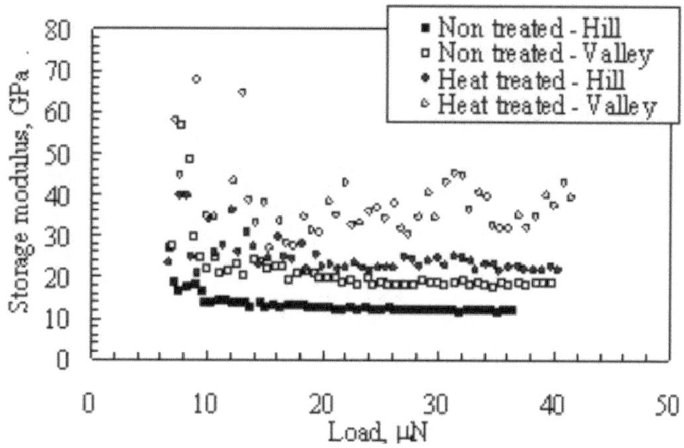

(a) Storage modulus

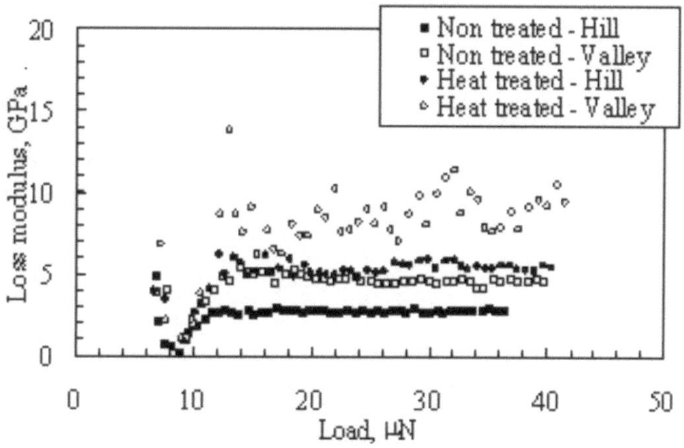

(b) Loss modulus

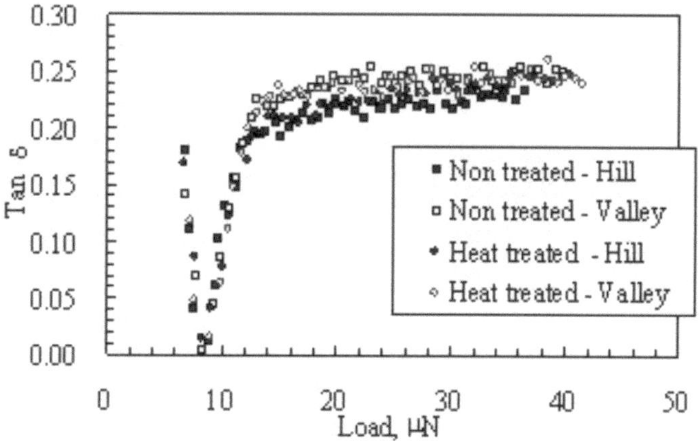

(c) Tan δ

Fig. 26. Viscoelastic properties of untreated and heat-treated disks.

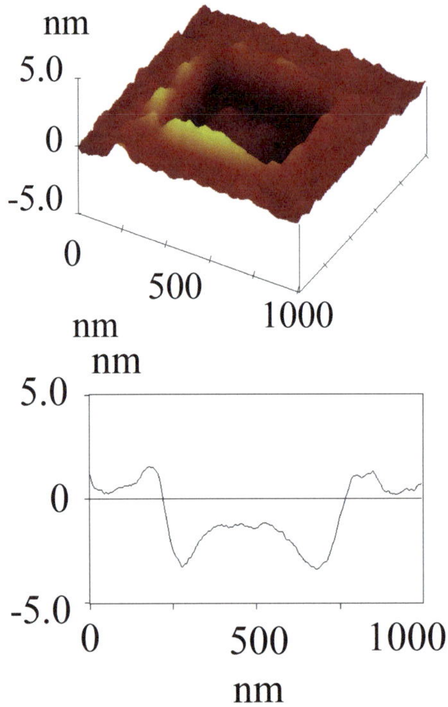

(a) Non treated

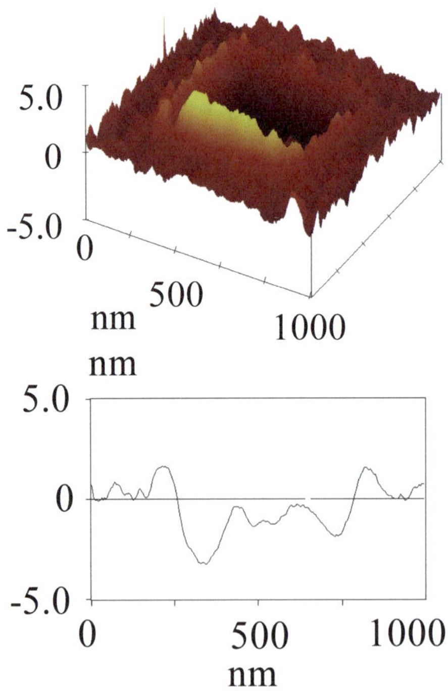

(b) Heat treated

Fig. 27. Surface and cross-sectional profiles of untreated and heat treated disk surfaces.

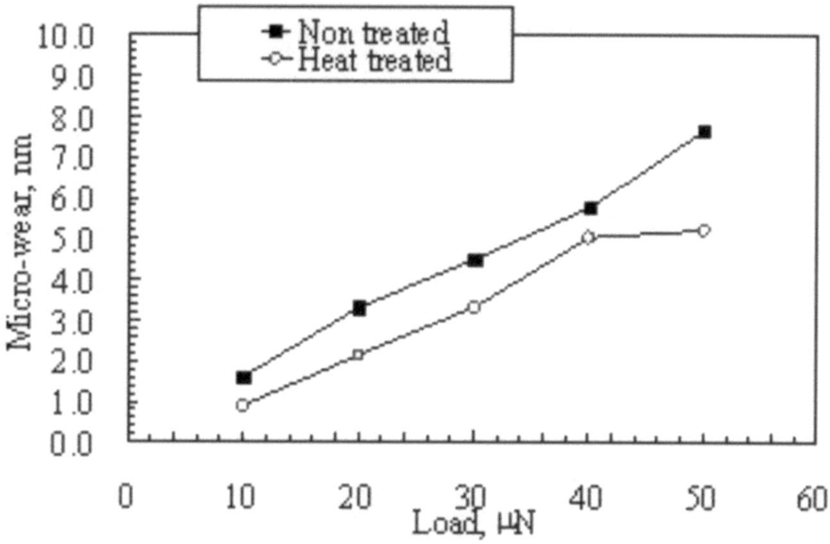

(a) Wear depth

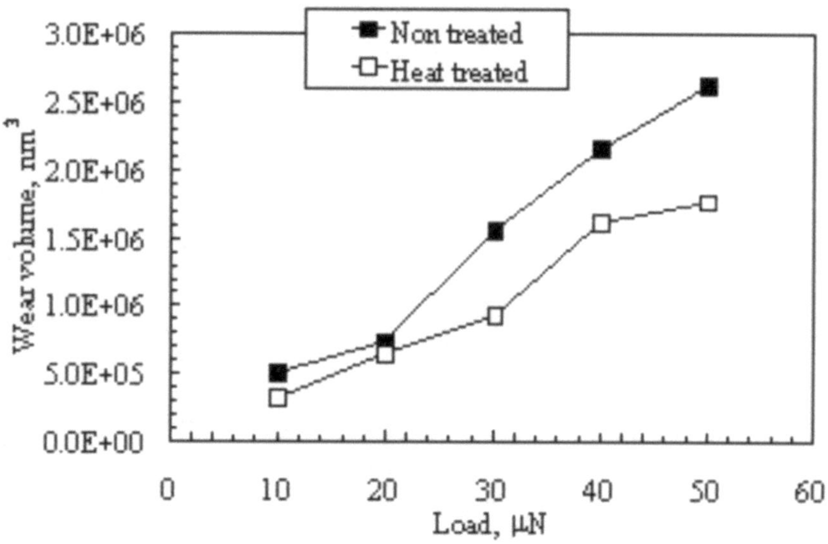

(b) Wear volume

Fig. 28. Dependence of microwear properties on applied load.

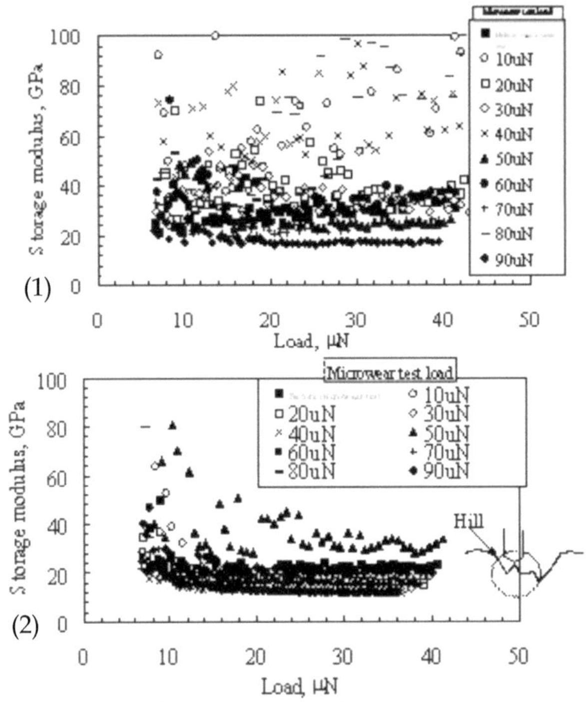

(a) Hill, (1) Untreated, (2) Heat-treated

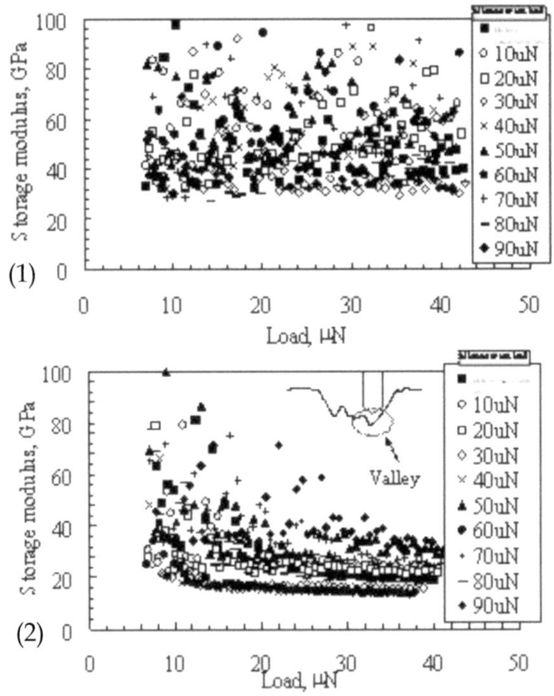

(b) Valley, (1) Untreated, (2) Heat-treated

Fig. 29. Storage moduli of wear and nonwear parts.

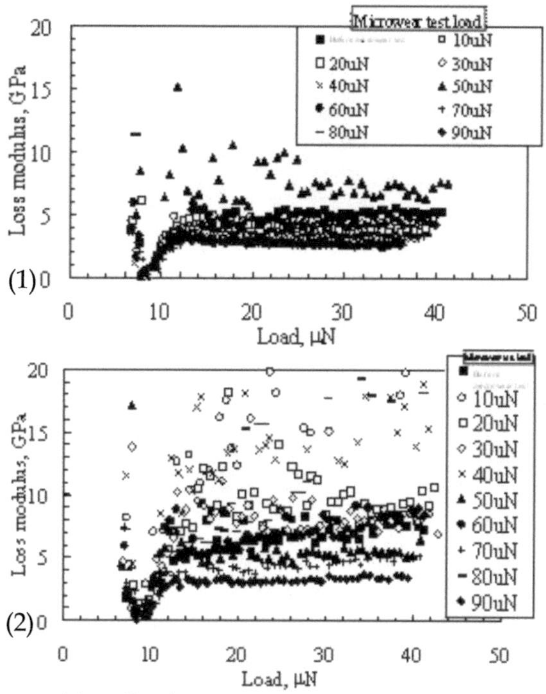

(a) Hill, (1) Untreated, (2) Heat-treated

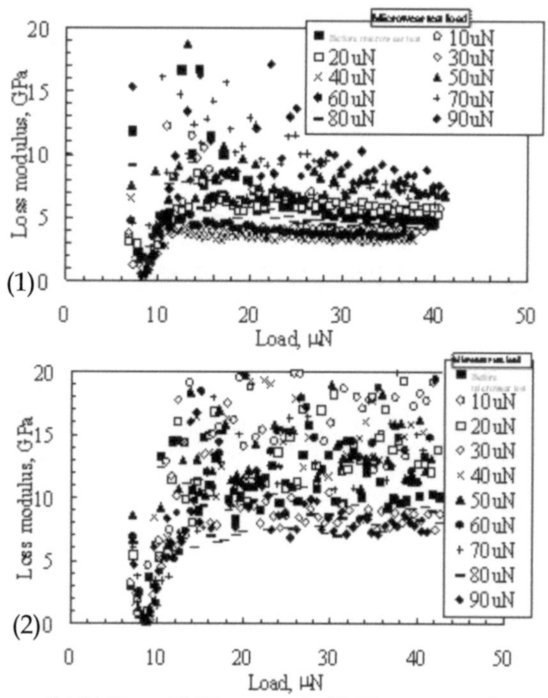

(b) Valley, (1) Untreated, (2) Heat-treated

Fig. 30. Loss moduli of wear and nonwear parts.

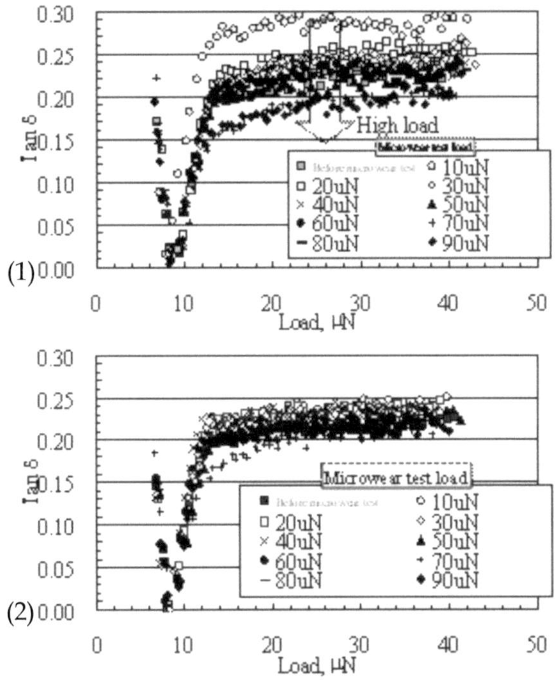

(a) Hill, (1) Untreated, (2) Heat-treated

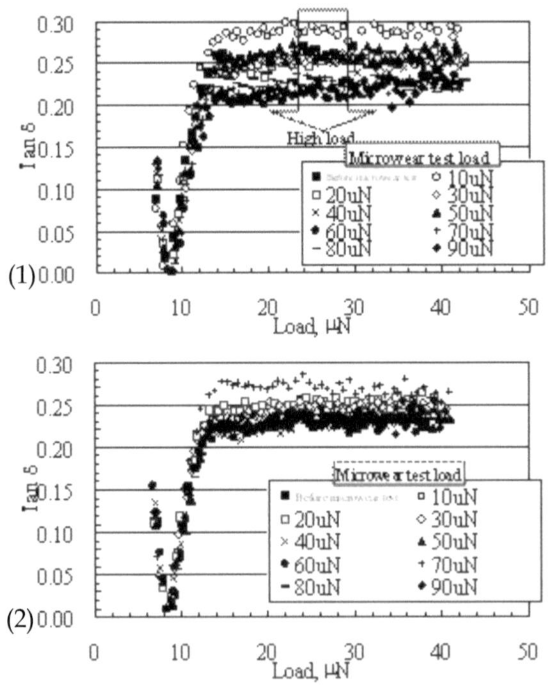

(b) Valley, (1) Untreated, (2) Heat-treated

Fig. 31. Tan δ for untreated and heat-treated disk surfaces after microwear test.

4. Conclusion

In this chapter we reported the present status of research on the nanomechanical evaluation of ultrathin films using AFM for hard disk applications. As stated, nanomechanical evaluation involves the evaluation of nanoindentation hardness, surface deformation properties, nanoscale wear properties and nanotribological properties.

Extremely thin nitrogen-containing carbon (C-N) films with thicknesses of 3, 6 and 9 nm were prepared as a protective overcoat on CoCrTa magnetic disks and their microtribological and micromechanical properties were evaluated by AFM. It is clear that both the nanoindentation hardness and microwear resistance of complex C-N films were significantly improved by complex treatment and He addition. The 3-nm-thick complex C-N films exhibited excellent wear resistance compared with other films, resulting from their graded composition. The effect of the deterioration of the layer generated by plasma irradiation on the magnetic spacing was negligible since a plasma-irradiated layer had a thickness of less than 1 nm. In contrast, the effect of the complex treatment on the mechanical properties of C-N films was significant.

According to the results of AFM, the nanoindentation hardnesses of the 100-nm-thick DLC films deposited by FCVA and ECR-CVD were 57 and 25 GPa, respectively, at a 40 μN load. Hardly any difference between the nanoindentation curves of extremely thin DLC films deposited by FCVA-CVD and ECR-CVD was detected, although it was difficult to evaluate the hardness of the DLC films with a thickness of a few nm by this nanoindentation test. In contrast, nm wear tests were successfully used to evaluate the surface of the extremely thin DLC films after wear and to clarify their mechanical properties. The wear depths of 1- and 2-nm-thick FCVA-DLC films were extremely low, less than 1 nm even after 20 sliding cycles. However, the wear depth of the 0.8-nm-thick FCVA-DLC film increased rapidly with the number of sliding cycles similarly to the result for a Si substrate. The wear depths of the 0.8-, 1.0- and 2.0-nm-thick ECR-CVD-DLC films were nearly 1 nm after one sliding cycle and exceeded the film thickness after a few sliding cycles. These results reveal the variation of wear resistance among these extremely thin DLC films and the superior wear resistance of FCVA-DLC thin films.

The microtribological properties of heat-treated disks were investigated using the force modulation method with SPM. In the quasi-static nanoindentation hardness test, the hardness of the hard disk was found to increase after heat treatment. The evaluation of the viscoelastic properties showed that tan δ for the valley part was higher than that for the hill part. The microwear depth and volume of the disks were decreased by heat treatment. This is consistent with the result of the quasi-static nanoindentation hardness test of the heat-treated disk. Lubricant present on the heat-treated disk was removed by sliding under high loads. AFM plays a vital role in the nanomechanical evaluation of ultrathin films used for hard magnetic disks. It is possible to evaluate the micromechanical and microtribological properties of such ultrathin films and to apply these evaluation methods in future works.

5. Acknowledgment

The research work was partly supported by the Storage Research Consortium (SRC).

6. References

Asif, S.A.S.; Wahl, K.J. & Colton, R.J. (1999). Nanoindentation and contact stiffness measurement using force modulation with a capacitive load-displacement transducer, *Rev Sci Inst.*, Vol. 70, pp. 2408–2413. ISSN 0034-6748

Bhushan, B. (1995). *Handbook of Micro/Nanotribology*, CRC Press, Inc. 1997, pp.321-322. ISBN 0-8493-8401-X, USA

Cutinogco, E.C.; Li, D.; Chung, Y.; & Bhatia, S.C. (1996). Tribological behavior of amorphous carbon nitride overcoats for magnetic thin-film rigid disks, *ASME Trans. J. Tribology*, Vol 118, 7, pp. 543-548. ISSN 0022-2305

Doerner, M.F.; Gardner, D.S. & Nix, W.D. (1986). Plastic properties of thin films on substrates as measured by submicron indentation hardness and substrate curvature techniques, *J. Mater. Res.*, Vol. 1, 6, pp. 845-851. ISSN 0884-2914

Farhat, Z.N.; Ding, Y.; Northwood, D.O. & Alps, A.T. (1997). Nanoindentation and friction studies on Ti-based nanolaminated films, *Surf. Coat. Technol.*, Vol. 89, pp. 24-30. ISSN 0257-8972

Ferrari, A.C. (2002). Determination of bonding in diamond-like carbon by raman spectroscopy, *Diamond and Related Materials*, Vol. 11, pp. 1053-1061. ISSN 0925-9635

Hyodo, H.; Yamamoto, T. & Toyoguchi, T. (2001). Properties of tetrahedral amorphous carbon film by filtered cathodic arc deposition for disk overcoat, *IEEE Trans. on Magn.*, Vol. 37, pp. 1789.-1791. ISSN 0018-9464

Kaneko, R.; Oguchi, S.; Miyamoto, T.; Andoh, Y. & Miyake, S. (1990). Micro-tribology for magnetic recording, *STLE Publication*, SP-27, 1990, pp. 31-34

Lemoine, P.; Quinn, J.P.; Maguire, P. & McLaughlin, J.A. (2004). Comparing hardness and wear data for tetrahedral amorphous carbon films, *Wear*, Vol. 257, pp. 509-522. ISSN 0043-1648

Lemoinea, P.; Quinna, J.P.; Maguirea, P.D.; Zhaob,J.F. & McLaughlina, J.A. (2007). Intrinsic mechanical properties of ultra-thin amorphous carbon layers, *Applied Surface Science*, Vol. 253, 14, pp. 6165-6175. ISSN 0169-4332

Liu, A.Y. & Cohen, M.L. (1989). Prediction of new low compressibility solid, *Science*, Vol. 245, pp. 841-842. ISSN 0036-8075

Liu, A.Y. & Cohe, M.L. (1990). Structure properties and electronic structure of low-compressibility materials: β-Si_3N_4 and hypothetical β-C_3N_4. *Physical Review*, Vol. B41, pp. 10727-10734. ISSN 1050-2947

Miyake, S.; Kaneko, R.; Kikuya, Y.; & Sugimoto. I. (1991). Micro-tribological studies on fluorinated carbon film, *Trans. ASME J. Tribology*, Vol. 113, pp. 384-389. ISSN 0742-4787

Miyake, S.; Kaneko, R. & Miyamoto, T. (1992). Micro- and macro-tribological improvement of CVD carbon film by the inclusion of silicon, *Diamond Films Technol.*, Vol. 1, 4, pp. 205-217. ISSN 0917-4540

Miyake, S.; Miyamoto, T. & Kaneko, R. (1993). Microtribological improvement of carbon film by silicon inclusion and fluorination, *Wear*, Vol. 168, pp. 155-159. ISSN 0043-1648

Miyake, S. (1994). Microtribology of Carbonaceous films, approach to atomic scale zero wear, *Trans. of the Mater. Res. Soci. of Japan*, , 1994, 15B, pp. 919-922. ISSN 1382-3469

Miyake, S.; Watanabe, S.; Miyazawa, H.; Murakawa, M.; Kaneko, R. & Miyamoto, T. (1994). Micro scratch hardness increasing of ion plated carbon film by nitrogen inclusion evaluated by atomic force microscopy, *Appl. Phys. Lett.*, Vol. 65, 9, pp. 3206-3208. ISSN 0003-6951

Miyake, S.; Watanabe, S.; Miyazawa, H.; Murakawa, M.; Kaneko, R. & Miyamoto, T. (1997). Modification of nanometer scale wear of nitrogen-containing carbon film due to ion implantation, *Nuclear Inst. and Methods in Phys. Res.*, Vol. B 122, pp. 643-649. ISSN 0168-9002

Miyake, S.; Sekine, Y. & Watanabe, S. (2002). Surface modification of magnetic recording layer by low energy beam irradiation (in Japanese), *J. Surf. Finishing Soc. Jpn.*, Vol. 53, pp. 939–944. ISSN 0915-1869

Miyake, S. (2003). Improvement of mechanical properties of nanometer period multilayer films at interfaces of each layer, *J. Vac. Sci. Tech.*,Vol. B21, 2, pp. 785-789. ISSN 0734-2101

Miyake, S. & Wang, M. (2004). Mechanical properties of extremely thin B-C-N protective layer deposited with helium addition, *Jpn. J. Appl Phys.*, Vol. 43, 6A, pp. 3566-3571. ISSN 0021-4922

Miyake, S.; Wang, M.; Saitoh, T.; & Watanabe, S. (2005). Microtribological properties of B-C-N extremely thin protective films deposited on plasma pretreated magnetic layers, *Surf. Coat. Technol.*, Vol. 195, 2-3, pp. 214-226. ISSN 0257-8972

Miyake, S.; Saito, T.; Wang, M. & Watanabe, S. (2006). Tribological properties of extremely thin protective carbon nitride films deposited on magnetic disk by complex treatment, *Proc. IMechE Vol. 220 Part J: J. Engineering Tribology*, JET124 , IMechE, pp. 587-595. ISSN 1350-6501

Miyake, S.; Kurosaka, W. & Oshimoto, K. (2009). Nanometer scale mechanical properties of extremely thin diamond-like carbon films, *Tribology*, Vol 3, No. 4, pp. 158-164. ISSN 0742-4787

Robertson, J. (2008). Ultrathin carbon coating for magnetic storage technology, *Int. J. Product Development*, Vol. 5, 3-4, pp. 321-338. ISSN 1477-9056

Saitoh, T. & Miyake, S. (2003). Dynamic deformation properties of the PFPE coated hard disk evaluated by force modulation method, (in Japanese). *J Surf Finishing Soc Japan, Vol.* 54, pp. 471–476. ISSN 0915-1869

Saperstein, D.D. & Lin, J.L. (1990). Improved surface adhesion and coverage of perfluoropolyether lubrication following far-UV irradiation, *Langmuir*, Vol. 6, pp. 1522–1524. ISSN 0743-7463

Shi, X.; Tay, B.K.; Tan, H.S.; Zhong, L.;Tu, Y.Q.; Silva, S.R.P. & Milne, W.I. (1996). Properties of carbon ion deposited tetrahedral amorphous carbon films as a function of ion energy, *J. Appl. Phys.* Vol. 79, 9, pp. 7234-7240. ISSN 0021-8979

Sugimoto, I. & Miyake, S. (1989). Optical emission studies on interaction between halogenated carbon species and noble gas during fuoropolymer sputtering, *J. Appl. Phys.*, Vol. 65, 12, pp. 4639-4645. ISSN 0021-4922

Tani, H. (1999). Observation of PFPE lubricant film on magnetic disk surface by atomic force microscopy. *IEEE Trans Magn.*, Vol. 35, pp. 2397–2399. ISSN 0018-9464

White, R.L.; Bahatia, S.S.; Meek, S.W.; Friedenberg, M.C. & Mate, C.M. (1996). Tribology of contact/near-contact recording for ultrahigh density magnetic storage, *ASME Tribology*, Vol. 6, pp. 33-41. ISSN 0022-2305

Yamamoto, T.; Toyoguchi, T. & Honda, F. (2000). Ultrathin amorphous C:H overcoats by PCVD on thin film media, *IEEE Trans. Magn.*, Vol. 36, pp. 115-119. ISSN 0018-9464

Yamamoto, T.; Hyodo, H.; Tsuchitani, S. & Kaneko, R. (2003). Ultrathin amorphous carbon overcoats by filtered cathodic arc deposition, *IEEE Trans. on Magn.*, Vol. 39, pp. 2201-2204. ISSN 0018-9464

9

Estimation of Grain Boundary Sliding During Ambient-Temperature Creep in Hexagonal Close-Packed Metals Using Atomic Force Microscope

Tetsuya Matsunaga and Eiichi Sato
Institute of Space and Astronautical Science, Japan Aerospace Exploration Agency
Japan

1. Introduction

Grain-boundary sliding is one of important deformation mechanisms during creep, which is generally activated by a diffusion process. The apparent activation energy (Q) of the phenomenon is similar to that of dislocation-core diffusion (Frost & Ashby, 1982), and affects creep at high temperatures. However, the present authors observed another type of grain-boundary sliding activated by a remarkably low Q value of about 20 kJ/mol for ambient-temperature creep in hexagonal-close packed metals and alloys. The Q value is about one-fourth of that of grain boundary diffusion of each material (Matesunaga et al., 2009a).

Heretofore, grain boundary sliding has been measured using scribe lines or micro-grid on the specimen surface which was fabricated by a small needle (Harper et al., 1958), focused ion beam (Koike et al., 2003; Rust and Todd, 2011), or stencil method (Parker and Wilshire, 1977). However, atomic force microscope is much more convenient method to measure a vertical component of grain-boundary sliding than the other method because of no advance preparation. In this chapter, to reveal a detailed role of grain-boundary sliding on ambient-temperature creep, atomic force microscopy was conducted to measure a travel distance of it.

Ambient-temperature creep was observed about a half century ago in pure Ti (Adenstedt, 1949; Kiessel & Sinnott, 1953; Luster et al., 1953). Since that time, several studies of this phenomenon have been carried out using Ti alloys. Several Ti alloys were investigated in the 1960s and 70s such as Ti-5Al-2.5Sn (Kiefer & Schwartzberg, 1967; Thompson & Odegard, 1973) and Ti-6Al-4V (Reiman, 1971; Odegard & Thompson, 1974; Imam & Gilmore, 1979). Later, Mills' group intensively studied this phenomenon using Ti-6Al and Ti-6Al-2Sn-4Zr-2Mo (Suri et al., 1999; Neeraj et al., 2000; Deka et al., 2006). Among these experimental studies restricted to Ti alloys, the present authors found recently that all hexagonal close-packed metals and alloys show creep behavior at ambient temperature below their 0.2% proof stresses, but no cubic metals or alloys show it under the same condition (Sato et al., 2006). The former group included commercially pure Ti, magnesium, zinc, Ti-6Al-4V, Zircaloy-4, and AZ31, and the latter group included pure iron, 5052 Al, and Ti-15V-3Cr-3Sn-3Al. In addition, the authors introduced a new region into the Ashby-type deformation mechanism map of commercially pure titanium, the ambient-temperature creep region (Tanaka et al., 2006).

In the ambient-temperature creep region, the authors identified dominant intragranular deformation as a planar slip with no tangled dislocation, which lies on a single slip plane, by transmission electron microscopy (Matesunaga et al., 2009b). The reason why only one slip system is activated inside of a grain is low crystalline symmetry of hexagonal close-packed structure. Its creep parameters were, then, obtained that a stress exponent (n) of 3.0 and a grain-size exponent (p) of 1.0 despite the creep is a dislocation creep. According to these experimental results, we proposed a constitutive relation of ambient-temperature creep as

$$\dot{\varepsilon}_s = AD_0 \exp\left(-\frac{Q}{RT}\right)\frac{Gb}{kT}\left(\frac{\sigma}{E}\right)^n\left(\frac{b}{d}\right)^p , \tag{1}$$

where A is a dimensionless constant, D_0 is a frequency factor, G is a shear modulus, R is the gas constant, k is Boltzmann's constant, b is a Burgers vector, σ is applied stress, E is a Young's modulus, and d is a grain size.

The positive p value means that grain boundaries function as a barrier to dislocation motion and lattice dislocations pile up at grain boundaries. On the other hand, the high-temperature dislocation creep possesses zero p value, which means dislocations can propagate to an adjacent grain easily. In addition, backflow is observed after complete stress-drop tests which mean that the recovery rate controls ambient-temperature creep (Matsunaga et al, 2009b). Results in the previous paper also claimed that grain boundary must accommodate piled-up dislocations to continue and to rate-control deformation because no obstacles exist against dislocation motion inside of a grain.

Describing above, the intragranular deformation mechanism have been studied and discussed well, whereas a role of grain boundary has not been revealed despite of the positive p value of unity. To reveal detailed intergranular deformation mechanism, atomic force microscopy was effective to measure the travel distance of grain-boundary sliding. In addition, electron back-scattered diffraction pattern analysis was conducted to observe grain-boundary structure. Combining of these technics, actual influence of grain boundary on deformation was evaluated.

2. Experimental procedure

The sample was rolled sheets of pure zinc with d of 210 μm and its purity was 99.995 mass%, which shows considerable creep at ambient temperature, as do other hexagonal close-packed materials. No impurity was detected using inductively coupled plasma optical emission spectroscopy (ICPS-8000, Shimadzu). Zinc has also been used for the study on grain-boundary deformation based on a displacement-sift-complete dislocation model above 0.7 T_m (Bollmann, 1967; Schober et al., 1970). Therefore, the features between low- and high-temperature deformation modes at grain boundaries can be compared.

Tensile tests were performed to evaluate 0.2% proof stress ($\sigma_{0.2}$) at ambient temperature with a constant crosshead speed corresponding to an initial strain rate of 1×10^{-3} s^{-1} using an Instron-type machine (AG-100KGN, Shimadzu). Tensile creep tests were then conducted using a dead-load creep frame with loads below $\sigma_{0.2}$ up to 4.3×10^5 s. Then, some tests were conducted with $0.8\sigma_{0.2}$ (19 MPa) at 300 K; they were interrupted after 0.18×10^5 s, 0.86×10^5 s, 2.6×10^5 s and 4.3×10^5 s, respectively. In the both tests, the loading direction corresponded to the rolling direction. Strain was measured using strain gauges with a resolution of 3x10^{-6} mounted directly on the specimen surfaces. By reducing creep of the strain gauge itself, a strain rate of 1x10^{-10} s^{-1} was measured directly. These creep curves were fitted by the logarithmic creep equation (Garofalo, 1963) to evaluate steady state creep rates:

$$\varepsilon = \varepsilon_i + \varepsilon_p \ln\left(1 + \beta_p \mathrm{t}\right) + \dot{\varepsilon}_s \mathrm{t} \tag{2}$$

where ε is true strain, ε_i is instantaneous strain, ε_p and β_p are parameters characterizing primary creep region, and *t* is the elapsed time. Temperature dependent of Young's modulus was calculated from the data of Ashby's textbook (Frost & Ashby, 1982); Poisson's ratio of zinc equals 0.244 (Chen & Sundman, 2001).

Optical microscopy (VHX-600, Keyence), electron back-scattered diffraction pattern analyses (OIM, TSL) were performed before and after the creep tests. The specimens were polished mechanically using colloidal silica before the microscopy. Electron back-scattered diffraction pattern analyses revealed no distortion in a grain before the tests and lattice-rotation distribution after those.

After mechanical tests, vertical components of grain-boundary sliding were measured at more than 40 grain boundaries using an atomic force microscope (VN-8000, Keyence) with a scan speed of 1.11 Hz on contact mode with scanning area of 100x100 μm for each observation. In this condition, an accuracy of an amount of grain-boundary sliding is 0.1 nm. The flatness of surface at grain boundary before the tests was also confirmed by this technique.

The contributions of grain-boundary sliding to the total creep strain, ε_{GBS}, were calculated for individual grains using

$$\varepsilon_{GBS} = \frac{u}{d} \tag{3}$$

$$= \frac{kv}{d} \tag{4}$$

where *u* is the travel distance parallel to the tensile direction and *v* is that perpendicular to the sample as well as to the tensile direction, *k* is a geometrical factor with the value of 1.1 (Bell & Langdon, 1967). In this study, the authors evaluated *v* by atomic force microscopy.

3. Experimental result

Figures 1a and 1b are optical microscope images of specimen surfaces taken after creep tests of 0.18x10^5 s and 2.6x10^5 s, respectively, with the load of 19 MPa at 300 K . Flat surface was confirmed before the tests, whereas Figs. 1a and 1b show slip lines on the surface and grain boundaries become apparent with increasing the test time. In addition, they show that only one slip line was activated inside each grain. This description coincides with the results of previous transmission electron microscopy (Matsunaga et al., 2009 b). Figure 1c is an 3-D profile image with aspect ratio of 10:1 taken after a creep test of 4.3x10^5 s by atomic force microscopy. It is a joined picture of five observations. This image also shows slip lines and

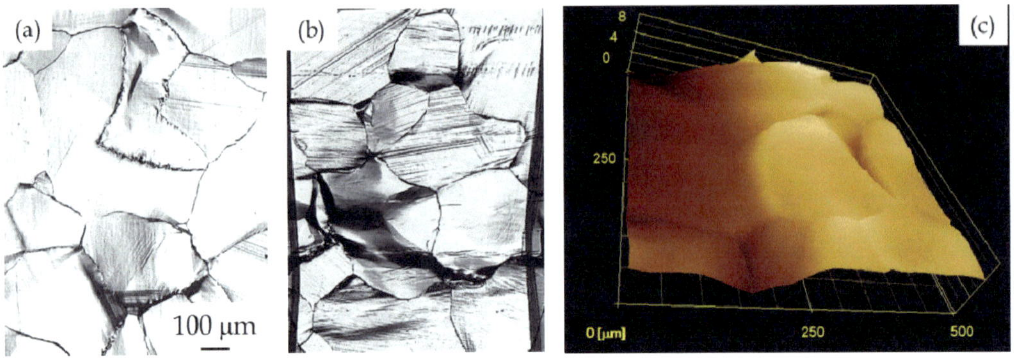

Fig. 1. Optical microscope images of specimen surfaces taken after creep test of (a) 1.8×10^5 s, and (b) 2.6×10^5 s. (c) 3-D surface profile taken after creep test of 4.3×10^5 s. Traces of grain boundaries and slip lines become significant with increasing the test time.

steps at grain boundaries clearly. An average height of the steps for the observed grain boundaries is about 1.0 μm.

Figure 2a shows a creep curve (solid line) and a time dependency of ε_{GBS} (closed circles) under the same condition with that in Fig. 1. A broken line is a fitting curve of ε_{GBS} using equation (2). Creep behavior was observed significantly for both creep strain ($\varepsilon_{creep}=\varepsilon-\varepsilon_i$) and ε_{GBS}. The strain rate by grain-boundary sliding at the infinite period is evaluated as 2.1×10^{-10} s^{-1}, which is only 6% of the steady-state creep rate on the same condition. Moreover, ε_{GBS} yielded most of the creep strain immediately after loading, but $\varepsilon_{GBS}/\varepsilon_{creep}$ decreased by 33% at 4.3×10^5 s as shown in Fig. 2b. It indicates that main part of the strain is yielded by the dislocation motion inside of grains in ambient-temperature creep region. (Matsunaga et al., 2010)

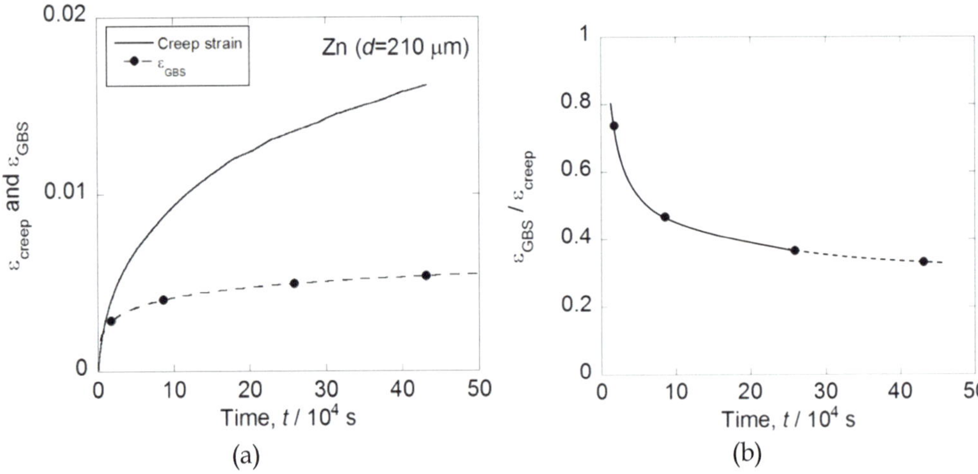

Fig. 2. (a) Creep strain and ε_{GBS} with 19 MPa at 300 K (Matsunaga et al., 2010). (b) Time dependency of $\varepsilon_{GBS}/\varepsilon_{creep}$ under the same condition. ε_{GBS} increase with lasting time, but contribution of ε_{GBS} on ε_{creep} decreases.

Figure 3a is an optical microscope image taken after a creep test for 4.3x10^5 s. Atomic force microscopy was conducted along a white line in Fig. 3a, and a surface profile shows a displacement of 0.77 μm at the grain boundary as shown in Fig. 3b. Figures 3c and 3d portrays results of electron back-scattered diffraction pattern analyses in a white flame in Fig. 3a before and after the creep test. They depict crystal orientation maps and the red point in Fig. 3c was the base point of orientation. Figure 3c was colored blue all around, which means that there were no change of crystal orientation near the grain boundary. On the other hand, Fig. 3d showed gradations with widths of 20 μm near the grain boundary, which implies that the change of crystal orientation of about five degree exists. Orientation gradient near grain boundaries is introduced by lattice dislocations which distort crystal lattice by piled-up dislocations. Figures 3c and 3d demonstrate that recrystallisation did not occur because the shape of the grain boundary was maintained during the creep test (Matsunaga et al., 2010).

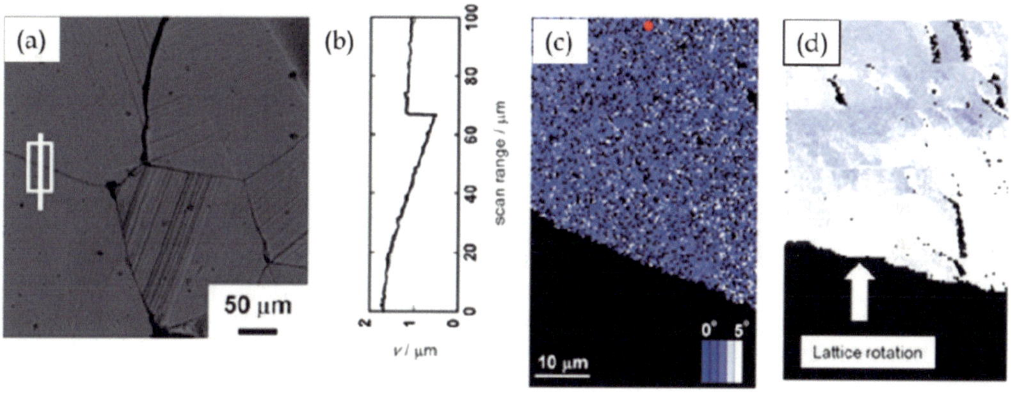

Fig. 3. (a) Optical microscope image taken after the creep test of 4.3x10^5 s. (b) Surface profile at the white line in (a) showing v=0.77 μm. Cristal orientation maps taken before (c) and after (d) the creep test with 19 MPa (Matsunaga et al., 2010). They imply that the change of crystal orientation of about five degree after the test.

A grain-boundary-structure dependency of ε_{GBS} in the ambient-temperature creep region was then analyzed in Fig. 4. The data of grain boundaries having a common axis of $<10\bar{1}0>$ were picked up. Grain boundaries with Σ number slide slightly compared to random grain boundaries. It is similar to the results described by Watanabe et al. (Watanabe et al., 1979, 1984).

4. Discussion

Because hexagonal close-packed materials have low crystal symmetry, strain compatibility is not satisfied at grain boundaries. Although five independent slip systems are necessary for bringing about creep in polycrystals (the von Mises condition), basal or prismatic slip system in the structure has only two or three independent slip systems, respectively. This necessitates some accommodation mechanisms acting at grain boundaries. Because the observed Q value (20 kJ/mol) is much smaller than those of diffusion processes, they are not activated. A possible mechanism of grain-boundary sliding of ambient temperature is slip-induced grain-boundary sliding (Valiev & Kaibyshev, 1977; Mussot et al., 1985; Koike et al., 2003).

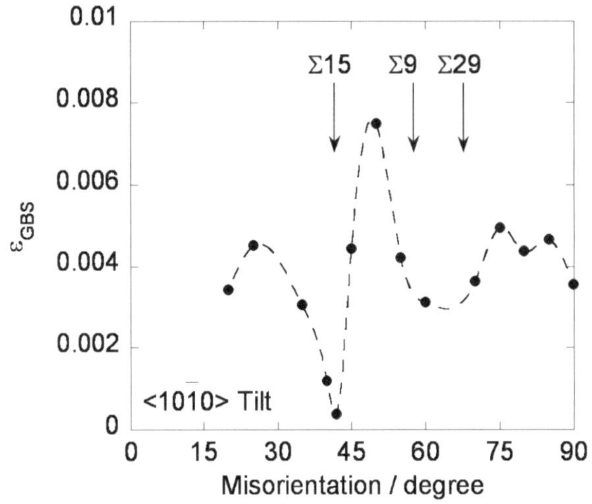

Fig. 4. Grain-boundary-structure dependency of ε_{GBS} of grain boundaries having a common axis of $< 10\bar{1}0 >$. Σ-numbered grain boundaries slide more slightly than random grain boundaries (Matsunaga et al., 2010).

Figure 5 describes a model of slip-induced grain-boundary sliding as follows; intragranular glide is activated first (Fig. 5a); plastic compatibility is responsible for internal stresses, which might be sufficiently strong to activate grain-boundary sliding. This activation is an incompatible process that is responsible for a new internal stress field, which modifies the plastic and total strain distribution near the sliding grain boundary by absorption of piled-up dislocations (Fig. 5b), so that the final intragranular and intergranular total strain distributions are compatible with grain-boundary sliding (Fig. 5c). Magnesium alloy (AZ31) had shown the grain-boundary sliding during tensile deformation in near room temperature with an estimated Q value of 15 kJ/mol (Koike et al., 2003). Features of the slip-induced grain-boundary sliding satisfy the condition of ambient-temperature creep of hexagonal close-packed metals.

On the other hand, grain-boundary sliding at coherent grain boundaries was described by the displacement-sift-complete dislocation (Bollmann, 1964; Schober, 1970) using zinc and aluminum bicrystals at temperatures higher than 0.7 T_m (Rae & Smith, 1980; Smith & King, 1981; Takahashi & Horiuchi, 1985). In this model, grain boundaries slide through the movement of grain boundary dislocations absorbed at a grain boundary with a diffusion process. It is showed that the amount of grain-boundary sliding varied by grain boundary structure (Watanabe et al, 1979, 1984).

In this study, the grain-boundary-structure dependency of grain-boundary sliding in ambient-temperature creep in zinc was then analyzed (Fig. 5). Σ-numbered grain boundary only slides slightly compared to random grain boundaries similar to the behavior of high-temperature grain-boundary sliding (Watanabe et al., 1979, 1984). And, it is considered that grain boundaries having a stable structure (a Σ number) hardly absorb piled-up dislocations. However, in this temperature region, it is believed that diffusion processes do not function, and that a different process of grain-boundary sliding is activated during the ambient-temperature creep.

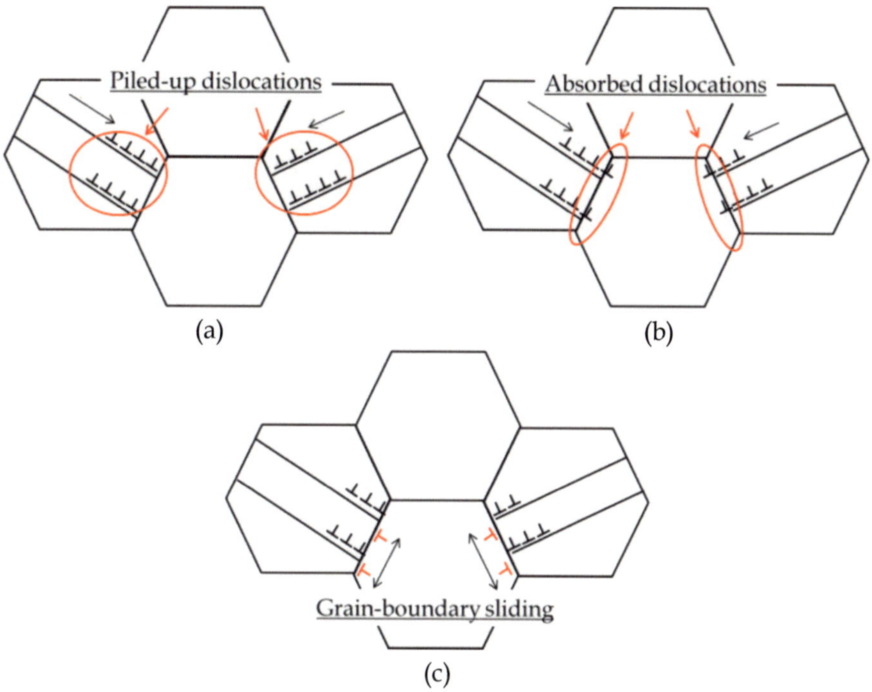

Fig. 5. Schematic drawing of slip-induced grain-boundary sliding: (a) pile-up lattice dislocations at grain boundaries, (b) which absorb these dislocations, (c) and thereby generating grain-boundary sliding.

Based on the discussion presented above, a constitutive relationship for the ambient temperature creep is considered. Because of a barrier against dislocation motion inside of each grain, the creep is controlled by recovery at grain boundaries. Therefore, a steady state creep rate is expressed as

$$\dot{\gamma} = \frac{r}{h}, \tag{5}$$

where r (=dτ/dt) is the recovery rate and h (=dτ/dγ) is the work-hardening rate (Orowan, 1934). In this condition, the stress applied to the leading dislocation piled up at the grain boundary, τ, is expressed as $NGb/\pi L$ where N denotes the number of piled-up dislocations (=ρL^2, ρ is the dislocation density), and L is the pile-up distance. The shear strain γ is related with the dislocation density as $sb\rho$ where s is the displacement of dislocations. Because h can be explained as (dτ/dρ)(dρ/dγ),

$$h = A\frac{G^2 b}{s\tau}. \tag{6}$$

Since the slip-induced grain-boundary sliding with shuffling proceeds through the dislocation absorption, the climbing rate based on the theory of superplasticity could be applied as a time-dependent change of dislocation density near the grain boundary, where is limited to b^2d. Thus, dρ/dt is expressed as following equation:

$$\frac{d\rho}{dt} = \frac{AD}{b^2}\left(\frac{Gb}{kT}\right)\left(\frac{b}{d}\right)^3\left(\frac{\tau}{G}\right)^2 , \tag{7}$$

where D is the diffusivity. Therefore, since r can be expressed as $(d\tau/d\rho)(d\rho/dt)$,

$$r = AD\left(\frac{G^2 b}{kT}\right)\left(\frac{b}{d}\right)^2\left(\frac{\tau}{G}\right)^2 . \tag{8}$$

Substituting equations (3) and (5) into equation (2) and interpreting s as d because there is few obstacles on dislocation motion inside of grains, the steady state creep rate is expressed by the equation (9);

$$\dot{\varepsilon} = AD_0\left(\frac{Gb}{kT}\right)\left(\frac{\sigma}{E}\right)^3\left(\frac{b}{d}\right)\exp\left(-\frac{Q}{RT}\right). \tag{9}$$

The effective diffusivity (D_{eff}) (Springarn et al., 1979) is given by the diffusivity of body diffusion (D_{v}), dislocation-core one (D_{d}) and shuffling (D_{s}):

$$D_{\mathrm{eff}} = D_{\mathrm{p}} + \alpha\left(\frac{\tau}{G}\right)^2 D_{\mathrm{d}} + \left(\frac{b}{d}\right)D_{\mathrm{s}} , \tag{10}$$

where α is a constant. Using the effective diffusion, all dislocation creep regions are expressed by an identical constitutive relation at all temperatures. Considering at very low temperature where the third term becomes dominant, equation (1) is obtained.

Using the constitutive relation, i.e., equation (9), the conventional deformation mechanism maps of hexagonal-close packed metals are rewritten. These maps of metals and alloys are often plotted in Ashby-type (Frost & Ashby, 1982) or Langdon-type (Langdon & Mohamed, 1978) maps. Each map is composed of modulus-compensated applied stress, homologous temperature and strain rate, and gives a particular deformation mechanism as a function of these parameters. Therefore, the maps are used to select a structural material in a design condition. In this paper, the Langdon-type map is used because each deformation region is split with a straight line.

The creep parameters of zinc are listed in Table 1 which includes the values of conventional creep regions represented by Ashby's textbook (Frost & Ashby, 1982). Figures 6a and 6b are conventional and modified deformation mechanism maps of zinc with d of 100 μm using the parameters in Table 1. The ambient-temperature creep region appears at low temperatures in the modified map. By means of this experiment, the detailed deformation mechanism maps of hexagonal close-packed metals are proposed and the new creep mechanisms are revealed.

Ultra-fine grained metals might show the similar deformation mechanism at low temperatures because they do not have any spaces generating cell structure inside of grains, which means that dislocation-grain-boundary interaction, i.e., slip-induced grain boundary sliding, becomes significant comparing with coarse-grained metals, bringing about the creep. Therefore, the kind of metals shows creep similar to the ambient-temperature creep of hexagonal close-packed metals. To reveal the above assumption, mechanical tests and some microscopy are in action using ultra-fine grained aluminum.

	n	p	Q [kJ/mol]
Ambient-temperature creep	3.4	1.2	18
Low-temperature dislocation creep	6.5	0	60
High-temperature dislocation creep	4.5	0	92
Coble creep	1.0	3.0	60
Nabarro-Herring (N-H) creep	1.0	2.0	92

Table 1. Creep parameters of zinc

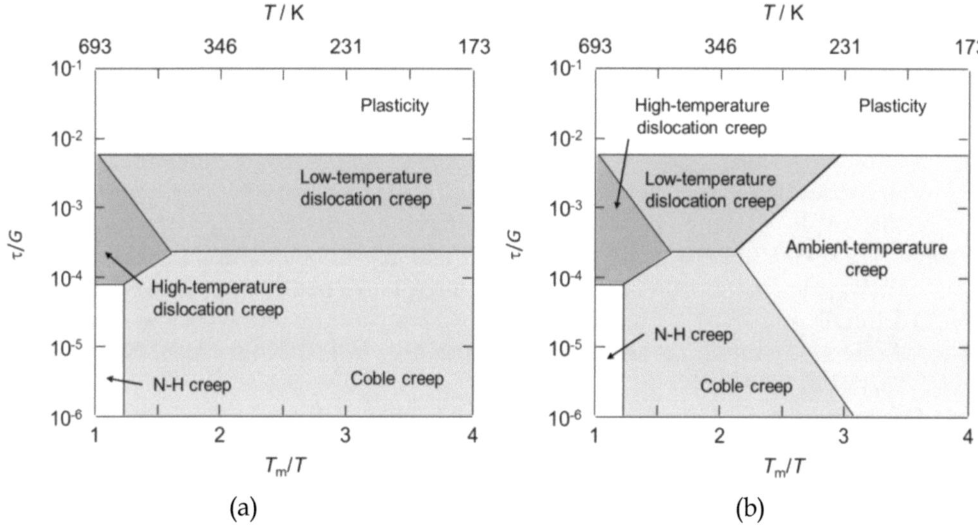

Fig. 6. (a) Conventional and (b) modified deformation mechanism maps of zinc. Using the parameters in Table1, the ambient-temperature creep region appears at low temperatures.

5. Conclusions

The role of the intergranular deformation mechanism was investigated to reveal an accommodation process in ambient-temperature creep of hexagonal close-packed metals using zinc by atomic force microscopy and electron back-scattered diffraction pattern analyses before and after the creep tests. Then, it evaluated the extension of grain-boundary sliding and revealed the role of grain boundary. These experiments produced the following important results:

1. Lattice rotation of about five degree is observed after a creep test, which shows that lattice dislocations piled up at grain boundaries.
2. Dislocations do not pass through grain boundaries because of the lack of equivalent slip systems in the hexagonal close-packed structure.
3. Grain boundary sliding depends on the grain boundary structure. The Σ-numbered grain boundary does not slide in ambient-temperature creep.
4. Strain by grain boundary sliding contributes to about 30% of the entire creep strain.

Therefore, the general representation of the ambient-temperature creep is described in Figs. 5 and 6 as a new creep mechanism.

6. Acknowledgment

The authors sincerely appreciate Prof. Takeshi Ogawa, Aoyama Gakuin University, for atomic force microscopy. The authors also appreciate funding and support from a Grant-in-Aid for Scientific Research and a Fellowship (08J00141) from the Japan Society for the Promotion of Science.

7. References

Adenstedt H. (1949). Creep of titanium at room temperature, *Metal Progress*, Vol.56, (November 1949), pp.658–660.

Bell, R.L. & Langdon T.G. (1967). An investigation of grain-boundary sliding during creep, *Journal of Materials Science*, Vol.2, No.4, (May 1967), pp.313-323, ISSN 0022-2461.

Blum, W. & Zeng, X.H. (2009). A simple dislocation model of deformation resistance of ultrafine-grained materials explaining Hall–Petch strengthening and enhanced strain rate sensitivity, *Acta Materialia*, Vol.57, No.6, (April 2009), pp.1966-1974, ISSN 1359-6454.

Bollmann, W. (1967). On the geometry of grain and phase boundaries I. General theory, *Philosophical magazine*, Vol.16, No.140, (March 1967), pp.363-381, ISSN 0031-8086.

Chen, Q & Sundman B. (2001). Calculation of debye temperature for crystalline structures – a case study on Ti, Zr, and Hf, *Acta Materialria*, Vol.60, No.6, (April 2001), pp.947-961, ISSN 1359-6454.

Deka, D.; Joseph, D.S.; Ghosh, S. & Mills, M.J. (2006). Crystal plasticitymodeling of deformation and creep in polycrystalline Ti-6242, *Metallurgical and Materials Transactions A*, Vol.37, No.5, (May 2006), pp.1371-1388, ISSN 1073-5623.

Frost, H.J. & Ashby, M.F. (1982). *DEFORMATION-MECHANISM MAPS*, Pergamon Press, ISBN 0-08-0293387, Oxford, England.

Furukawa, M.; Horita, Z.; Nemoto, M.; Valiev, R.Z. & Langdon, T.G. (1996). Microhardness measurements and the Hall-Petch relationship in an Al-Mg alloy with submicrometer grain size, *Acta Materialia*, Vol.44, No.11, (November 1996), pp.4619-4629, ISSN 1359-6454.

Garofalo, F. (1963). *Fundamentals of creep and creep-rupture in metals*, MacMillan, New York, USA.

Harper, J.G. Shepard, L.A. & Dorn, J.E. (1969). Creep of aluminum under extremely small stresses, *Acta Metallurgica*, Vol.6, No.7, (July 1958) pp.509-518, ISSN 0001-6160.

Imam, M.A. & Gilmore, C.M. (1979). Room temperature creep of Ti-6Al-4V, *Metallurgical Transactions A*, Vol.10, No.4, (April 1979), pp.419-425, ISSN 1073-5623.

Kiefer T.F. & Schwartzberg F.R. (1967). NASA-CR-92418.

Kiessel, W.R. & Sinnott, M.J. (1953). Creep properties of commercially pure titanium, *Journal of Metals*, Vol.5, No.2, (February 1953), pp.331-338, ISSN 0148-6608.

Koike, J.; Ohyama, R.; Kobayashi, T.; Suzuki, M. & Maruyama, K. (2003). Grain-Boundary Sliding in AZ31 Magnesium Alloys at Room Temperature to 523 K, *Materials Transactions*, Vol.44, No.4, (April 2003), pp.445-451, ISSN 1345-9678.

Langdon, T.G. & Mohamed, F.A. (1978). A simple method of constructing an Ashby-type deformation mechanism map, *Journal of Materials Science*, Vol.13, No.6, (June 1978), pp.1282-1290, ISSN 0022-2461.

Luster, D.R.; Wentz W.W. & Kaufman D.W. (1953). Creep properties of titanium, *Materials and Methods*, Vol.37, No.6, (June 1953), pp.100-103.

Matsunaga, T.; Kameyama, T.; Takahashi, K. & Sato. E. (2009a). Constitutive Relation for Ambient-Temperature Creep in Hexagonal Close-Packed Metals, *Materials Transactions*, Vol.50, No.12, (December 2009) pp.2858-2864, ISSN 1345-9678.

Matsunaga, T.; Kameyama, T.; Takahashi, K. & Sato. E. (2009b). Intragranular Deformation Mechanisms during Ambient-Temperature Creep in Hexagonal Close-Packed Metals, *Materials Transactions*, Vol.50, No.12, (December 2009) pp.2865-2872, ISSN 1345-9678.

Matsunaga, T.; Kameyama, T.; Ueda, S. & Sato. E. (2010). Grain boundary sliding during ambient-temperature creep in hexagonal close-packed metals, *Philosophical Magazine*, Vol.90, No.30, (October 2010), pp.4041-4054, ISSN 1478–6435.

Mussot, P.; Rey, C. & Zaoui, A. (1985). Grain boundary sliding and strain compatibility, *Res Mechanica: International Journal of Structural Mechanics and Materials Science*, Vol.14, No.1, (January-Mars 1985), pp.69-79, ISSN 0143-0084.

Neeraj, T.; Hou, D.H.; Daehn, G.S. & Mills, M.J. (2000). Phenomenological and microstructural analysis of room temperature creep in titanium alloys, *Acta Materialia*, Vol.48, No.6, (April 2000), pp.1225-1238, ISSN 1359-6454.

Odegard B.C. & Thompson A.W. (1974). Low temperature creep of Ti-6Al-4V, *Metallurgical Transactions A*, Vol.5, No.5, (May 1974), pp.1207-1213, ISSN 1073-5623.

Orowan, E. (1934). Zur Kristallplastizität. II - Die dynamische Auffassung der Kristallplastizität, *Zeitschrift für Physik A Hadrons and Nuclei*, Vol.89, No.9-10, (September 1934), pp.614-633, ISSN 0939-7922.

Parker, J.D. & Wilshire, B. (1977). A surface measurement study of grain-boundary sliding during creep of a two-phase, copper-cobalt alloy, *Materials Science and Engineering*, Vol.29, No.3, (August 1977) pp.219-225.

Rae, C.M.F. & Smith, D.A. (1980). On the mechanism of grain boundary migration, *Philosophical Magazine A*, Vol.41, No.4, (April 1980), pp.477-492, ISSN 0141-8610.

Reiman W.H. (1971). Room temperature creep in Ti-6Al-4V, *Journal of Materials*, Vol.6, No.4, (December 1971), pp.926-940.

Rust, M.A. & Todd, R.I. (2011). Surface studies of region II superplasticity of AA5083 in shear: Confirmation of diffusion creep, grain neighbor switching and absence of dislocation activity, *Acta Materialia*, Vol.59, No.13, (August 2011) pp.5159-5170, ISBN 1359-6454.

Sato, E.; Yamada, T.; Tanaka, H. & I. Jimbo (2006). Categorization of Ambient-Temperature Creep Behavior of Metals and Alloys on Their Crystallographic Structures, *Materials Transactions*, Vol.47, No.4, (April 2007), pp.1121–1126, ISSN 1345-9678.

Schober, T & Balluffi, R.W. (1970). Quantitative observation on misfit dislocation arrays on low and high angle twist grain boundaries, *Philosophical magazine*, Vol.21, No.169, (January 1970), pp.109-124, ISSN 0031-8086.

Smith, D.A. & King, A.H. (1981). On the mechanism of diffusion induced boundary migration, *Philosophical Magazine A*, Vol.44, No.2, (July 1981), pp.333-340, ISSN 0141-8610.

Spingarn, J.R.; Barnett, D.M. & Nix, W.D. (1979). Theoretical descriptions of climb controlled steady state creep at high and intermediate temperatures, *Acta Metallurgica*, Vol.27, No.9, (September 1979), pp.1549-1561, ISSN 0001-6160.

Suri, S.; Viswanathan, G.B.; Neeraj, T.; Hou, D.H. & Mills, M.J. (1999). Room temperature deformation and mechanisms of slip transmission in oriented single-colony crystals of an α/β titanium alloy, *Acta Materialia*, Vol.47, No.3, (February 1999), pp.1019-1034, ISSN 1359-6454.

Takahashi, T. & Horiuchi, R. (1985). Coupling of sliding and migration in coincidence boundaries of Zn and Al bicrystals and their DSC dislocation models, *Transactions of Japan Institute of Metals*, Vol.26, No.11, (November 1985), pp.786-794, ISSN 1345-9678.

Tanaka, H.; Yamada, T.; Sato, E. & Jimbo, I. (2006). Distinguishing the Ambient-Temperature Creep Region in a Deformation Mechanism Map of Annealed CP-Ti, *Scripta Materialia*, Vol.54, No.1, (January 2006), pp.121-124, ISSN 1359-6462.

Thompson A.W. & Odegard B.C. (1973). The influence of microstructure on low temperature creep of Ti-5Al-2.5Sn, *Metallurgical Transactions A*, Vol.4, No.4, (April 1973), pp.899-908, ISSN 1073-5623.

Valiev, R.Z. & Kaibyshev, O.A. (1977). Mechanism of superplastic deformation in a magnesium alloy-2. The role of grain boundary, *Physica Status Solidi (A)*, Vol.44, No.2, (December 1977), pp.477-484, ISSN 1862-6319.

Valiev, R.Z.; Korznikov, A.V. & Mulyukov, R.R. (1993). Structure and properties of ultrafine-grained materials produced by severe plastic deformation, *Material Science and Engineering A*, Vol.168, No.2, (August 1993), pp.141-148, ISSN 0921-5093.

Watanabe, T.; Yamada, M.; Shima, S. & Karashima, S. (1979). Misorientation dependence of grain boundary sliding in $<10\bar{1}0>$ tilt zinc bicrystals, *Philosophical Magazine A*, Vol.40, No.5, (November 1979), pp.667-683, ISSN 0141-8610.

Watanabe, T.; Kimura, S. & Karashima, S. (1984). The effect of a grain boundary structural transformation on sliding in $<10\bar{1}0>$-tilt zinc bicrystals, *Philosophical Magazine A*, Vol.49, No.6, (Jun 1984) pp.845-864, ISSN 0141-8610.

10

Elastic and Nanowearing Properties of SiO_2-PMMA and Hybrid Coatings Evaluated by Atomic Force Acoustic Microscopy and Nanoindentation

J. Alvarado-Rivera[1], J. Muñoz-Saldaña[2] and R. Ramírez-Bon[2]
[1]*Departamento de Investigación en Física, Universidad de Sonora*
[2]*Centro de Investigación y de Estudios Avanzados del IPN, Unidad Querétaro*
Mexico

1. Introduction

Organic-inorganic hybrid materials have been widely studied during the past two decades. The capability to combine materials which are completely different has expanded the number of new materials with unique properties that are possible to synthesize. Thus, hybrid materials are nanocomposites where at least one of the components has a domain size of at least several nm [Sanchez, 1994]. In 1985, the first hybrid material was synthesized by the sol-gel process, combining alkoxides with organic monomers or oligomers leading to a hybrid network [Huang et al., 1985; Schmidt, 1985]. At this respect, sol-gel is a feasible, simple and low-cost technique. The hybrid materials obtained are polydisperse in size and locally heterogeneous in chemical composition [Sanchez et al., 2005]. These hybrids are also inexpensive, versatile and present many interesting applications converting them into makertable products as films, powders or monoliths. Among the wide variety of hybrid materials, SiO_2-Poly(methylmethacrylate) (SiO_2-PMMA), hybrids have shown interesting properties as higher hardness than some common thermoplastics like poly(methyl methacrylate), poly(ethylene terpththalate) and poly(ethylene naphthalate); low surface roughness; low friction coefficient; and high optical transparency. These properties are attractive to be applied as protective coatings as well as in the fabrication of electronic devices on flexible substrates like OLED's used as both planarization and insulator coatings. Plastic surface roughness lower than 2 nm are required in order to start deposition of ceramic films to assemble the device. This ultra-smooth surface can be achieved by hybrid coatings due to the combination of components at a molecular level. Moreover, SiO_2-PMMA hybrids are flexible, a property conferred by the organic component, and thus they can undergo higher deformations before cracking if applied on plastic substrates.

Mechanical property evaluation using conventional mechanical testing is not suitable to evaluate mechanical properties of thin films, but instrumented indentation has provided a viable method. In nanoindentation, the surface of a material of unknown properties is indented at high spatial resolution with another material of known properties (such as diamond) while load and displacement is continuously recorded [Fischer-Cripps, 2004; Hay

& Pharr, 2000]. The load-displacement curves obtained from the test are used to determine elastic modulus and hardness by applying the method developed by Oliver and Pharr [Oliver & Pharr, 1992]. Moreover, nanoindentation also allows performance of wear tests at nanometric scale on thin films. Nanowells can be "machined" by the indenter applying a normal force. Then the wear resistance can be evaluated in terms of the wear loss volume, which is obtained by image analysis [Alvarado-Rivera et al., 2007, 2010].

As mentioned above, organic and inorganic phases are bonded at a molecular scale, and consequently the obtained films are optically transparent. Thus it is difficult to observe the distribution of both phases using conventional scanning electron microscopy. However, atomic force acoustic microscopy is a technique that offers the ability to observe a material's phase distribution at the surface by measuring the difference in elastic properties. Amplitude vibration of the surface is mapped by a cantilever (excited at a fixed frequency near its resonance). Depending on the variations of the local contact stiffness, resonance frequency will vary, causing the amplitude vibration of the work frequency to change, which will be reflected in contrast differences on the amplitude images [Rabe, 2006; Kopycinska-Müller et al., 2007]. In this chapter, the elastic and viscoelastic behaviour, hardness and nanowear characterization of SiO_2-PMMA hybrid coatings by atomic force acoustic microscopy (AFAM) imaging and nanoinentation are presented and discussed.

2. Wear behaviour of SiO_2-PMMA hybrid coatings reinforced with Al_2O_3 whiskers and nanoparticles

2.1 Materials preparation and experimental set-up

For the sol-gel synthesis of the hybrid coatings, tetraethylortosilicate (TEOS) and methyl methacrylate (MMA) were used as silica and polymethyl methacrylate (PMMA) precursors. To crosslink both phases, 3-Trimetoxisilylpropyl methacrylate was added to the precursor formulation. The TEOS:MMA:TMSPM molar ratio was chosen according to the enhanced mechanical properties of the hybrid coatings obtained with this precursor composition. In this study the effect of Al_2O_3 nanoparticles and nanowhiskers on the hybrid coatings was analysed. Four hybrid solutions with different content of Al_2O_3 nanoparticles and whiskers were prepared. Details of the combinations are shown in Table 1. The alumina nanoparticles (pAl_2O_3) and whiskers (wAl_2O_3) were provided as nanopowders. The nanoparticles have an average size of 50 nm with a superficial area of 33 m^2/g. The whiskers have a diameter of 2-4 nm and a length of 2800nm. This information was provided by the supplier, Sigma-Aldrich.

Sample	TEOS:MMA:TMSPM (Molar ratio)	Al_2O_3 Nanoparticles (wt. %)	Al_2O_3 Whiskers (wt.%)
SiO_2-PMMA	1:0.25:0.25	0	0
SiO_2-PMMA-0.05pAl_2O_3	1:0.25:0.25	0.05	0
SiO_2-PMMA-0.1pAl_2O_3	1:0.25:0.25	0.1	0
SiO_2-PMMA-0.1wAl_2O_3	1:0.25:0.25	0	0.1

Table 1. Details of the prepared solutions.

Al_2O_3 nanoparticles and whiskers were added in a beaker with deionized water and then placed into an ultrasonic bath for 20 minutes to aid dispersion. TEOS was mixed with the deionized water containing the nanoparticles or the whiskers, then ethanol and clorhidric acid (HCl) as a catalyst were added and mixed together for 30 minutes. The MMA precursor solution was prepared by adding benzoil peroxide to promote radical polymerization. The crosslinking agent was also mixed with deionized water and ethanol to hydrolyse the metoxy groups of the molecule. Afterwards, the three precursor solutions were mixed together in a molar ratio of TEOS:MMA:TMSPM of 1:0.25:0.25 and then the solution was stirred for 30 minutes. The solutions were left to age for one day. Sets of Corning glass and commercial acrylic substrates were coated with the four different hybrid solutions by dip coating. Finally, the coated substrates were dried in a conventional oven at 70°C for 6 hours.

2.1.1 Nanoindentation testing

Nanoindentation or instrumented indentation is a technique, which basically consists of indenting a material of unknown properties and continuously recording the applied load and displacement with high spatial resolution [Hay & Pharr, 2000]. In this technique it is essential to have a material of known properties for the calibration of the measurement system prior to the analysis. All the nanoindentation tests and nanowearing testing were performed on a Hysiton Ubi-1 nanoindenter (Minneapolis, MN) equipped with a Berkovich diamond tip. In Fig. 1 a schematic representation of the nanoindentation set-up is presented. The equipment is provided with a piezoelectric actuator, which precisely moves the tip before and after the test. The system is able to record images of the sample surface before and after the indentation and perform scratch testing using the same diamond tip used for indentation. The force/displacement transducer (shown in Fig. 1) consists of an arrangement of three capacitive plates, with the tip attached to a beam that is fixed to the central plate. The force is administered by applying a voltage to the bottom capacitor, creating an electrostatic attraction between the bottom and central plate, and moving the central plate towards the bottom. The magnitude of the force is calculated from the applied voltage. The load-displacement curves (*P-h*) obtained from the indentation test (an example is depicted in Fig. 1) are used to extract the hardness and elastic modulus of the indented material using the method introduced by Oliver and Pharr [Oliver & Pharr, 1992]. A simple load-unload cycle was used to perform the indentation test on the hybrid coatings. Several indents were applied varying the load from 50 μN to 9 mN . Since the hardness calculated from the *P-h* curves is the hardness of the substrate/film system, the work-of-indentation model was applied to extract the film hardness [Korsunsky et al., 1998)]. This method describes the behaviour of the hardness over a wide range, and it can be applied to systems with either fracture of plastic deformation. The method is described by the following equation:

$$H_c = H_s + \frac{H_f - H_s}{1 + k\beta^2} \quad (1)$$

where H_c is the composite hardness (substrate/film system); H_s and H_f are the substrate and film hardness, respectively; and k is a constant. In order to apply this equation to fit the experimental data, it is necessary to plot H_c as a function of the relative indentation depth, $\beta = h/t$, which is the penetration depth (h) divided by the film thickness (t). The film thickness of all hybrid coated glass and acrylic were determined using a Dektak II profilometer.

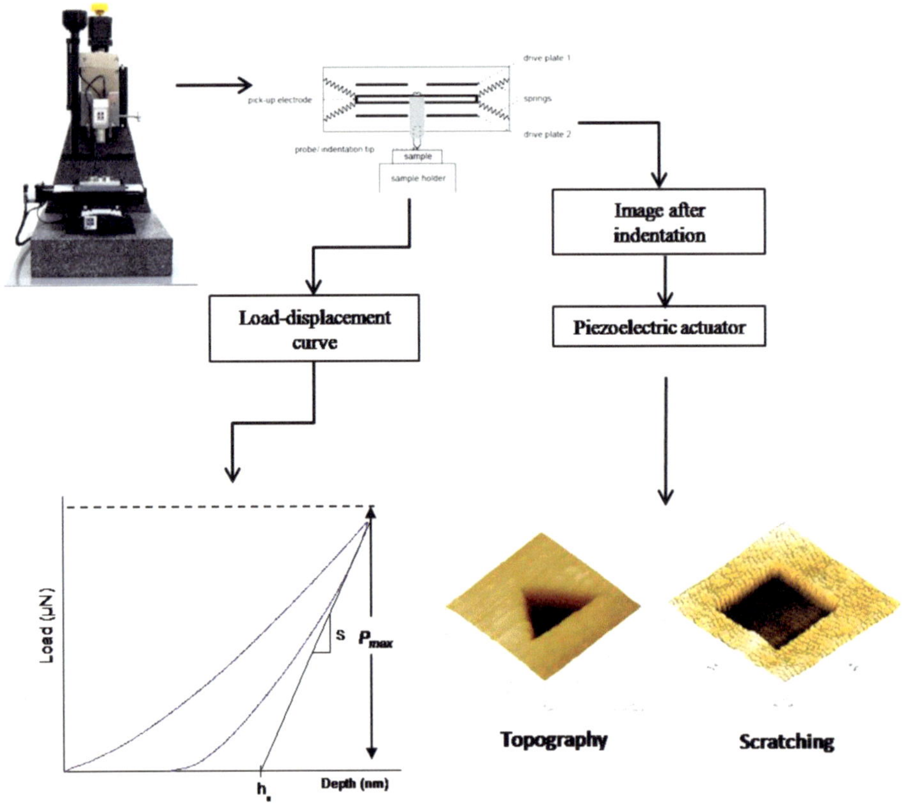

Fig. 1. Schematic representation of the Hysitron-Ubi-1 nanoindentation system.

2.1.2 Sliding life test

Sliding life tests were carried out on a pin-on disc tribometer (CSM Instruments) using a 10 mm diameter steel pin and applying a normal force of 1 N without the aid of lubricant. The test consists of the steel pin sliding on the sample surface at a constant velocity, and the instantaneous friction coefficient is recorded as a function of the covered distance. For the hybrid coatings a velocity of 1 cm/s was used.

2.1.3 Nanowear

Nanoscratch testing was performed with the Hysitron Ubi-1 nanoindenter by machining nanometric cavities on the coating surface using the Berkovich diamond tip. A fixed area of 6 x 6 µm was scanned ten times applying a normal load of 70 µN at a frequency of 1Hz. In Fig. 2 a schematic representation of the nanoscratch test is presented. The diamond tip moves along the x-axis removing material; after the scratching is finished an image of the surface is then taken. Afterwards, the images of the nanowells were analyzed using WSxM (software) to determine the material wear loss volume [Horcas et al., 2007]. The borders of the wells are irregular due to the tip geometry, thus with image analysis is possible to determine the real borders of the hole and the volume of the wells using the command flooding, which can detect holes and hills on the image establishing a reference height.

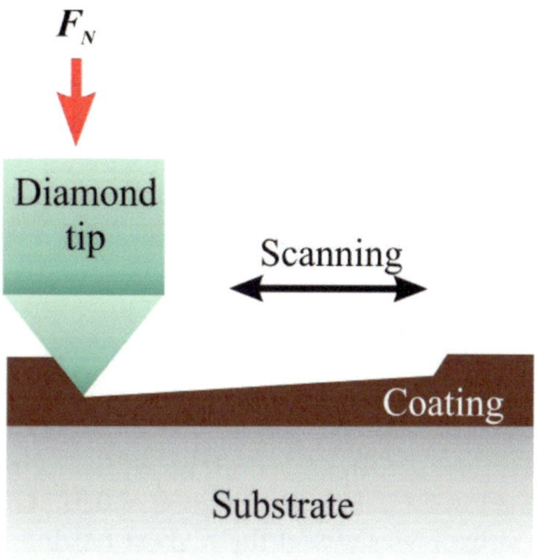

Fig. 2. Schematic representation of nanoscratch testing on the sample surface performed by the nanoindenter diamond tip.

2.2 Hybrid optical characterization

Fig. 3 shows the UV-Vis optical transmission of the studied hybrid coatings and acrylic substrate for comparison. It is observed that the acrylic substrate has about 92 %, on average, optical transmission in the visible range. All hybrid coatings clearly show higher optical transmittance than the substrate in the visible region. There is no evidence of light dispersion due to the incorporated nanoparticles or whiskers. The oscillations observed in the spectra of the hybrid layers are due to interference produced because the coating thickness is of the order of the incident wavelength. The high optical transparency of the hybrid coatings evidences that there is no phase separation in the hybrid materials, and therefore they are homogeneous, showing incorporated organic and inorganic phases.

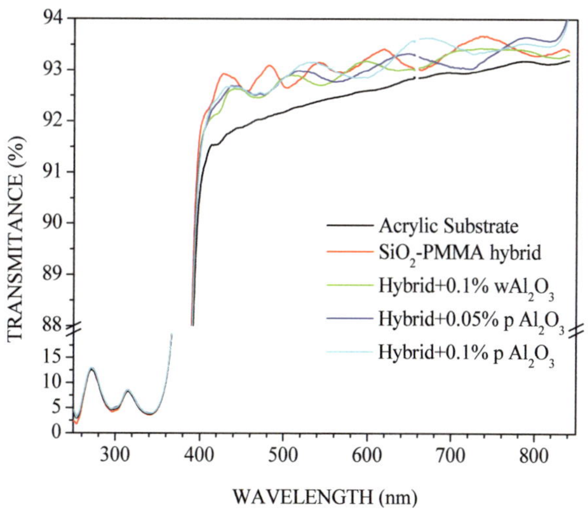

Fig. 3. Optical transmission spectra of the hybrid coatings and the acrylic substrate.

2.3 Nanoindentation hardness

A series of at least 25 indentations in the range load of 30 to 6000 μN were performed on each type of hybrid coating, and uncoated acrylic substrate was also tested. For comparison a Corning glass slice was coated with the hybrid solution with 0.1 wt.% alumina whiskers. The composite hardness (H_c) as a function of the relative indentation depth of the hybrid coatings is displayed in Fig. 4. In all cases the curves show the characteristic form of a hard coating on a soft substrate system. In Fig. 4 a) the coatings with alumina nanoparticles and the hybrid coating without alumina are compared. The continuous line for each set of data corresponds to the work of indentation model fitting to estimate the film hardness (H_f). The hardness of the coatings with 0.05 and 0.1 wt.% were similar to that of the SiO_2-PMMA coating. However these values are higher than that of the acrylic substrate of 0.17 GPa, also measured by nanoindentation. The hybrid coating with alumina nanowhiskers on the acrylic substrate showed the lowest values of hardness of 0.512 GPa (Table 2). However, the same coating on the glass substrates showed the highest hardness value of 0.814 GPa. This difference can be attributed to the substrate itself, since glass is much harder than PMMA. Thus, when the coating/acrylic substrate system is under load the coating deformation volume also includes the substrate, and its elastic properties contribute to the final measurement of the penetration depth, and consequently the hardness and reduced modulus. In the case of the coating on glass, the substrate has a high elastic modulus and is harder than the coating, and thus the deformation volume under the indenter is confined within the coating [Fischer-Cripps, 2004].

2.4 Sliding life testing

All the hybrid coatings were tested to determine their sliding life and friction coefficient. The results for the coatings with better performance are show in Fig. 5; the others failed in early stages of the test. The coatings that showed more resistance for the test were those with alumina nanowhiskers with a friction coefficient lower than that of the acrylic substrate. The coating with the best performance was SiO_2-PMMA-0.1wAl_2O_3 on glass, while the same coating on acrylic started to fail after 4 m of sliding. This can be explained in terms of the differences in elastic properties of the substrates in the same way that occurs for the hardness explained in the previous section. For this test, a pin with a 10 mm of radius applying a normal load of 1N is in contact with the coating surface. The size of the pin and the applied load are high enough to generate a large deformation volume within the substrate, affecting the coating which is also being subjected to wear. It is worth to mention that this coating on acrylic resisted more than the hybrid without alumina whiskers also on acrylic, thus alumina whiskers provide reinforcement in terms of enhancing the wear resistance of the SiO_2-PMMA hybrid material.

The hybrid-whisker coating on glass resisted the test without failing after 40 m of sliding, but it showed some abrupt increments of the friction coefficient. As the sliding contact between the surfaces continues the material adhered to the pin is detached and transferred to the coating surface by cold welding causing a drop in μ values. Optical photomicrographs of the wear zone on the hybrid surfaces were taken and they are displayed in Fig. 6. For hybrid coatings both with and without whiskers on acrylic substrate, zones with removed material, caused by fracture of the coating surface, are visible. The generated debris promotes abrasive wear between the sliding surfaces (increasing friction) and, thus, the

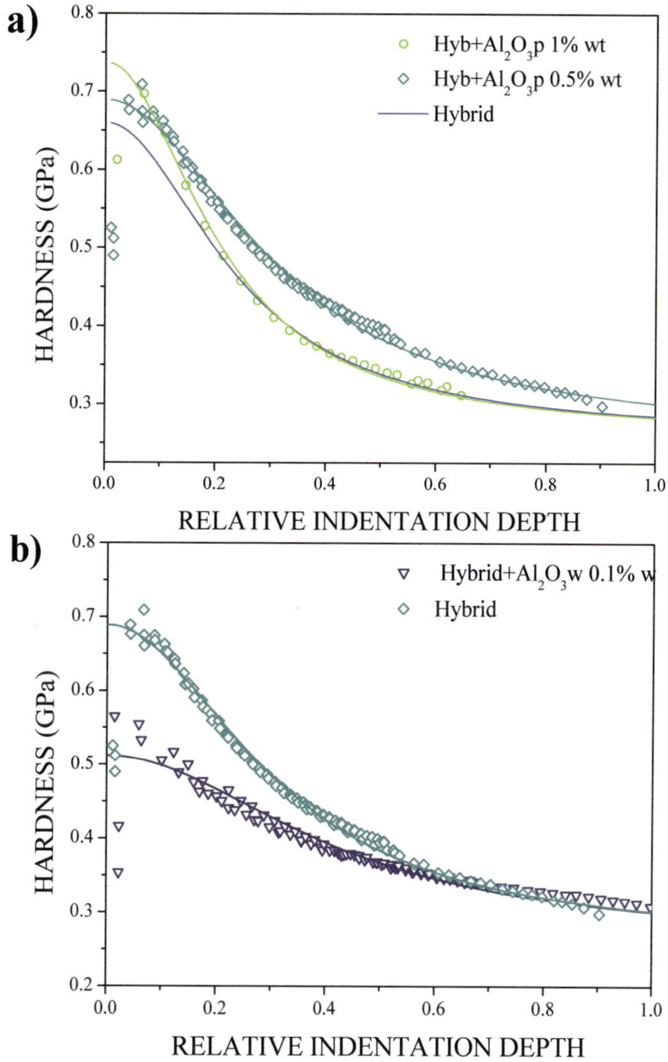

Fig. 4. Composite hardness as a function of the relative indentation depth of the hybrid coatings on acrylic: a) Al_2O_3 nanoparticles and b) Al_2O_3 whiskers.

Sample	**Film Hardness (GPa)**
Hybrid SiO_2-PMMA	0.689
Hybrid+0.05% pAl_2O_3	0.66
Hybrid+0.1% pAl_2O_3	0.738
Hybrid+0.1% wAl_2O_3on Acrylic	0.512
Hybrid+0.1%wAl_2O_3on glass	0.814

Table 2. Film hardness of all hybrid coatings.

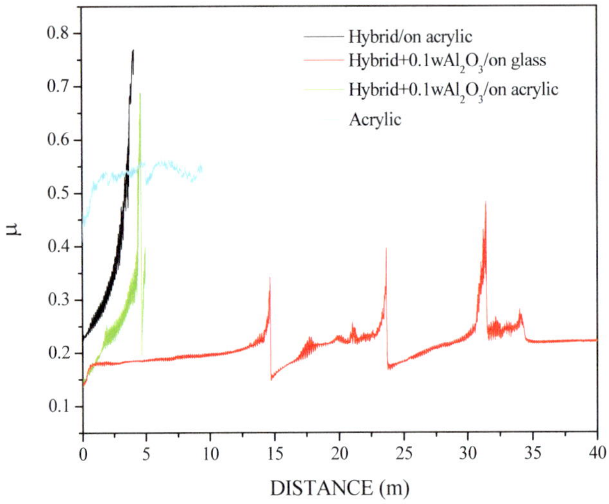

Fig. 5. Sliding life test of the hybrid coatings with and without reinforcement.

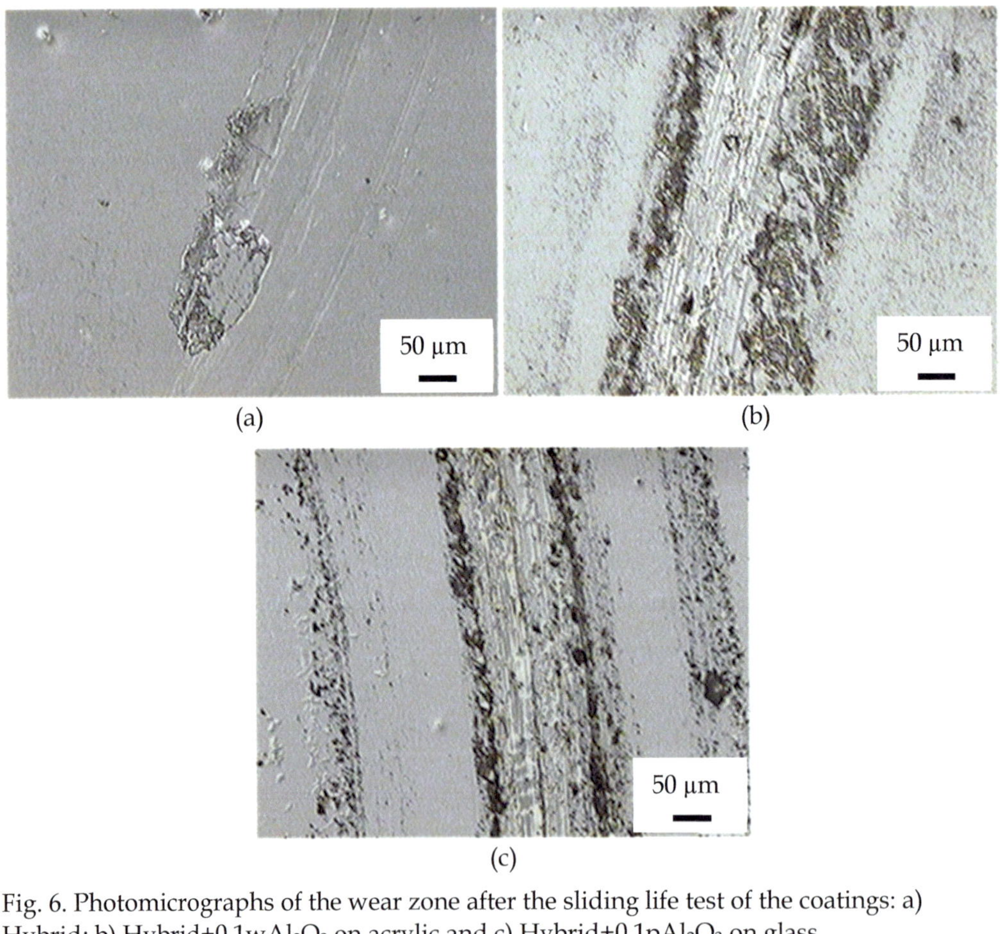

Fig. 6. Photomicrographs of the wear zone after the sliding life test of the coatings: a) Hybrid; b) Hybrid+0.1wAl_2O_3 on acrylic and c) Hybrid+0.1pAl_2O_3 on glass.

failure of the coating. As mentioned before, the coating on glass did not fail under sliding conditions, but in Fig. 6c, one can see on the sliding path, some material encrusted due to cold welding of previously removed coating pieces. Comparing the performance of both types of reinforcement, nanoparticles increase friction between the sliding surfaces when the cracking of the coating starts because alumina is harder than the coating and thus the cracking extends until the film completely fails. However, this does not occur with whiskers, which can be attributed to the shape and size of the materials. Whiskers are short discontinuous fibres with a diameter substantially smaller than the length (the whiskers used in this study have a diameter of 2-4 nm and a length of 2800 nm). The length provides more resistance to applied normal forces and instead of cracking only plastic deformation occurs. In this process some material was removed and adhered to the pin due to the local heating caused by friction, making the hybrid material more fluid.

2.5 Nanowear testing

Fig. 7 shows scaningg probe microscopy (SPM) images of the wear cavities on the coating surfaces after wear testing together along with height profiles of the transverse section. The wear-loss volume was estimated from the depth of the cavities using WSxM software as explained earlier. The profile graphs show that the hybrid material tends to form pile-ups on the cavity edges, generated by the material flow towards the coating surface during the scratch testing (due to the normal and lateral forces applied by the indenter). This is also related to the viscoelastic nature of the hybrid, which is a contribution of the PMMA component; this behaviour will be further explained in a subsequent section. The SiO_2-PMMA hybrid coating depth profile shows that the edges and walls of the well are more defined than those of the other coatings. The hybrid is a porous material, so when scratched it will tend to compress under the action of the applied normal force during the scanning. If we compare the SiO_2-PMMA-0.05pAl_2O_3 coating with the SiO_2-PMMA, it is possible to observe that the depth of the cavity is smaller than that of the coating without alumina nanoparticles. Another feature that can be observed from the SPM images is that no trace of debris was detected in any of the coatings. The diamond tip does not remove material from the surface; the hybrid is plastically deformed instead. Thus, the nanoparticles at the surface are "displaced" by the tip during scratching, as can be observed for the SiO_2-PMMA-0.05pAl_2O_3 coating (Fig. 7-b). The alumina nanoparticles are harder than the coatings, so when the normal force is applied the nanoparticles move within the hybrid film while it is plastically deformed. The SiO_2-PMMA-wAl_2O_3 showed a wear-loss volume higher than that of the SiO_2-PMMA-0.1pAl_2O_3. These results can be related to the coating hardness, where the coating with nanowhiskers had the lowest.

Values of wear-loss volume of all hybrid coatings and the acrylic substrate as a function of alumina concentration are presented in Fig. 8. As can be observed, the wear-loss volume of the coatings diminishes as the concentration of alumina nanoparticles increases. At higher contents of nanoparticles the material has more resistance to being compressed by both applied normal and lateral forces. In nanoindentation testing to evaluate hardness, a normal force is applied to the surface of a magnitude necessary to plastically deform the material. In scratch test when lateral and normal forces are applied the material is compressed and flows to the surface, comparable with a hardness test in 2 dimensions. The wear-loss volume of the acrylic substrate without coating is larger by two orders of magnitude than that of all

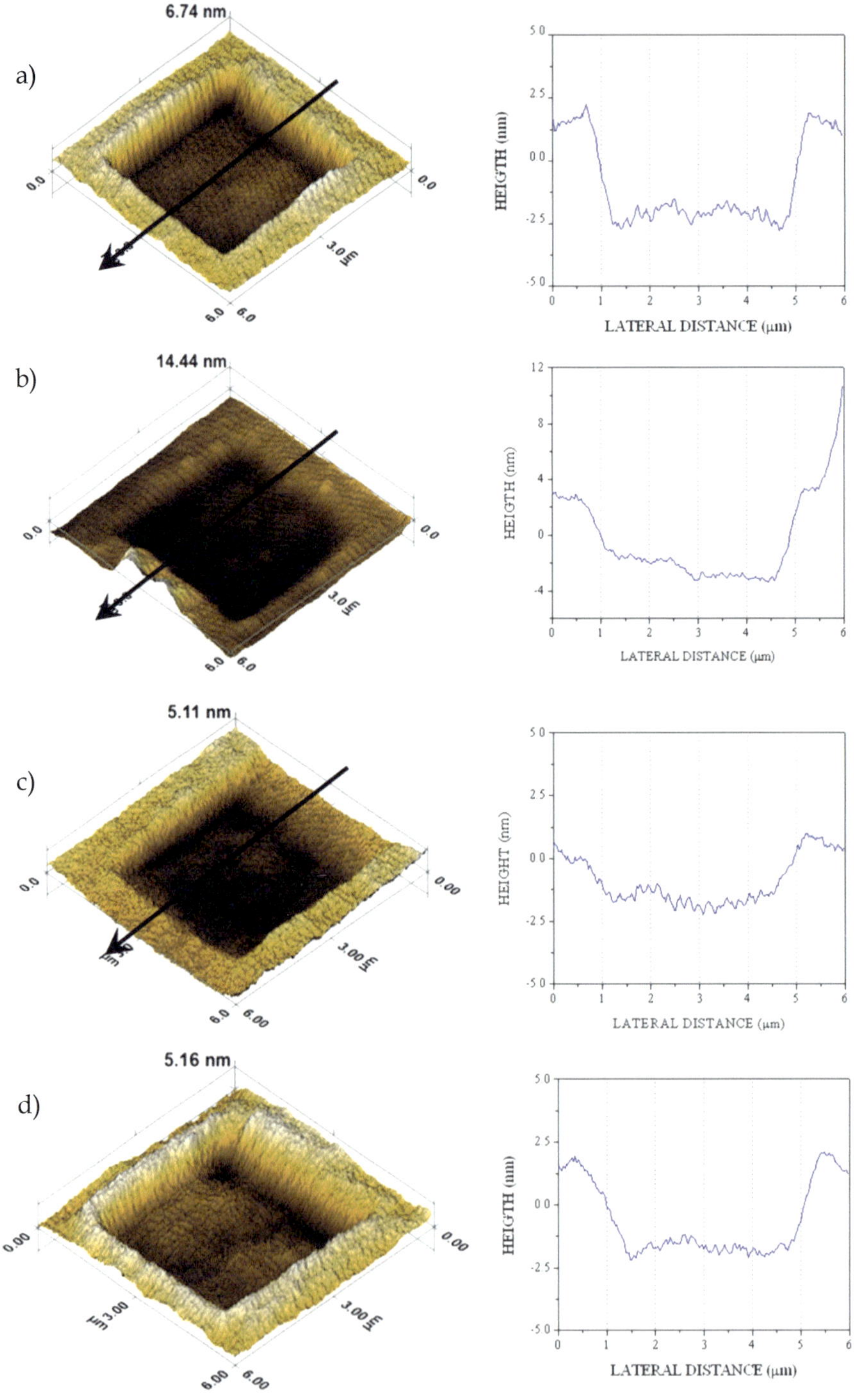

Fig. 7. SPM images of the dwells and the corresponding height profiles after the nanoscratch tests on the surface of: a) Hybrid; b) Hybrid+0.05pAl_2O_3, c) Hybrid+0.1pAl_2O_3 y d) Hybrid+0.1wAl_2O_3.

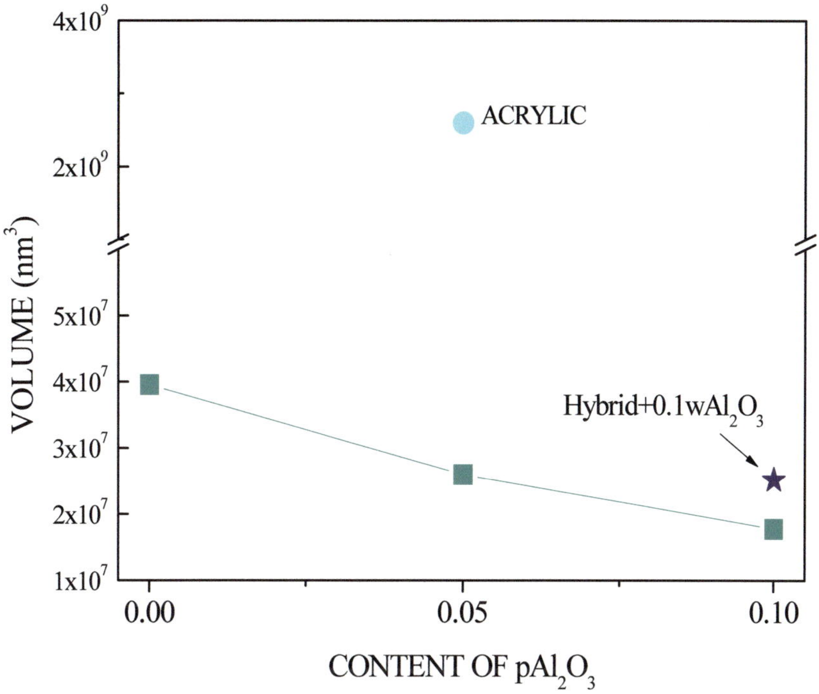

Fig. 8. Wear loss volume of the tested hybrid coatings.

hybrid coatings. The wear-loss volume of the SiO_2-PMMA-0.05pAl_2O_3 and SiO_2-PMMA-0.1wAl_2O_3 coatings decreases by 32-34%, while for the SiO_2-PMMA-0.1pAl_2O_3 coating it drops down by 55%. Therefore, the addition of alumina nanoparticles and whiskers effectively improves the abrasion resistance of the SiO_2-PMMA hybrid coatings.

3. Elastic properties of SiO_2-PMMA coatings measured by atomic force microscopy

3.1 Material preparation

The hybrid coatings were prepared using TEOS and MMA as precursors and TMSPM as cross linker. For this set of samples a fixed TEOS:MMA composition of 1:0.25 was used, only the TMSPM content was varied from 0.05 to 0.2 molar ratio with respect to TEOS. Corning glass slices were dip coated with the hybrid solutions, and the wet films were dried in an oven. Two different temperatures, 80 and 90°C, were used, along with two drying times, 3 and 6 h. Additional coatings were prepared on silicon substrates for infrared spectra measurements to monitor the hybrid formation. In Table 3, the composition, drying conditions and acronyms to designate each coating are presented.

FT-IR measurements were performed in a Perkin Elmer Spectrum GX. Film thickness of the coatings was measured with a Dektak II Profilometer. Nanoindentation testing was performed on a Hysitron Ubi-1 nanoindenter (Minneapolis, MN). Images of the surfaces to evaluate roughness were taken in a Nanoscope IV Atomic Force Microscope in tapping mode using a silicon rectangular cantilever.

Hybrid composition	Drying conditions		Acronym
	Temperature (°C)	Time (h)	
1:0.25:0.05	80	3	5803
	80	6	5806
	90	3	5903
	90	6	5906
1:0.25:0.2	80	3	20803
	80	6	20806
	90	3	20903
	90	6	20906

Table 3. Characteristics of the synthesized hybrid coatings.

3.2 Hybrid characterization

In Fig. 9 the FT-IR spectra of all hybrid coatings are presented. For all coatings a broad band at 1070 cm^{-1} is present and corresponds to the Si-O-Si bonds, characteristic of silica. This band shows a shoulder at ~1170 cm^{-1}, formed by the absorption bands at 1144 and 1200 cm^{-1}, corresponding to the C=O and C-O-C bonds, respectively, and indicating the presence of PMMA. It is worth noticing that this shoulder increases in intensity at the highest content of TMSPM in the hybrid solution, thus it can be assumed that the cross linker promotes the formation of PMMA chains. Moreover, the presence of the bands at 1640 cm^{-1} and near 1700 cm^{-1} corresponding to the C=C and C=O, respectively, are present, confirming the presence of PMMA in the hybrid material. The band at 940 cm^{-1} corresponds to uncondensed silanol groups and also the broad band at 3370 cm^{-1} characteristic of the presence of OH^- groups.

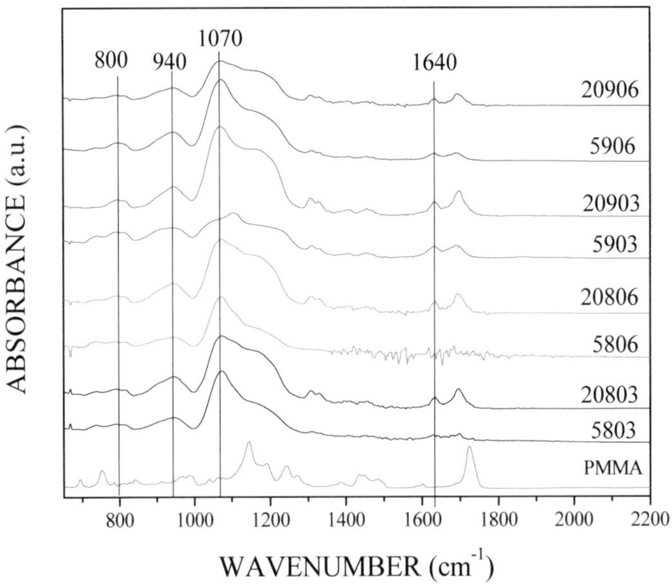

Fig. 9. Infrared spectra of the synthesized hybrid coatings. The spectrum of pure PMMA is also included for comparison.

To evaluate the surface quality of the hybrid coatings, atomic force microscopy (AFM) images in tapping mode were taken in a 5 x 5 μm area. The images and the corresponding values of RMS roughness are presented in Fig. 10. For all measured coatings smooth surfaces were revealed with ultra-low roughness values of less than 1 nm. This feature is of great interest in the fabrication of electronic devices, as extremely flat surface are required as a base for further growth of the thin films that compose the electronic devices.

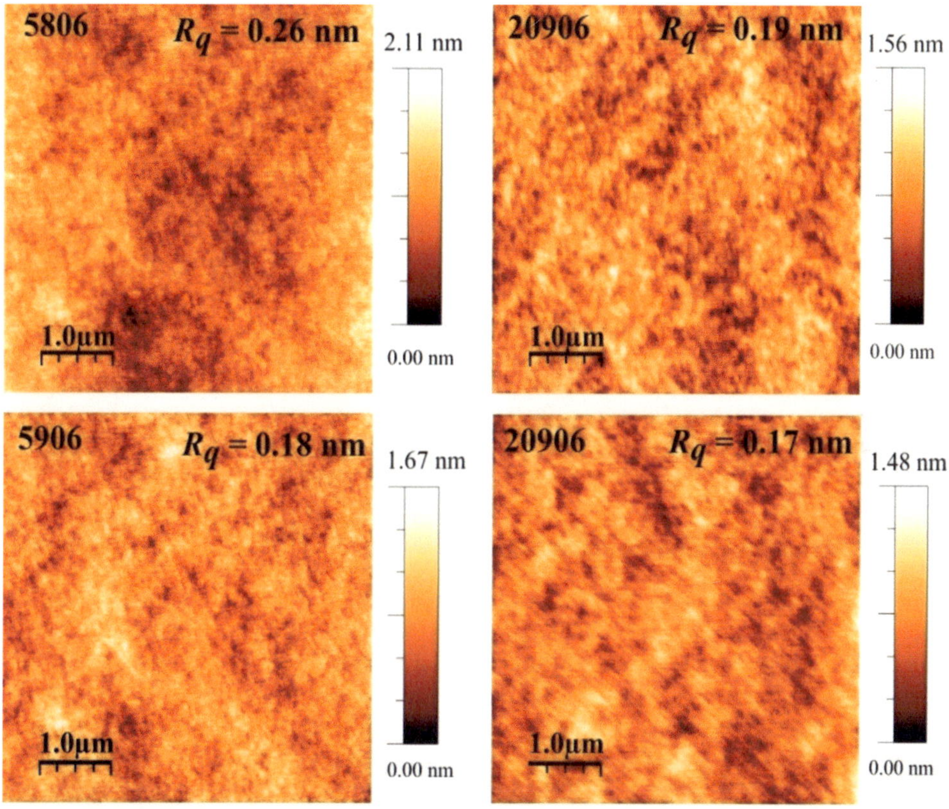

Fig. 10. AFM images of the surface of hybrid coatings.

3.3 Nanoindentation testing

One of the characteristics of the hybrid coatings that can be evaluated by nanoindentation is the viscoelastic behaviour, which is defined as follows. A material under an applied stress deforms with a combined behaviour of an elastic solid and a viscous flow. Thus, in viscoelastic materials, the stress-strain relationship depends on time or frequency [Lake, 2004]. Viscoelastic behaviour is also related with materials that have a glass transition temperature similar to thermopolymers such as PMMA. The load-displacement curve of a material with time-dependent response will show a "nose" at the beginning of the unloading curve, as can be seen in Fig. 11a. The material continues to flow under the indenter tip when it is under load; this phenomenon is called creep [Tweedie & Van Vliet, 2006]. A strategy to avoid this type of unloading curve shape is to maintain the maximum load for a certain time, allowing the material to flow viscously until it deforms permanently,

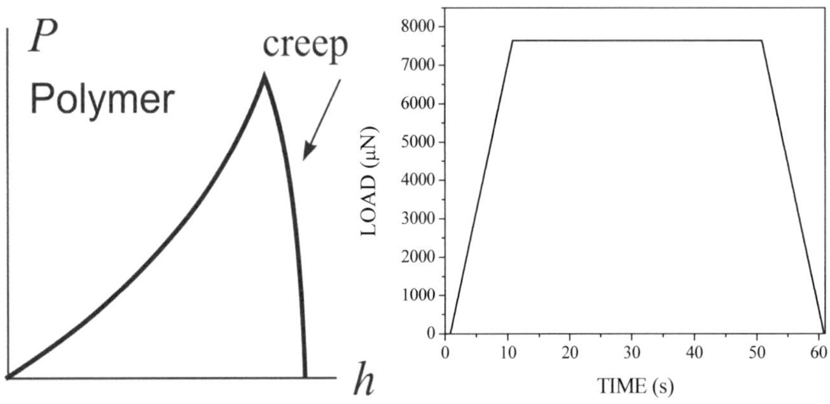

(a) (b)

Fig. 11. a) Typical shape of a *P-h* curve of a polymer [Fischer-Cripps, 2004] and b) Load-time cycle used for the nanoindentation testing on the hybrid coatings.

and then unload. Then, the Oliver and Pharr method can be applied to extract hardness and reduced modulus. The load-time indentation cycle has a trapezoidal shape (Fig. 11b), in the case of the hybrid coatings the maximum load was maintained for 50 seconds.

3.3.1 Hardness and reduced elastic modulus

Film hardness was estimated using the work-of-indentation model [Korsunsky et al., 1998]. In Table 4 the obtained values of H_f are presented. All the coatings showed film hardness higher than 1.2 GPa, which is several times higher than that for commercial acrylic (~260 MPa, measured by nanoindentation). The hardness values of coatings 20806 and 5903 were not possible to estimate due to an unexpected behaviour of the composite hardness. In a typical nanoindentation test of a coated system, two different behaviours can be observed. When the coating is harder than the substrate, the H_c values will decreases to values closer than that of H_s. When the coating is softer than the substrate, the H_c values will increases near to that of the substrate as the indenter approaches values of β=1 [Korsunsky et al., 1998]. The same phenomena occur for reduced modulus [Fischer-Cripps, 2004]. If the hybrid coatings are softer than the glass substrate, then H_c increases gradually as the indenter goes deeper into the film, as can be seen for the coatings 5803 and 5806 in Fig. 12. The continuous line is the work-of-indentation model fitting. However, for 20806 and 5903 samples, after reaching a maximum, H_c starts to decrease. This behaviour can be associated with the presence of one or two internal soft layers, thus the hybrid coatings have a gradient of hardness through the thickness. Regarding the hybrid samples (Fig. 12, 20806 and 5903 samples) the data suggests the presence of an external layer on top of a harder layer, causing an increment in the composite hardness. Finally, after the hard layer another soft layer is present and then H_c decreases. Furthermore, the substrate is harder than the coating, so the hardness values will increase again. However, to observe this behaviour, employment of loads greater than 9000 μN is recommended which it was not possible with the nanoindenter used in this study.

Coating	Film thickness (nm)	Hardness (GPa)
5803	500	1.86
20803	767	1.3
5806	513	1.94
20806	718	--
5903	705	--
20903	744	1.27
5906	446	1.96
20906	1065	1.4

Table 4. Film hardness values of the hybrid coatings obtained by the work-of-indentation model.

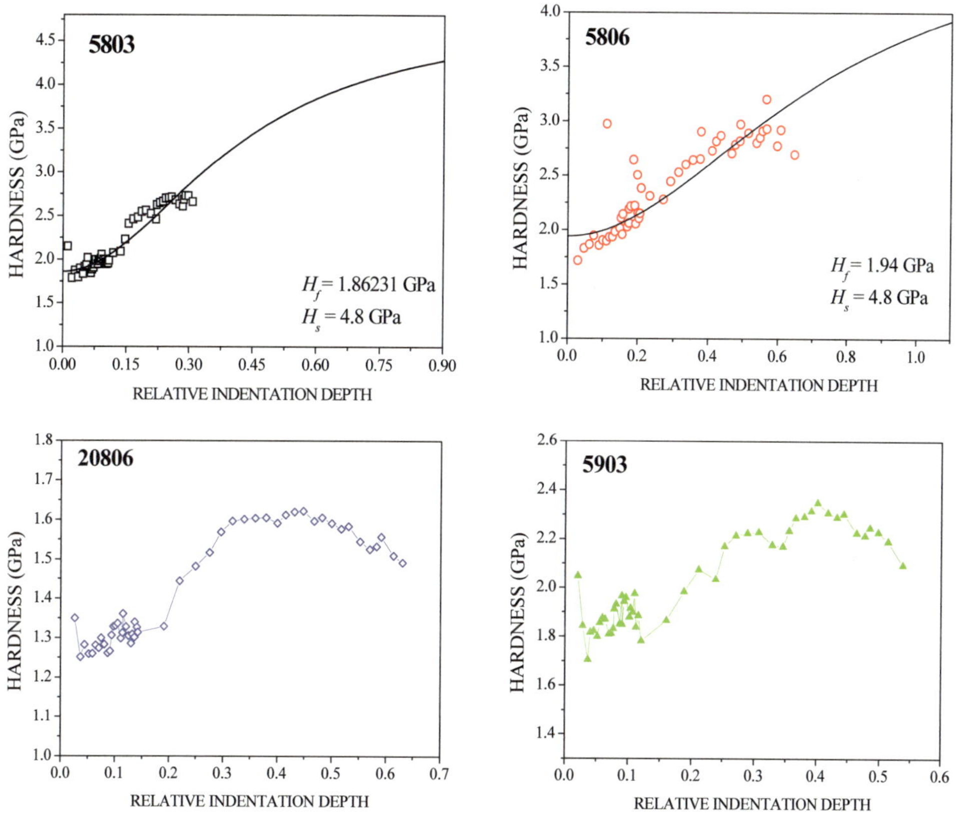

Fig. 12. Composite hardness as a function of the relative indentation depth for some of the hybrid coatings.

In Fig. 13 the reduced modulus as a function of the relative indentation depth is displayed. In the case of the film/substrate system, values of Er as a function of the relative indentation depth show the same tendency as the 20806 and 5903 films showed for hardness. The hybrid

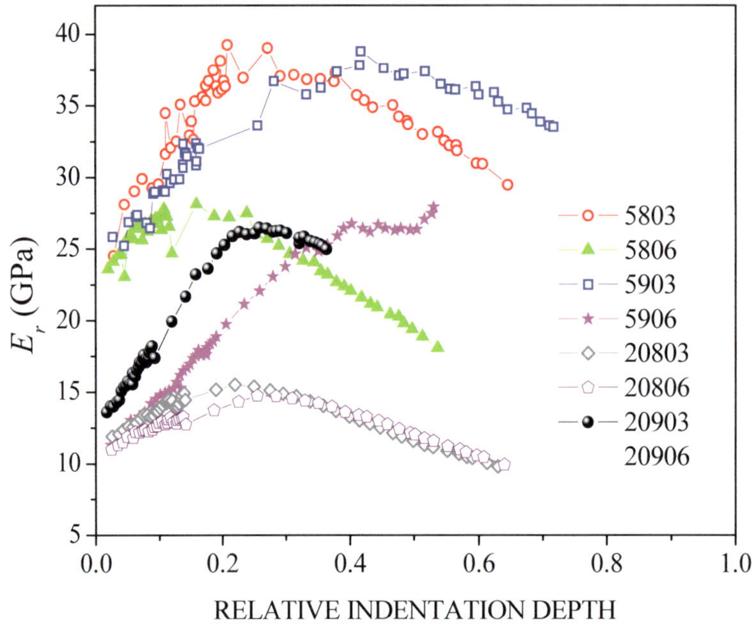

Fig. 13. Reduced modulus of the coating-substrate system as a function of the relative indentation depth.

coatings have elastic modulus gradients through the film thickness. As a result, it was not possible to apply a model to extract the film elastic modulus. Nevertheless, it is possible to observe overall that the films have high values of Er, which are higher than that of the PMMA (3.6 GPa) [Cardarelli, 2008]. As with the hardness, the films with higher values of Er are those with the lowest content of TMSPM.

3.3.2 Creep and stress relaxation

As mentioned above, viscoelastic materials creep under an applied load. This capacity to flow is known as a transitory property, which shows a response in a certain time frame. Creep, creep compliance and stress relaxation are transitory properties of viscoelastic materials [Lake, 2004]. Creep is the time-dependant response to an applied constant stress, and creep compliance is defined as the change in strain as a function of time under an applied constant stress. On the other hand, stress relaxation monitors the change in stress under an applied constant strain [Lake, 2004; Tweedie & Van Vliet, 2006]. To evaluate creep and stress relaxation of the hybrid coatings by nanoidentation, the ISO 14577 standard was employed [Fischer-Cripps, 2004]. This standard makes use of several aspects in instrumented indentation in different penetration depth intervals at macro, micro and nanometric scales and also includes coated systems. In this work we applied the suggested analysis for materials with time-dependant response in order to perform creep and stress relaxation tests.

Creep of a specimen can occur under indentation loading and manifests itself as a change in depth when the applied load is held constant. The relative change in the penetration depth is referred to as the material creep and its value, C_{IT}, is expressed as:

$$C_{IT} = \frac{h_2 - h_1}{h_1} 100 \quad (2)$$

Where h_1 is the depth at which the maximum applied load begins to be maintained constant, and t_1 is the corresponding time; h_2 is the depth that has been reached at the time, t_2, when the unloading starts (Fig. 14). As an example of how C_{IT} is reported: C_{IT}0.5/10/50 = 2.5 which means that a creep of 2.5% was determined in a test applying a load of 0.5 N in a time of 10 seconds and maintained constant for 50 seconds.

Stress relaxation R_{IT} is the relative change in force under an applied constant displacement, thus instead of a constant maximum load, a constant displacement or penetration depth is maintained t while the change in force is measured. Stress relaxation is given by:

$$R_{IT} = \frac{F_1 - F_2}{F_1} 100 \quad (3)$$

This equation is very similar to Eqn. (2). F_1 is the force at which the maximum displacement is reached and kept constant and F_2 is the measured force value at time t_2, the end of the period at which the displacement is maintained constant. A typical curve of displacement as a function of time is depicted in Fig. 14.

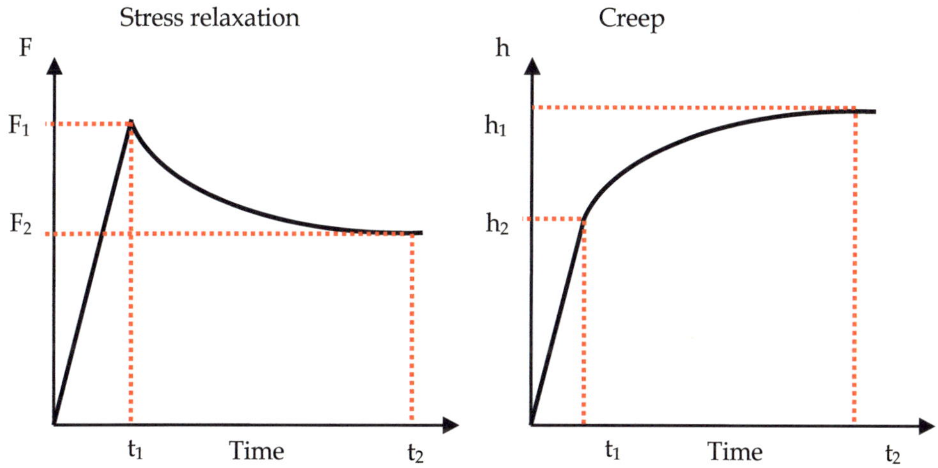

Fig. 14. Schematic representation of the curves of displacement and force as a function of time for creep and stress relaxation tests, respectively.

The results of C_{IT} and R_{IT} for the hybrid coatings on glass with 0.2 and 0.05 TMSPM content and standard PMMA as reference are presented in Fig. 15. A series of several indentations varying the maximum load and displacement were performed on all coatings, then the creep and stress relaxation ratios were calculated from all the indentations performed on each coating. For both C_{IT} and R_{IT} the hybrid coatings showed values lower than that of the PMMA, which is at least twice as high. However, there is no a marked difference between the different contents of TMSPM or drying temperatures and time. The creep values are more dispersed than those for stress relaxation (i.e., the error bars are smaller for stress

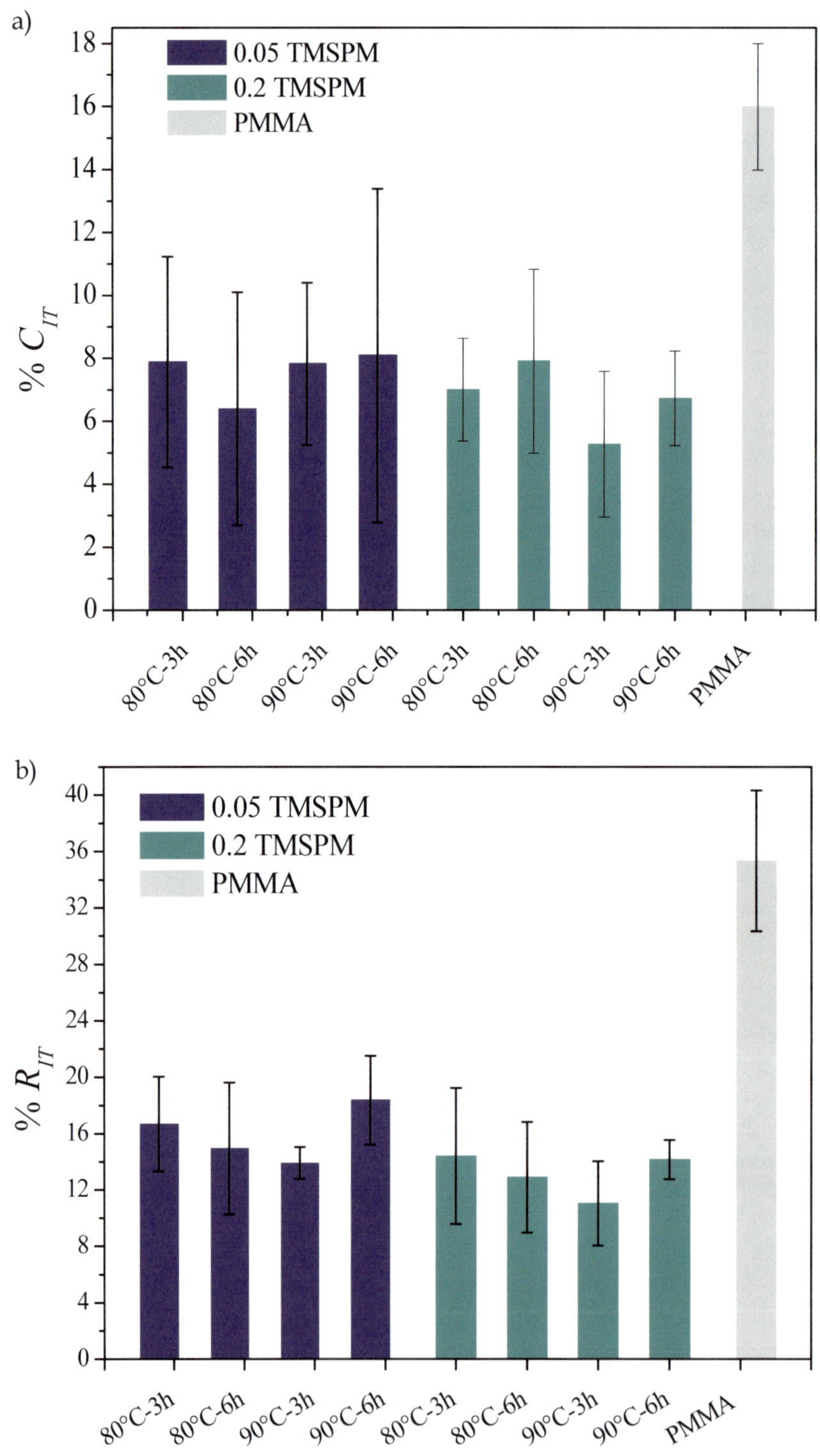

Fig. 15. a) Creep and b) Stress relaxation results of the hybrid coatings on glass.

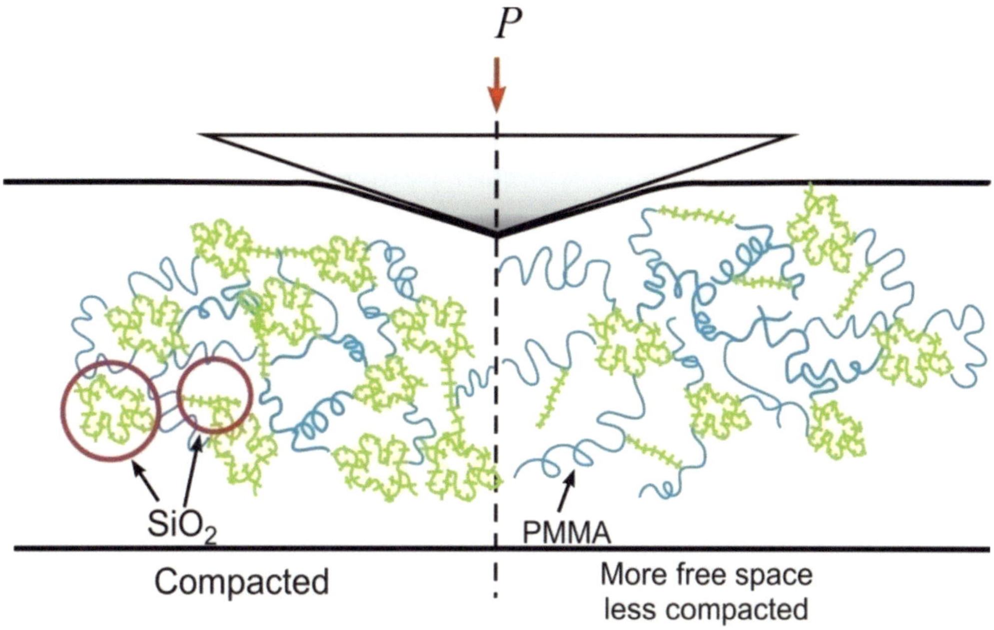

Fig. 16. Schematic representation of the hybrid conformation and distribution of phases when a load is applied.

relaxation). This distinctive behaviour can be associated with localized phenomenon of the hybrid structure or phase distribution. Since creep is the material's capacity to flow, this value will depend upon the chain coiling and available space to the chain to unroll when it is under load. In Fig. 16 a schematic representation of the possible phase distribution of the organic and inorganic phases which are responsible of the viscous flow of the hybrid coating is presented.

3.4 Atomic Force Acoustic Microscopy (AFAM)

During the last decades acoustic force microscopy has been employed to image differences in elastic properties and for detection of defects, applied in a wide range of fields, including physics, non-destructive testing and medicine. This technique is based on transmission and reflection of ultrasonic waves [Rabe, 2006]. The acoustic microscope is a confocal system, this is that focus occurs when both acoustic waves travel through the specimen and are detected by the lenses. Contrast in the image depends on the acoustic impedance and consequently in the elastic constants of the specimen. This technique has a restricted lateral resolution of about the half of the value of the wavelength [Briggs, 1985]. With the invention of atomic force microscopy (AFM), other techniques that combined its characteristic with ultrasonic imaging methods were developed, such as ultrasonic force microscopy (UFM), scanning atomic force microscopy (SAFM), ultrasonic atomic force microscopy (UAFM) and atomic force acoustic microscopy (AFAM). The advantage of combining AFM with ultrasonic techniques is that the probe has a tip with a radius less than 100 nm, allowing high image resolution. Thus tip contact radius is several orders of magnitude lower than the acoustic wavelength, which defines the local resolution [Briggs, 1985]. In AFAM a transducer placed under the sample sends longitudinal waves through the sample causing out-of plane ultrasonic vibrations of the surface. These vibrations are coupled with the AFM cantilever tip generating flexural vibrations of the probe [Kopycinska-Müller, et al., 2007]. This technique can be used to obtain images by mapping the vibration amplitude of the sample surface. In this case, the probe is excited at a fixed frequency near to its resonance frequency. Depending on the local contact stiffness the resonance frequency will change and consequently the vibration amplitude of the work frequency will change, which will also be reflected in contrast differences in amplitude images. These images provide qualitative information about differences in stiffness in regions of the sample surface [Kopycinska-Müller, et al., 2007; Rabe, 2006].

3.4.1 AFAM spectroscopy mode

For AFAM measurements, the information of the contact resonance can be collected in either, step-by-step or sweep mode. In the former, the wave generator changes its output frequency from an initial set value to a final one and this type of measurement is used to analyze a single point on the surface sample. In sweep mode, a frequency interval is scanned in 0.5 seconds, producing a great number of spectra. Contact resonance frequencies are measured as a function of the cantilever static deflection, which is affected by the tip geometry. If the tip has a different geometry from that of a flat indenter, then the increment in static force will lead to an increment in the contact area between the tip and the sample and therefore in the contact stiffness. Thus, the resonance frequency will change to higher values [Kopycinska-Müller et al., 2007].

In our analysis, the tests were performed on a modified Dimension 3000 atomic force microscope in the Fraunhofer Institut for Non-destructive Testing in Saarbruecken, Germany. A diagram of the microscope and associated equipment is presented in Fig. 17. This set-up is used to excite and detect AFAM contact-resonance frequencies in order to measure the local elastic constants of the material. The sample is placed on an ultrasonic transducer that emits longitudinal waves and the produced out-of-plane surface vibrations are detected by the cantilever beam when it is in contact with the surface. These vibrations are the contact resonance frequencies, and they are a consequence of the tip-sample interactions that modify the boundary conditions for the vibrating cantilever. The tip-sample interactions depend on the static force $F_C = k_C \times \Delta z$ applied to sensor tip by the cantilever deflection Δz and on the attractive tip-sample forces, such as electrostatic and adhesion forces [Rabe et al., 2002]. In this set-up the contact-resonance frequencies are measured as a function of the static deflection of the cantilever.

Since the tip of the cantilever probe is in contact with the surface applying a certain load, only a small volume of the sample determines the elastic contact forces. According to the Hertzian model [Johnson, 1985] a contact area with a radius of:

$$a = \sqrt[3]{\frac{3F_C R}{4E^*}} \tag{4}$$

is formed when an isotropic sphere of radius R contacts an elastic isotropic flat surface. Here, E^* is the reduced elastic modulus and is given by:

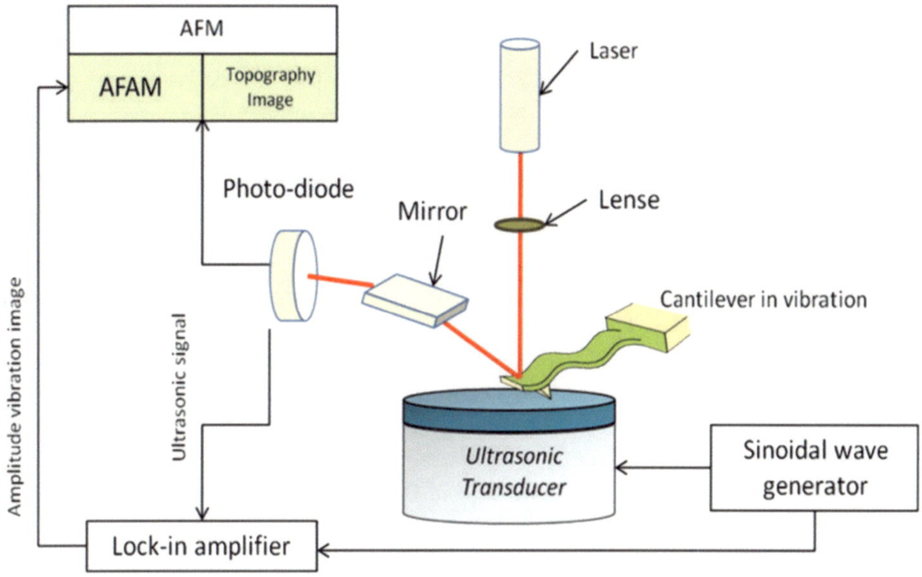

Fig. 17. Experimental set-up of AFM for the acoustic spectroscopy [Rabe, 2006].

$$\frac{1}{E^*} = \frac{\left(1-\nu_S^2\right)}{E_S} + \frac{\left(1-\nu_T^2\right)}{E_T} \tag{5}$$

where E_S, E_T, ν_S, and ν_T are the Young´s moduli, the Poisson's ratios of the surface and the tip, respectively. At small vibration amplitudes the tip-sample forces can be linearized and are represented by a contact stiffness k*

$$k^* = \sqrt[3]{6E^* RF_C}$$

However, the tip shapes often deviate from that of a sphere [Kopycinska-Müller et al., 2007; Rabe et al., 2002]. In the case of a flat punch, the radius of the punch R_p is equal to the contact radius a_C, and the contact stiffness k^*, which is no longer load-dependent, is given by:

$$k^* = 2R_p E^* \tag{6}$$

For anisotropic solids an indentation modulus M is introduced and it is calculated from the single-crystal elastic constants. Then the equation for the reduced elastic modulus E^* can be replaced by:

$$\frac{1}{E^*} = \frac{1}{M_S} + \frac{1}{M_T} \tag{7}$$

were M_S and M_T are the indentation modulus of the sample and the tip, respectively. The tip shape can be characterized by evaluating the contact resonances for reference samples with known indentation modulus. The elastic properties can be evaluated using the following equation by comparative measurements:

$$E_s^* = E_r^* \left(k_r^* / k_s^* \right)^m \tag{8}$$

Here, r and s refer to the reference and the unknown sample, respectively, and m describes the tip geometry. For a flat punch $m = 1$ and for a spherical tip $m = 3/2$.

3.4.2 Determination of the resonance frequencies of the hybrid coatings

The set of samples analyzed by AFAM in spectroscopy mode were 5806, 20806, 5906 and 20906 together with standard samples of fused silica and PMMA for comparison and calibration. NCL silicon probes from Nanosensors with rounded tips and spring constants k_c ranging from 33 to 34 N/m were used. The free resonance frequencies of this cantilever were 159.2 KHz, 989 KHz and 2710.5 KHz for the first, second, and third flexural mode, respectively. The contact-resonances were taken at a static cantilever deflection p_f = 40 nm, which means a static force of 1360 nN was applied to the hybrid surfaces. The resonance frequencies recorded for hybrid sample 5806 are presented in Fig. 18. The standards were used to estimate the shape and elastic moduli of the tip. The first, second and third contact resonances were obtained for this purpose. For the analysis of the results two Labview programs were used. The first allows calculation of the tip position and the contact stiffness with or without considering the lateral forces from two flexural modes of contact resonances. The second Labview program determines the tip position using two flexural modes of contact resonance from two reference samples. It calculates the contact stiffness from the unknown sample using only one contact resonance. For the experiment a sequence for measurement was established as follows: fused silica, PMMA, 5806, 20806, 5906, 20906, fused silica and PMMA. The results for the contact resonance frequencies of the first and second flexural mode for each sample are presented in Fig. 19. These values were used to estimate the contact stiffness of the hybrid coatings. It can be seen that sample 5906 shows the highest values of all hybrid coatings, and it is also close to that of the fused silica for both flexural modes. In the case of sample 20806, the values of the second flexural mode are considerably different from each other, thus only the first flexural mode contact resonance frequency values were used to estimate the contact stiffness of this sample.

In order to calculate the contact stiffness it is necessary to determine the position of the cantilever tip along the length of the probe. For this, the measured values of the contact resonance frequencies of the first and second mode of the fused silica and PMMA were used. The estimated position was $L_1/L = 0.949$, where L is the probe length and L_1 is the tip position. With this value the contact stiffness of the hybrid coatings surfaces was determined, and the results are presented in Table 5.

As was mentioned earlier, coating 5906 showed the highest value of contact resonance frequency, and the contact stiffness is also the highest of all samples. It is worth mentioning that the contact stiffness of fused silica is $k^* = 2639$ N/m, thus in the case of this coating, the surface must be constituted primarily of a dense silica layer. This coating also showed the highest measured value of hardness in nanoindentation testing. The k^* values presented in Table 5 were obtained using both first and second flexural mode contact resonance frequencies. For sample 20806 the contact stiffness was measured considering only the first flexural mode, and the calculated value was 949.6 N/m, which is similar to that obtained for the sample with the same TMSPM content but dried at 90°C (Sample 20906). The hybrid

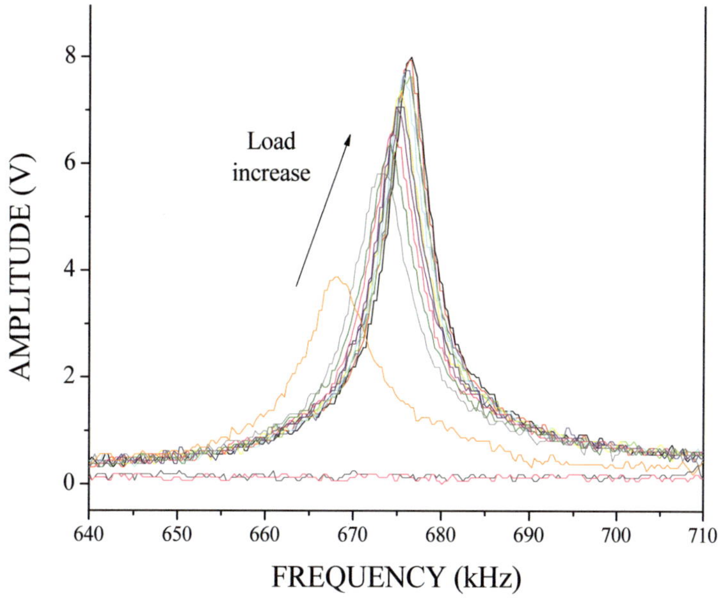

Fig. 18. Recorded contact resonance spectra of the 5806 coating during cantilever loading.

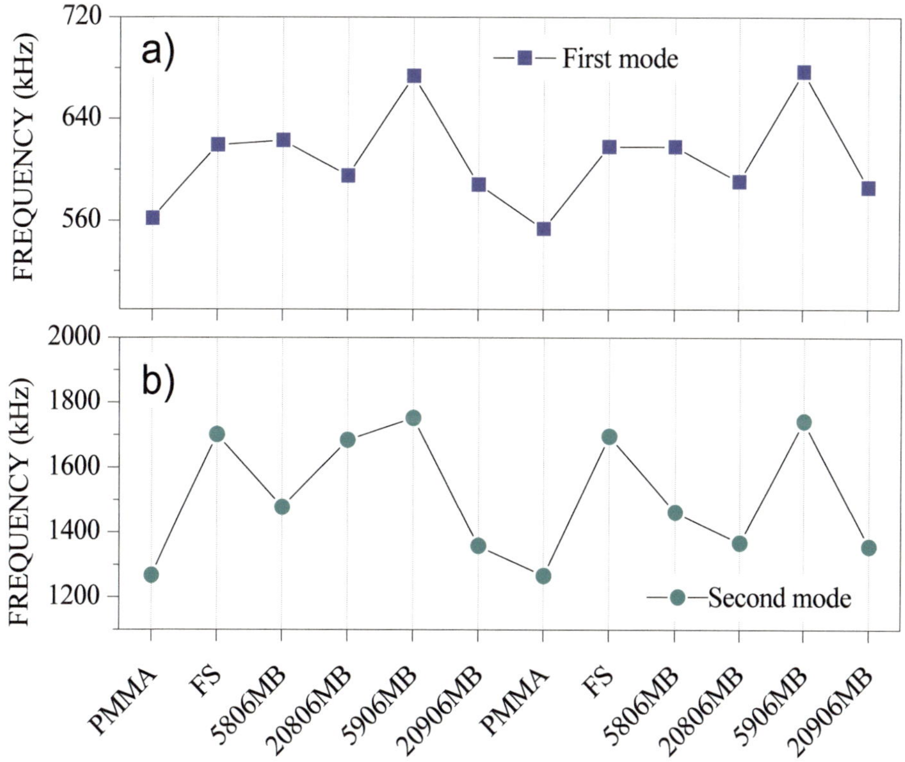

Fig. 19. Variation of the contact resonance frequencies of a) first and b) second flexural modes as a function of measurement order.

Sample	Contact stiffness, k^* N/m
5806	1328
20806	--
5906	2593.5
20906	964

Table 5. Contact stiffness values of the analyzed hybrid samples.

with 0.2 molar ratio of TMSPM showed lower values of contact stiffness. This result can be associated to the more PMMA chains in the surface. According to FT-IR spectra, a major content of the crosslinker promotes the formation of PMMA. Thus, the hybrid coating will show elastic properties similar to that of the polymer.

3.4.3 AFAM imaging mode

AFAM images of the hybrid coatings were taken in a modified Nanoscope IV Dimension 3000. A silicon cantilever coated with Cr/Pt (Budget sensors), with a spring constant of 3 N/m and a tip radius of 20-25 nm, was used. The samples measured were those analyzed by AFAM spectroscopy. Images of a 1x1 µm area of the hybrid film surfaces were taken and are presented in Fig. 20. This is a complementary analysis the spectroscopy. Even though it is a qualitative analysis it is very useful to observe differences in stiffness of the sample surface. The first step is to find the local contact resonance frequency of the surface and then tune the cantilever near this frequency. Afterwards, the vibration amplitude of the surface is scanned, and the changes in stiffness will be represented as contrast between dark and bright zones. When the surface has lower stiffness than the measured local stiffness, it will show as dark zones; otherwise when the stiffness is higher it will show as bright zones. Amplitude images show details that are not always perceptible in the topography images. The phase images show the different components of the sample surface, which for the case of hybrid coatings are silica and PMMA. Then the phase and amplitude images reveal structural details regarding the local stiffness of the analyzed zones.

As mentioned earlier when discussing the AFM topography images, the hybrid coatings have an ultra-low surface roughness and a smooth surface. The resonance frequencies used to tune the cantilever for each sample are presented in Table 6, these values were taken measuring the local contact resonance frequency at a single point on the coating surface. Then with AFAM structural details of the PMMA and silica component distribution were revealed. Samples 5806 and 20806 phase images show distinct distribution of bright and dark regions in the phase images and some black spots, which can be pores. Sample 5906 does not show a clear distribution of shapes either in the phase or amplitude image, probably because of a more random distribution of silica and PMMA. In the case of sample 20906, zones with different contrast are visible showing a particular morphology of elongated "beans" whit sizes of 30 to 80 nm. These images demonstrate the formation of a material where both phases are distributed in regions of less than 100 nm. This provides the material with its high transparency, ultra low roughness and low friction coefficient. Further studies on coatings with different contents of PMMA will be interesting to observe if the materials arranges itself in a specific shape and size.

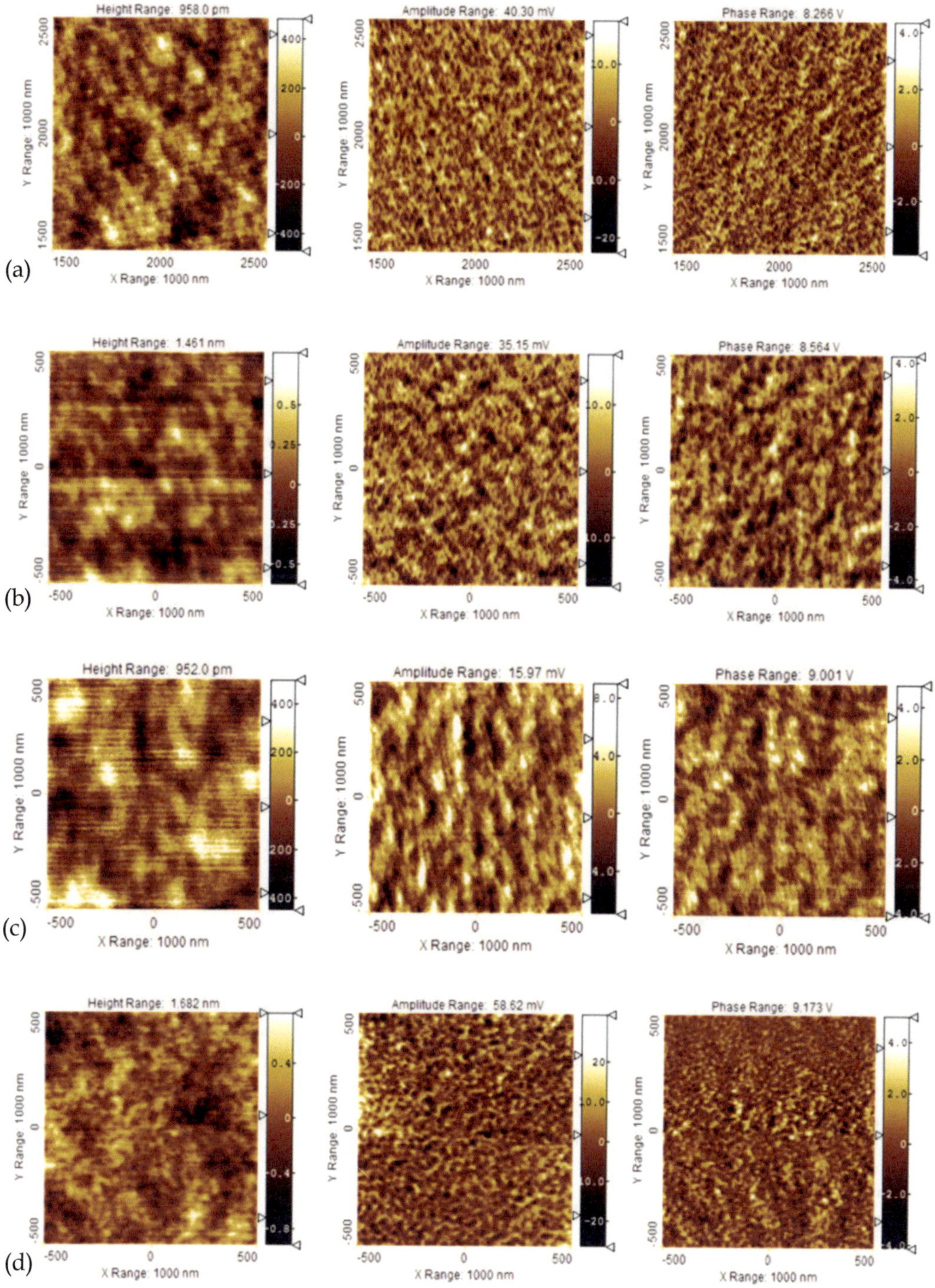

Fig. 20. Height, amplitude and phase images of the coatings: a) 5806, b) 20806, c) 5906 and d) 20906.

Sample	Cantilever Excitation frequency (kHz)
5806	339.84
2806	338.86
5906	340.82
20906	339.84

Table 6. Frequency values used to tune the cantilever for AFAM imaging.

4. Conclusion

The addition of Al_2O_3 nanoparticles and whiskers improve the wear resistance behaviour of the hybrid material even though the hardness of the coatings does not show an increase. The hybrid coating SiO_2-PMMA-0.1wAl_2O_3 was the one with the best performance in the sliding life test resisting the whole test without failing. In general, in the nanoscratch testing all the coatings have better wear resistance than that of the acrylic substrate, showing values of wear loss volume two orders of magnitude lower than that of the substrate. Moreover, the concentrations of nanoparticles and whiskers used in this study improved the wear resistance behaviour and transparency of the films was maintained.

According to the AFAM spectroscopy mode results, the TMSPM content and drying temperature of the hybrid coatings have an important effect on the contact stiffness of the hybrid coating surfaces. The hybrid coating with a TMSPM content of 0.5 and dried at 90°C for 6 hours showed the highest value of contact stiffness, near to that of the fused silica standard. This result is in good agreement with the nanoindentation hardness, in which the same coating showed the highest hardness value. In general, the coatings with a TEOS:TMSPM molar ratio of 1:0.5 have higher values than those of the coatings with a TMSPM content of 0.2. According to the FT-IR spectra results, these results are associated with the capability of the TMSPM to promote the formation of PMMA chains. In this respect, the AFAM imaging testing showed that the silica and PMMA phases are homogeneously distributed, forming nanometric domains of each component. This can be related to the transparency of the film and the smooth surfaces of the hybrid coatings with roughness values of less than 2 nm.

5. Acknowledgments

Authors thank Prof. Dr. W. Arnold and Dr. U. Rabe from Fraunhofer Institut for Non-Destructive Testing in Saarbruecken, Germany for their support in AFAM spectroscopy mode measurements and Dr. Francisco Javier Espinoza Beltrán from Cinvestav-Unidad Querétaro for his help in the AFAM imaging testing.

6. References

Alvarado-Rivera, J., Muñoz-Sadaña, J., & Ramírez-Bon, R. (2010). Nanoindentation testing of SiO_2-PMMA hybrid films on acrylic substrates with variable coupling agent content. *Journal of Sol-Gel Science and Tehcnology, Vol. 54, No. 3* , pp. 312-318, ISSN 0928-0707.

Alvarado-Rivera, J., Muñoz-Saldaña, J., Castro-Beltrán, A., Quintero-Armenta, J. M., Almaral-Sánchez, J., & Ramírez-Bon, R. (2007). Hardness and wearing properties of SiO_2-PMMA hybrid coatings reinforced with Al_2O_3 nanowhiskers. *Physica Status Solidi C, Vol. 14, No. 11* , pp. 4254-4259, ISSN 1862-6351.

Briggs, A. (1985). *An Introduction to Scanning Acoustic Microscopy, Royal Microscopical Society Handbook 12.* Oxford University.

Cardarelli, F. (2008). *Materials Handbook: a concise desktop reference, 2nd Edition.* New York, NY: Springer, ISBN-13 9781846286681.

Fischer-Cripps, A. (2004). *Nanoindentation.* New York, United States of America: Springer-Verlag, ISBN 0-387-22045-3.

Hay, J., & Pharr, G. (2000). Instrumented Indentation Testing. In H. Kuhn, & D. Medlin, *ASM Handbook Vol.8 Mechanical Testing and Evaluation* (pp. 232-243). Ohio, United States of America: ASM International, ISBN 978-0871703897.

Horcas, I., Fernández, R., Rodríguez, J., Colchero, J., Gomez-Herrero, J., & Baro, A. (2007). WSXM: A software for scanning probe microscopy and a tool for nanotechnology. *Review of Scientific Instruments, vol. 78, No. 1* , 013705, ISSN 0034-6748.

Huang, H.-H., Orler, B., & Garth, L. (1985). Ceramers: Hybrid Materials Incorporating Polymeric/Oligomeric Species with Inorganic Glasses by a Sol-Gel Process 2. Effect of Acid Content on the Final Properties. *Polymer Bulletin, Vol. 14* , pp. 557-564, ISSN 557-564.

Johnson, K. (1985). *Contact Mechanics.* Cambridge: Cambridge University Press.

Kopycinska-Müller, M., Caron, A., Hirsekon, S., Rabe, U., Natter, H., Hempelmann, R., et al. (2007). quantitative Evauation of Elastic Properties of Nano-Crystalline Nickel Using Atomic Force Acoustic Microscopy. *Zeitschrift für Physikalische Chemie, Vol. 222, No. 2-3* , pp. 471-498, ISSN 004-3336.

Korsunsky, A., McGurk, M., Bull, J., & Page, T. (1998). On the hardness of coated systems. *Surface and Coatings Technology, Vol. 99, No. 1-2* , pp. 171-183, ISSN 0257-8297.

Lake, R. (2004). Viscoelastic measurements thecniques. *Review of Scientific Instruments, Vol. 75, No. 4* , 797-810, ISSN 0034-6748.

Oliver, W., & Pharr, G. (1992). An improved technique for determining hardness and elastic moduus using load and displacement sensing indentation experiment. *Journal of Materials Research, Vol. 7, No. 6* , pp. 171-183, ISSN 0884-2914.

Rabe, U. (2006). Atomic Force Acoustic Microscopy. In B. Bhushan, & H. Fuchs, *Applied Scanning Probe Methods II* (pp. 37-90). Germany: Springer-Verlag, ISBN 978-3-540-26242-8.

Rabe, U., Amelio, S., Kopycinska, M., Hirsekorn, S., Kempf, M., Göken, M., et al. (2002). Imaging and measurement of local mechanical material properties by atomic force acoustic microscopy. *Surface and Interface Analysis, Vol.33, No.1* , 65-70, ISSN 1096-9918.

Sanchez, C., & Ribot, F. (1994). Design of hybrid organic-inorganic materials synthesized via Sol-Gel chemistry. *New Journal of Chemistry, Vol. 18, No. 10* , pp. 1007-1047, ISSN 0959-9428.

Sanchez, C., Julián, B., Belleville, P., & Popall, M. (2005). Applications of hybrid organic-inorganic nanocomposites. *Journal of Materias Chemistry, Vol. 15, No. 35-36* , pp. 3559-3592, ISSN 0959-9428.

Schmidt, H. (1985). New type of non-crystalline solids between inorganic and organic materials. *Journal of Non-Crystalline Solids, vol. 73, No. 1-3* , pp. 681-691, ISSN 0022-3093.

Tweedie, C., & Van Vliet, K. (2006). Contact creep compliance of viscoelastic materials via nanoindentation. *Journal of Materials Research, Vol. 21, No. 6* , 1576-1589, ISSN 0884-2914.